HSC Year 12
PHYSICS

ADAM SLOAN

+ summary notes
+ revision questions
+ detailed sample answers
+ study and exam preparation advice

STUDY NOTES

A+ HSC Physics Study Notes
1st Edition
Adam Sloan
ISBN 9780170465304

Publisher: Cathy Beswick-Davison
Series editor: Catherine Greenwood
Copyeditor: Marta Veroni
Reviewer: Martin Barkl
Series text design: Nikita Bansal
Series cover design: Nikita Bansal
Series designer: Cengage Creative Studio
Artwork: MPS Limited
Production controller: Karen Young
Typeset by: Nikki M Group Pty Ltd

For product information and technology assistance,
in Australia call **1300 790 853**;
in New Zealand call **0800 449 725**

For permission to use material from this text or product, please email **aust.permissions@cengage.com**

ISBN 978 0 17 046530 4

Cengage Learning Australia
Level 7, 80 Dorcas Street
South Melbourne, Victoria Australia 3205

Cengage Learning New Zealand
Unit 4B Rosedale Office Park
331 Rosedale Road, Albany, North Shore 0632, NZ

For learning solutions, visit **cengage.com.au**

Printed in China by 1010 Printing International Limited.
1 2 3 4 5 6 7 26 25 24 23 22

CONTENTS

HOW TO USE THIS BOOK vi
A+ DIGITAL viii
PREPARING FOR THE END-OF-YEAR EXAM ix
ABOUT THE AUTHOR xiv

CHAPTER 1

MODULE 5: ADVANCED MECHANICS

Module summary	2
1.1 Projectile motion	5
1.1.1 Using horizontal and vertical components	5
1.1.2 Trajectory of projectiles	6
1.1.3 Analysing projectile motion with equations	6
1.2 Circular motion	7
1.2.1 Features of uniform circular motion	7
1.2.2 Special cases of uniform circular motion	8
1.2.3 Rotation and torque	10
1.3 Motion in gravitational fields	10
1.3.1 Gravitational forces and gravitational fields	10
1.3.2 Satellite motion	11
1.3.3 Kepler's laws	12
1.3.4 Satellite energy	13
Glossary	15
Exam practice	16

CHAPTER 2

MODULE 6: ELECTROMAGNETISM

Module summary	31
2.1 Charged particles, conductors, and electric and magnetic fields	33
2.1.1 Charged particles in electric fields	33
2.1.2 Motion of charged particles in uniform electric fields	34
2.1.3 Charged particles in magnetic fields	35
2.1.4 Moving charged particles in electric and magnetic fields	37
2.2 The motor effect	38
2.2.1 Force on a current-carrying conductor in a magnetic field	38
2.2.2 Forces between parallel current-carrying wires	38
2.3 Electromagnetic induction	40
2.3.1 Magnetic flux	40
2.3.2 Processes of electromagnetic induction	40
2.3.3 Transformers	42

CONTENTS

9780170465304

2.4 Applications of the motor effect 44
2.4.1 DC motors 44
2.4.2 Generators 46
2.4.3 AC induction motors 47
2.4.4 Lenz's law – DC motors and electromagnetic braking 48
Glossary 49
Exam practice 51

CHAPTER 3

MODULE 7: THE NATURE OF LIGHT

Module summary 67
3.1 Electromagnetic spectrum 69
3.1.1 Maxwell's theory of electromagnetism 69
3.1.2 The speed of light 70
3.1.3 Spectra 71
3.1.4 Using spectra 71
3.2 Light: wave model 73
3.2.1 Diffraction of light 74
3.2.2 Interference of light 74
3.2.3 Competing models of light 76
3.2.4 Polarisation of light 77
3.3 Light: quantum model 77
3.3.1 Black body radiation and quanta 77
3.3.2 Photoelectric effect and the wave model of light 78
3.3.3 Photoelectric effect and the particle model of light 79
3.4 Light and special relativity 80
3.4.1 Postulates of special relativity 80
3.4.2 Time dilation 82
3.4.3 Length contraction 83
3.4.4 Evidence for special relativity 83
3.4.5 Relativistic momentum 84
3.4.6 $E = mc^2$ and the equivalence of mass and energy 84
Glossary 85
Exam practice 87

9780170465304

CHAPTER 4

MODULE 8: FROM THE UNIVERSE TO THE ATOM

Module summary 103

4.1 Origins of the elements 105

4.1.1 The Big Bang theory and an expanding Universe 105

4.1.2 Classification of stars from stellar spectra 107

4.1.3 The Hertzsprung–Russell diagram and stellar evolution 107

4.1.4 Nucleosynthesis in stars and mass–energy equivalence 110

4.2 Structure of the atom 112

4.2.1 Cathode rays 112

4.2.2 Thomson's charge-to-mass experiment 112

4.2.3 Millikan's oil drop experiment 113

4.2.4 Geiger–Marsden experiment and Rutherford's atomic model 114

4.2.5 Chadwick's discovery of the neutron 115

4.3 Quantum mechanical nature of the atom 116

4.3.1 The emission spectrum of hydrogen 116

4.3.2 Bohr's atomic model 116

4.3.3 de Broglie's matter waves 118

4.3.4 Schrödinger's contribution to the current atomic model 119

4.4 Properties of the nucleus 119

4.4.1 Spontaneous nuclear decay 119

4.4.2 Half-life 122

4.4.3 Nuclear fission 123

4.4.4 Energy from nuclear reactions 124

4.5 Deep inside the atom 126

4.5.1 Subatomic particles and the Standard Model of matter 126

4.5.2 Evidence from particle accelerators 127

Glossary 129

Exam practice 131

SOLUTIONS....................................144

DATA SHEET....................................215

FORMULAE SHEET..............................216

PERIODIC TABLE OF THE ELEMENTS...............218

CONTENTS

HOW TO USE THIS BOOK

The A+ HSC Physics resources are designed to be used year-round to prepare you for your HSC Physics exam. *A+ HSC Physics Study Notes* includes topic summaries of all key knowledge in the NSW HSC Physics syllabus that you will be assessed on during your exam. Each chapter of this book addresses one module. This section gives you a brief overview of each chapter and the features included in this resource.

Module summaries

The module summaries at the beginning of each chapter give you a high-level overview of the essential knowledge and key science skills you will need to demonstrate during your exam.

Concept maps

The concept maps at the beginning of each chapter provide a visual summary of each module outcome.

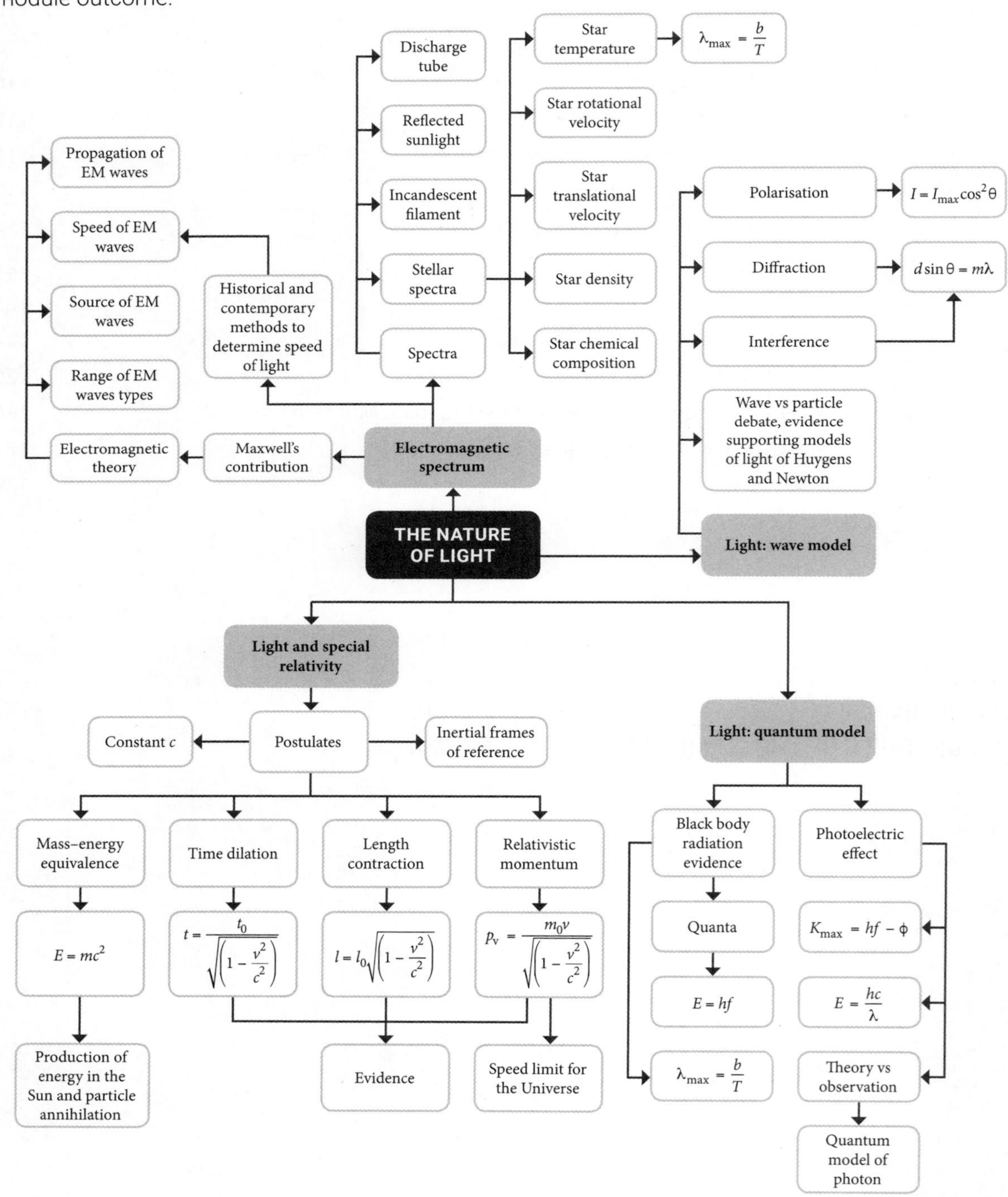

Inquiry question summaries

All of the dot points under each inquiry question are summarised sequentially throughout inquiry section summaries.

Exam practice

Exam practice questions appear at the end of each chapter to test you on what you have just reviewed in the chapter. These are written in the same style as the questions you will find in the actual HSC Physics exam. There are some official past exam questions in each chapter.

Multiple-choice questions

Each chapter has approximately 30 multiple-choice questions.

Short-answer questions

There are approximately 20 short-answer questions in each chapter, often broken into parts. These questions require you to apply your knowledge across multiple concepts. Mark allocations have been provided for each question.

Solutions

Solutions to practice questions are supplied at the back of the book. They have been written to reflect a high-scoring response and include explanations of what makes an effective answer.

Explanations

The solutions section includes explanations of each multiple-choice option, both correct and incorrect. Explanations of written response items include what a high-scoring response looks like and indications of potential mistakes.

SOLUTIONS

CHAPTER 1 MODULE 5 ADVANCED MECHANICS

Multiple-choice solutions

1 D

$u = 40\,\text{m s}^{-1}$

$\theta = 20°$

Therefore, $u_y = u \sin\theta = 40 \sin 20°\,\text{m s}^{-1}$.

$v_y = 0$ at maximum height

$a = -9.8\,\text{m s}^{-2}$

$s_y = ?$

Choose equation from data.	$v_y^2 = u_y^2 + 2as_y$
Substitute known values.	$0 = (40 \sin 20°)^2 + 2 \times -9.8 \times s_y$
Rearrange.	$s_y = -\frac{(40 \sin 20°)^2}{-19.6}$
Calculate the value.	$s_y = 9.549\,\text{m} = 9.5\,\text{m}$ (to two sig. fig.)

Short-answer solutions

30 a Just as masses have energy as a consequence of their position in a gravitational field, charged particles have energy as a consequence of their position in an electric field. This energy is electrical potential energy (analogous to gravitational potential energy). Because the particle is free to move, the potential energy is converted to kinetic energy.

Mark breakdown

- 1 mark: response identifies the energy as electrical potential energy or kinetic energy
- 2 marks: response clearly connects the force exerted by the electric field and the object's position within the field to the electrical potential energy it has or the kinetic energy it gains (or both)

HOW TO USE THIS BOOK

Icons

The following icons occur in the summaries and exam practice sections of each chapter.

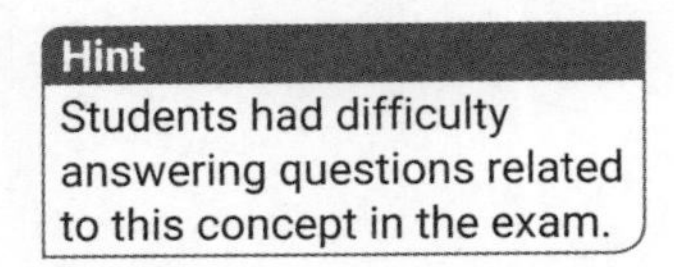

Hint and note boxes appear throughout to provide additional tips and support.

This icon appears with official past NESA questions.

These icons indicate whether the question is easy, medium or hard.

A+ HSC Physics Practice Exams

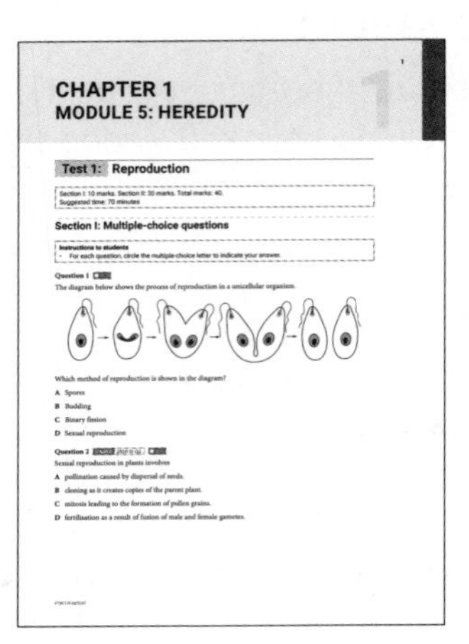

A+ HSC Physics Study Notes can be used independently or alongside the accompanying resource *A+ HSC Physics Practice Exams*. *A+ HSC Physics Practice Exams* features topic tests comprising original HSC-style questions and official HSC questions. Each topic test includes multiple-choice and short-answer questions and focuses on one inquiry question of the NSW HSC Physics syllabus. There are two complete practice exams following the tests. As in *A+ Physics Study Notes*, detailed solutions are included at the end of the book, demonstrating and explaining how to craft high-scoring exam responses.

A+ DIGITAL

Just scan the QR code or type the URL into your browser to access:

- A+ Flashcards: revise key terms and concepts online
- Revision summaries of all concepts from each inquiry question.

Note: You will need to create a free NelsonNet account.

PREPARING FOR THE END-OF-YEAR EXAM

Exam preparation is a year-long process. It is important to keep on top of the theory and consolidate often, rather than leaving work to the last minute. You should aim to have the theory learned and your notes complete so that by the time you reach STUVAC, the revision you do is structured, efficient and meaningful.

Effective preparation involves the following steps.

Study tips

To stay motivated to study, try to make the experience as comfortable as possible. Have a dedicated study space that is well lit and quiet. Create and stick to a study timetable, take regular breaks, reward yourself with social outings or treats, and use your strengths to your advantage. For example, if you are great at art, turn your Physics notes into cartoons, diagrams or flowcharts. If you are better with words or lists, create flashcards or film yourself explaining tricky concepts and watch the videos.

Another strong recommendation is to engage with the Performance Band Descriptors published by NESA. Clear information is provided as to what is expected of a student performing at band 6, band 5 and so on mapped against the knowledge and understanding and working scientifically course outcomes. Have an honest conversation with yourself as to what level you are currently performing at. This will in turn provide you with guidance on what you need to do to improve.
For example, a band 6 student:

- demonstrates an extensive knowledge and understanding of scientific concepts, including complex and abstract ideas
- communicates scientific understanding succinctly, logically and consistently using correct and precise scientific terms and application of nomenclature in a variety of formats and wide range of contexts
- designs and plans investigations to obtain accurate, reliable, valid and relevant primary and secondary data, evaluating risks, mitigating where applicable, and making modifications in response to new evidence
- selects, processes and interprets accurate, reliable, valid and relevant qualitative and quantitative primary or secondary data, and represents it using a range of scientific formats to derive trends, show patterns and relationships, explain phenomena and make predictions
- designs solutions to scientific problems, questions or hypotheses using selected accurate, reliable, valid and relevant primary and secondary data and scientific evidence, by applying processes, modelling and formats
- applies knowledge and information to unfamiliar situations to propose comprehensive solutions or explanations for scientific issues or scenarios

Revision techniques

Here are some useful revision methods to help information '**STIC**'.

Spaced repetition	This technique helps to move information from your short-term memory into your long-term memory by spacing out the time between your revision and recall flashcard sessions. As the time between retrieving information is slowly extended, the brain processes and stores the information for longer periods.
Testing	Testing is necessary for learning and is a proven method for exam success. If you test yourself continually before you learn all the content, your brain becomes primed to retain the correct answer when you learn it. As part of this process, engage with the marking criteria provided to help decide on areas where improvement is needed.
Interleaving	This is a revision technique that sounds counterintuitive but is very effective for retaining information. Most students tend to revise a single topic in a session, and then move on to another topic in the next session. With interleaving, you choose three topics (1, 2, 3) and spend 20–30 minutes on each topic. You may choose to study 1-2-3 or 2-1-3 or 3-1-2, 'interleaving' the topics and repeating the study pattern over a long period of time. This strategy is most helpful if the topics are from the same subject and are closely related.
Chunking	An important strategy is breaking down large topics into smaller, more manageable 'chunks' or categories. Essentially, you can think of this as a branching diagram or mind map where the key theory or idea has many branches coming off it that get smaller and smaller. By breaking down the topics into these chunks, you will be able to revise the topic systematically.

These strategies take cognitive effort, but that is what makes them much more effective than re-reading notes or trying to cram information into your short-term memory the night before the exam!

Note: Year 11 knowledge is assumed (Year 11 equations are still on Data and Formulae sheets etc.) and does get tested within Year 12 focused questions. Remember to plan for how your could revise Year 11 knowledge too.

Time management

It is important to manage your time carefully throughout the year. Make sure you are getting enough sleep, that you are getting the right nutrition, and that you are exercising and socialising to maintain a healthy balance so that you don't burn out.

To help you stay on target, plan out a study timetable. Here are some steps to help you do this.

1. Assess your current study time and social time. How much are you dedicating to each?
2. List all your commitments and deadlines, including sport, work, assignments etc.
3. Prioritise the list and re-assess your time to ensure you can meet all your commitments.
4. Decide on a format, whether it be weekly or monthly, and schedule in a study routine.
5. Keep your timetable somewhere you can see it.
6. Be consistent.

Studies suggest that 1-hour blocks with a 10-minute break is most effective for studying, and remember you that can interleave three topics during this time! You will also have free periods during the school day you can use for study, note-taking, assignments, meeting with your teachers and group study sessions. Studying does not have to take hours if it is done effectively. Use your timetable to schedule short study sessions often.

The exam

The examination is held at the end of the year and contributes 50% to your HSC mark. You will have 180 minutes plus 5 minutes of reading time. You are required to attempt 20 multiple-choice questions (Section I), and short-answer questions (Section II), covering all areas of study in modules 5–8.
The following strategies will help you prepare for the exam conditions.

Practise using past papers

To help prepare, download the past papers from the NESA website and attempt as many as you can in the lead-up to the exam. These will show you the types of questions to expect and give you practice in writing answers. It is a good idea to make the practice exams as much like the real exam as possible (conditions, time constraints, materials etc.). You can also use *A+ HSC Physics Practice Exams*.

Use trial papers, school-assessed coursework, and comments from your teacher to pinpoint weaknesses, and work to improve these areas. Do not just tick or cross your answers; look at the suggested answers and try to work out why your answer was different. What misunderstandings do your answers show? Are there gaps in your knowledge? Did you provide sufficient detail?

Did you use scientific language appropriately and structure your response clearly, logically and cohesively? Read the examiners' reports to find out the common mistakes students make.

Make sure you understand the material, rather than trying to rote learn information. Most questions are aimed at your understanding of concepts and your ability to apply your knowledge to new situations.

Try different strategies

You don't need to do the exam in order. It may be worth practising alternative approaches. The multiple-choice questions are ordered from easy to hard. Some students complete the first 10 multiple-choice questions, then move on to other questions and come back to the last 10 multiple-choice questions at the end.

Some students like to start with the two longest response questions (usually 1 × 7 marks and 1 × 9 marks) before moving on to other areas. Some students like to do all the calculation questions first. The main idea is to start in a positive and confident way, which should set you up positively for the rest of the paper. Don't dwell for too long on a question you are not confident about or struggling with.

The day of the exam

The night before your exam, try to get a good rest and avoid cramming, as this will only increase stress levels. On the day of the exam, arrive at the venue early and bring everything you will need with you. If you must rush to the exam, your stress levels will increase, thereby lowering your ability to do well. Further, if you are late, you will have less time to complete the exam, which means that you may not be able to answer all the questions or may rush to finish and make careless mistakes. If you are more than 30 minutes late, you may not be allowed to enter the exam. Do not worry too much about exam jitters. A certain amount of stress is required to help you concentrate and achieve an optimum level of performance. If, however, you are feeling very nervous, breathe deeply and slowly. Breathe in for a count of 6 seconds, and out for 6 seconds until you begin to feel calm.

Important information from the syllabus

Sixty per cent of your school-based assessment will have addressed the skills required for working scientifically. This will have included a mandatory depth study. You are strongly encouraged to engage with the working scientifically syllabus outcomes as part of your study as they will also constitute a significant component of the HSC exam.

Outcome	Description
PH11/12-1	**Questioning and predicting** develops and evaluates questions and hypotheses for scientific investigation
PH11/12-2	**Planning investigations** designs and evaluates investigations in order to obtain primary and secondary data and information
PH11/12-3	**Conducting investigations** conducts investigations to collect valid and reliable primary and secondary data and information
PH11/12-4	**Processing data and information** selects and processes appropriate qualitative and quantitative data and information using a range of appropriate media
PH11/12-5	**Analysing data and information** analyses and evaluates primary and secondary data and information
PH11/12-6	**Problem solving** solves scientific problems using primary and secondary data, critical thinking skills and scientific processes
PH11/12-7	**Communicating** communicates scientific understanding using suitable language and terminology for a specific audience or purpose

Section I of the exam

Section I consists of a question book and an answer sheet. The answers for multiple-choice questions must be recorded on the answer sheet provided. A correct answer scores 1, and an incorrect answer scores 0. There is no deduction for an incorrect answer, so attempt every question. Read each question carefully and underline key words. If you are given a graph or diagram, make sure you understand the graphic before you read the answer options. You may make notes on the diagrams or graphs.

Section II of the exam

Section II consists of a question book with space to write your answers. The space provided is an indication of the detail required in the answer. Most questions will be broken down into several parts, and each part will be testing new information; so, read the entire question carefully to ensure you do not repeat yourself. Use correct physics terminology and make an effort to spell it correctly! Look at the mark allocation. Generally, if there are two or three marks allocated to the question, you will be expected to show several linked and logical steps in your analysis/reasoning or to show several clear steps in your working/problem solving process. If you make a mistake, cross out any errors but do not write outside the space provided; instead, ask for another booklet and re-write your answer. Mark clearly on your paper which questions you have answered where.

Make sure your handwriting is clear and legible and attempt all questions! Marks are not deducted for incorrect answers, and you will often get a mark for an entry level attempt to respond to the question by engaging with it and producing some concept or detail or equation that is not already included in the question. You will definitely not get any marks if you leave a question blank!

Do not be put off if you do not recognise an example or context; questions will always be about the concepts that you have covered. The exam aims to test your ability to use your knowledge, understanding and skills in contexts you are not familiar with.

Reading time

Use your time wisely! *Do not* use the reading time to try and figure out the answers to any of the questions until you have read the whole paper! The exam will not ask you a question testing the same knowledge twice, so look for hints in the stem of the question and avoid repeating yourself. Plan your approach so that when you begin writing you know which section, and ideally which question, you are going to start with. You do not have to start with Section I.

Strategies for answering Section I

Read the question carefully and underline any important information to help you break the question down and avoid misreading it. Read all the possible solutions and eliminate any clearly wrong answers. You can annotate or write on any diagrams or infographics and make notes in the margins. Fill in the multiple-choice answer sheet carefully and clearly. Every few questions, check that your question number matches the answer number you are filling in, check your answer and move on. Do not leave any answers blank.

Strategies for answering Section II

The examiners' reports always highlight the importance of planning responses before writing. Remember you have 3 hours to complete 100 marks. This means you have an average of 1.8 minutes per mark. For a 5-mark question this equals 9 minutes. You should spend a good proportion of this time planning your response.

To do this, **CUBE** the question:

Circle the verb (identify, describe, explain, evaluate).

Underline the key Physics concepts to be covered in your response (e.g. electromagnetic induction, Hertzsprung–Russell diagram, satellite motion, photoelectric effect).

Box important information.

Elaborate on depth required to answer question.

Many questions require you to apply your knowledge to unfamiliar situations, so it is okay if you have never heard of the context before. You should, however, know which part of the course you are being tested on and what the question is asking you to do. Plan your response in a logical sequence based on the level of detail required by the verb of the question.

Another useful acronym to remember is based on the **ALARM** scaffold (**A L**earning **A**nd **R**esponding **M**atrix) developed by Max Woods. We typically accumulate knowledge in a hierarchical nature. For example, before you can explain back emf in a motor, you must first be able to identify when electromagnetic induction occurs and describe the factors that determine its magnitude. Plan your responses by following the same logic and scaffold your responses using **IDEA/E**. If the questions require an assessment or evaluation, first **Identify,** then **Describe**, then **Explain** and finish with the **Assessment/Evaluation**. If the question requires an explanation, stop at IDE.

Rote-learned answers are unlikely to receive full marks, so you must relate the concepts of the syllabus back to the question and ensure that you answer the question that is being asked and *not* the question you think they are asking. Planning your responses to include the relevant information and the key terminology will help you avoid writing too much, contradicting yourself, or 'waffling on' and wasting time. If you have time at the end of the paper, go back and re-read your answers.

ABOUT THE AUTHOR

Adam Sloan

A passionate Physics teacher with 30 years' experience, Adam Sloan has taught extensively in New South Wales as well as the United Kingdom and New Zealand. He has been a Head of Year for much of that time and more recently Assistant Head of Department at Knox Grammar School. An experienced HSC marker and an author of Physics resources, Adam has a well-developed appreciation for the Physics curriculum and how it can be taught and learned.

CHAPTER 1 MODULE 5: ADVANCED MECHANICS

Module summary 2
1.1 Projectile motion 5
1.2 Circular motion 7
1.3 Motion in gravitational fields 10
Glossary 15
Exam practice 16

Chapter 1
Module 5: Advanced mechanics

Module summary

Outcomes

On completing this module, you should be able to:

- select and process appropriate qualitative and quantitative data and information using a range of appropriate media
- analyse and evaluate primary and secondary data and information
- solve scientific problems using primary and secondary data, critical thinking skills and scientific processes
- communicate scientific understanding using suitable language and terminology for a specific audience or purpose
- describe and analyse qualitatively and quantitatively circular motion and motion in a gravitational field, in particular, the projectile motion of particles.

Working Scientifically skills

In this module, you are required to demonstrate the following Working Scientifically skills:

- develop and evaluate questions and hypotheses for scientific investigation
- design and evaluate investigations in order to obtain primary and secondary data and information
- conduct investigations to collect valid and reliable primary and secondary data and information
- select and process appropriate qualitative and quantitative data and information using a range of appropriate media
- analyse and evaluate primary and secondary data and information
- solve scientific problems using primary and secondary data, critical thinking skills and scientific processes
- communicate scientific understanding using suitable language and terminology for a specific audience or purpose.

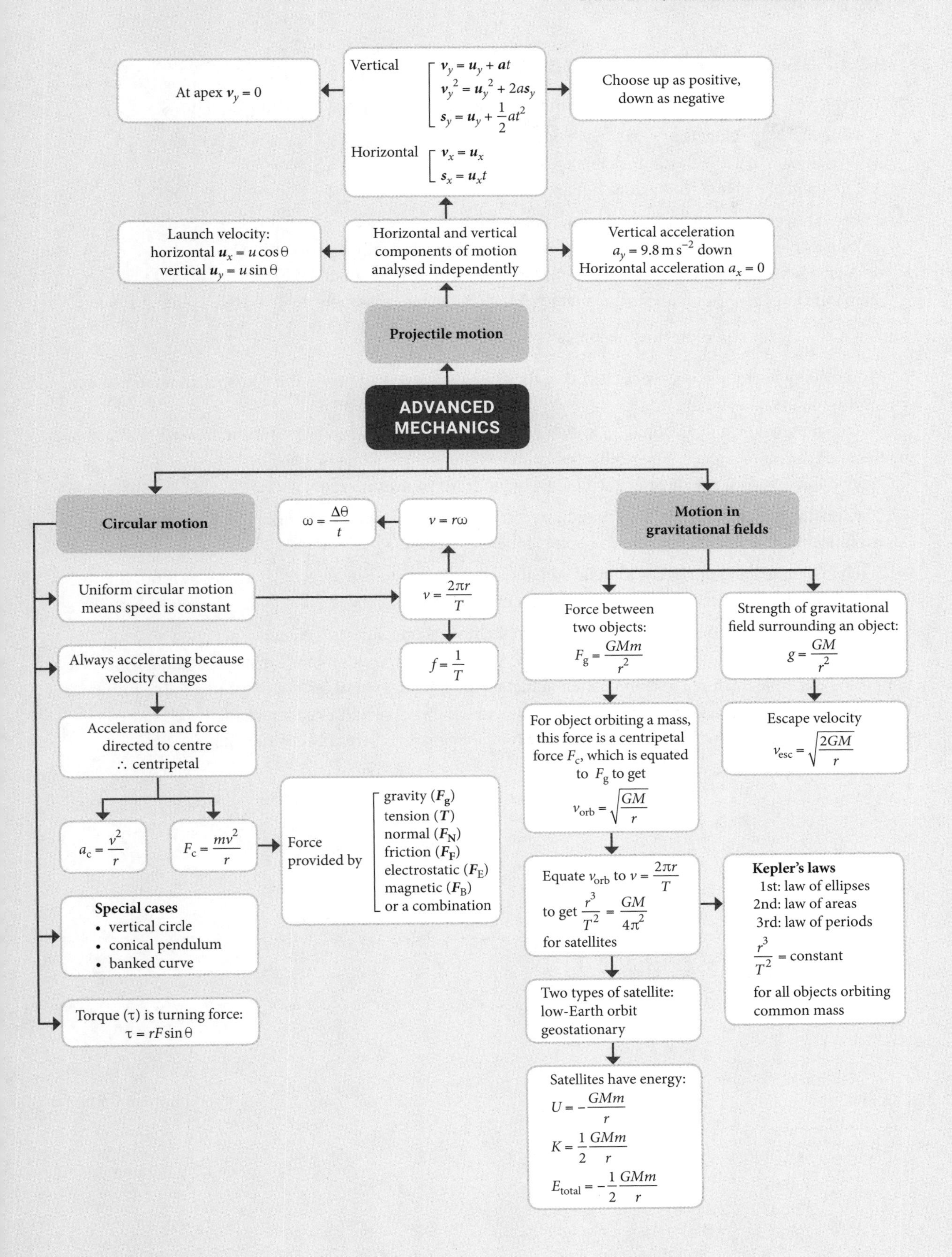

At apex $v_y = 0$
Vertical $v_y = u_y + at$, $v_y^2 = u_y^2 + 2as_y$, $s_y = u_y + \frac{1}{2}at^2$
Horizontal $v_x = u_x$, $s_x = u_x t$
Choose up as positive, down as negative
Launch velocity: horizontal $u_x = u\cos\theta$ vertical $u_y = u\sin\theta$
Horizontal and vertical components of motion analysed independently
Vertical acceleration $a_y = 9.8\text{ m s}^{-2}$ down Horizontal acceleration $a_x = 0$
Projectile motion
ADVANCED MECHANICS
Circular motion
$\omega = \frac{\Delta\theta}{t}$
$v = r\omega$
Motion in gravitational fields
Uniform circular motion means speed is constant
$v = \frac{2\pi r}{T}$
$f = \frac{1}{T}$
Always accelerating because velocity changes
Acceleration and force directed to centre ∴ centripetal
$a_c = \frac{v^2}{r}$
$F_c = \frac{mv^2}{r}$
Force provided by gravity (F_g) tension (T) normal (F_N) friction (F_F) electrostatic (F_E) magnetic (F_B) or a combination
Special cases
• vertical circle
• conical pendulum
• banked curve
Torque (τ) is turning force: $\tau = rF\sin\theta$
Force between two objects: $F_g = \frac{GMm}{r^2}$
Strength of gravitational field surrounding an object: $g = \frac{GM}{r^2}$
For object orbiting a mass, this force is a centripetal force F_c, which is equated to F_g to get $v_{orb} = \sqrt{\frac{GM}{r}}$
Escape velocity $v_{esc} = \sqrt{\frac{2GM}{r}}$
Equate v_{orb} to $v = \frac{2\pi r}{T}$ to get $\frac{r^3}{T^2} = \frac{GM}{4\pi^2}$ for satellites
Kepler's laws
1st: law of ellipses
2nd: law of areas
3rd: law of periods
$\frac{r^3}{T^2}$ = constant
for all objects orbiting common mass
Two types of satellite: low-Earth orbit geostationary
Satellites have energy:
$U = -\frac{GMm}{r}$
$K = \frac{1}{2}\frac{GMm}{r}$
$E_{total} = -\frac{1}{2}\frac{GMm}{r}$

Recall assumed knowledge

In Year 11, you analysed motion in one dimension (on a line) using the following equations:

$v = u + at$ Note that s does not appear.

$v^2 = u^2 + 2as$ Note that t does not appear.

$s = ut + \frac{1}{2}at^2$ Note that v does not appear.

The notes above help work out which equation to apply!

In one dimension, using vectors is easy. It is just a forward or reverse direction (+ or −).

In Year 12, you will need to analyse motion in two dimensions (in a plane).

Newton's laws of motion are encapsulated in the first equation above: $v = u + at$. Recall that $F = ma$, and so $a = \frac{F}{m}$. This equation then becomes $v = u + \frac{F}{m}t$.

By applying vectors, it becomes clear that the final velocity of a body is the result of an applied force changing the initial velocity.

In one dimension, the vectors $\vec{v}$, $\vec{u}$ and $\vec{F}$ are all on the same line, so $\vec{u}$ is either increased or decreased by the application of a force. The body continues to move on the same line as $\vec{u}$.

In two dimensions, the same equations apply but must be considered vectorially.

- If a constant force is applied and maintained at an angle to $\vec{u}$, then $\vec{v}$ changes in both magnitude and direction with t, and the path of the object follows a parabola.
- If a constant force is applied and maintained perpendicular to the object's changing motion, then $\vec{v}$ changes only in direction with t and the path of the object is a circle.

(Extra for experts! Look up 'conic sections' to see how these paths are related.)

Nature is complex. There are many variables! An approach to understanding is through analysis of simplified examples. Simplification is often achieved by ignoring variables that have negligible influence (e.g. friction) and/or considering special situations in which some variables are virtually constant (e.g. small changes in altitude within which the force of gravity is virtually constant).

9780170465304

1.1 Projectile motion

The motion of an object above Earth's surface, where it is moving in a uniform gravitational field and is therefore under the influence of gravity, is described as **projectile** motion. This field exerts a force straight down on the object, in the direction of the gravitational field lines. The principles at work in projectile motion are relatively simple to understand, but the applications can be more complex.

Typical examples of objects moving in projectile motion are a football being kicked, an arrow fired, a tennis ball struck, a marble rolling off a bench, a rubber ball being thrown, or an apple being dropped from a moving car.

1.1.1 Using horizontal and vertical components

An object launched into projectile motion can be described as having an **initial velocity**, u, and a **launch angle**, θ. Analysis of the motion of the projectile throughout its **trajectory** can be done by separating the motion into vertical and horizontal components. These two aspects of the motion will be independent of each other.

Resolving the initial velocity vector yields an **initial horizontal velocity** $u_x = u\cos\theta$ and an **initial vertical velocity** $u_y = u\sin\theta$.

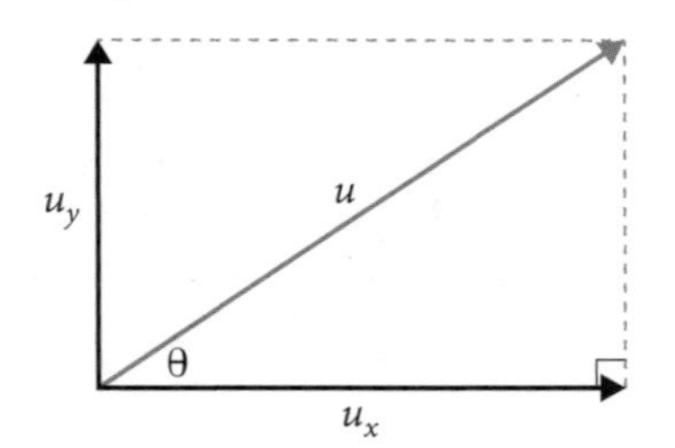

FIGURE 1.1 An object is launched with initial velocity u at an angle θ to the horizontal. Its motion is analysed using the components horizontal velocity u_x and vertical velocity u_y.

Air resistance is generally considered to be negligible and is ignored. There are no significant forces acting horizontally on the object and so its horizontal motion is constant – it will move forwards the same distance each second.

> **Note**
> Always ignore air resistance unless instructed otherwise.

Vertical motion is under the influence of Earth's gravitational field and, therefore, an object is accelerated by the gravitational force at a rate of $9.8\,\mathrm{m\,s^{-2}}$ downwards (referred to as g on the Data and Formulae sheets).

> **Note**
> Two objects set in motion simultaneously, one launched horizontally ($u_y = 0$) from a height and the other dropped ($u_y = 0$) from the same height, will land simultaneously, although at separate points, because their different horizontal velocities do not affect their vertical motions, which are identical.

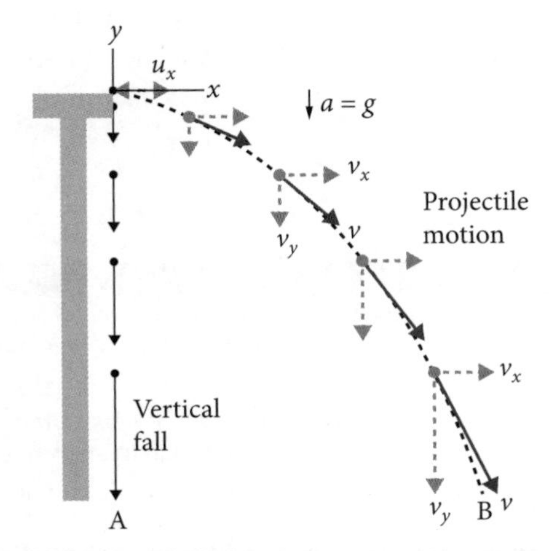

FIGURE 1.2 An object is launched with an initial horizontal velocity at the same time as another is dropped from the same location. They land at the same time but in different places.

1.1.2 Trajectory of projectiles

The sum of this constant horizontal motion and the accelerated vertical motion creates a trajectory that is some portion of a parabolic arc. The proportion and shape of this arc will depend on the launch velocity and launch angle and on where the projectile lands relative to the launch position.

There are five possible projectile motion situations that can be described under these conditions.

> **Note**
> There is symmetry in the motion of the projectile either side of the **maximum height** position in the first three situations only (Figure 1.3a–c), but the amount of symmetry is determined by the landing point relative to the launch point.

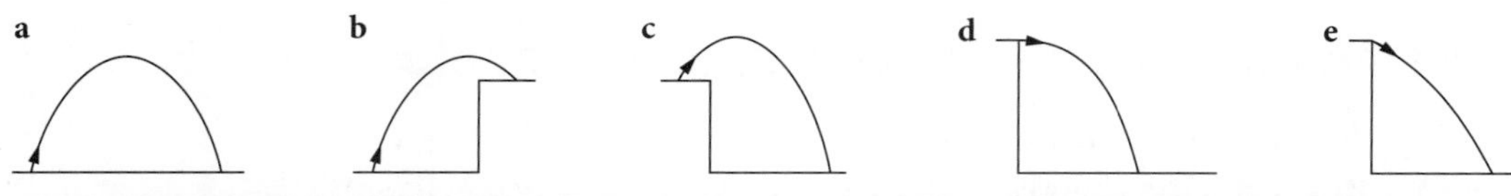

FIGURE 1.3 All the various forms of projectile motion can be grouped into one of five situation types: **a** launched upwards and lands at the same level; **b** launched upwards and lands at a higher level; **c** launched upwards and lands at a lower level; **d** launched horizontally and lands at a lower level; **e** launched downwards and lands at a level below.

1.1.3 Analysing projectile motion with equations

The three equations of motion from Year 11 (and found on the Formulae sheet) can be used to quantitatively analyse all aspects of the motion of a projectile:

$$v = u + at$$
$$v^2 = u^2 + 2as$$
$$s = ut + \frac{1}{2}at^2$$

Care must be taken to consider the way in which the equations are applied to achieve the required outcome.

Since horizontal motion (the x component) is constant ($a = 0$), the above equations can be rewritten for the horizontal aspect of the motion. A subscript x is usually used to indicate that these equations refer only to the horizontal motion.

$$v_x = u_x$$

and

$$s_x = u_x t$$

> **Note**
> The symbol s_x is usually referred to as the **range**.

For vertical motion (the y component), the projectile is accelerated constantly (where $a = g = 9.8\ \text{m s}^{-2}$ down), so the three equations of motion can be rewritten using a y subscript:

$$v_y = u_y + at$$
$$v_y^2 = u_y^2 + 2as_y$$
$$s_y = u_y t + \frac{1}{2}at^2$$

> **Note**
> The key to using the equations correctly is to decide which direction is positive and which direction is negative and ensure all data is consistently recorded according to this decision. Most people choose up as positive and down as negative.
>
> Many people mistakenly think that the journey of the projectile has to be quantitatively analysed in parts. This is an unnecessary complication because the equations are powerful enough to analyse the entire trajectory.
>
> At the highest point of the trajectory, the vertical velocity, v_y, is zero. If you are analysing only the second part of the motion (after the apex), then this means $u_y = 0$. If you are analysing only the first part of the motion (before the apex), then this means $v_y = 0$. This latter process is handy to determine maximum height.
>
> If substitution into the formula results in a quadratic equation, having used $s_y = u_y t + \frac{1}{2}at^2$, use the quadratic formula or just use the other two equations sequentially.
>
> Time of flight is determined by the acceleration due to gravity, the vertical displacement, and the initial vertical velocity – all y components of the motion – and so must be calculated from the set of equations with the y subscripts.

1.2 Circular motion

Many objects that are observed or considered in everyday life can be described as moving in circular motion (e.g. the Moon, a seat on a Ferris wheel, a pebble stuck in a cyclist's wheel).

1.2.1 Features of uniform circular motion

Uniform circular motion (UCM) describes the motion of an object following a circular path at constant speed. Although the speed is uniform, the velocity of the object is changing throughout its motion, because of the continuously changing direction. Hence, an object in UCM is accelerating and, from Newton's second law, this must be due to the action of a net force.

In fact, any object travelling in circular motion must do so under the influence of a net force acting towards the centre of the circle. It can be seen that both the net force acting upon the object, and the resultant acceleration, are, therefore, always perpendicular to the **tangential velocity** of the object. This 'pointing to the centre' nature is indicated by the term *centripetal*.

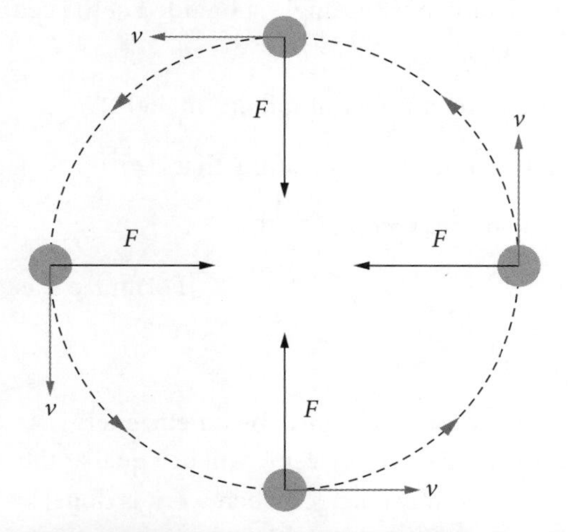

FIGURE 1.4 Wherever the object is in its uniform circular motion, there will be a net force directed towards the centre of the circle, which will always be perpendicular to the object's velocity.

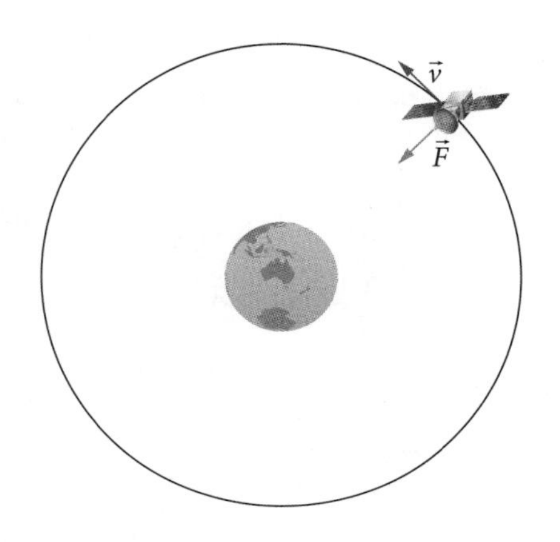

FIGURE 1.5 An orbiting satellite moves in uniform circular motion because gravity provides a centripetal force.

The net force acting on an object moving in UCM, called the **centripetal force**, can be the result of a single force (such as the gravitational force acting on a satellite orbiting Earth) or a combination of forces in more complex examples.

The motion of an object with mass *m* undergoing UCM can be described quantitatively using the terms listed in Table 1.1.

TABLE 1.1 Key quantities of UCM

Term	Symbol	Definition	Unit
Period	T	Time taken to complete one revolution	seconds (s)
Frequency	f	Number of revolutions completed per second	hertz (Hz) or per second (s^{-1})
Angular displacement	θ	Angle through which an object has moved	radian (rad)
Angular velocity	ω	Rate at which an object is moving through its arc	radians per second ($rad\,s^{-1}$)
Tangential velocity	v	Speed of an object at any point on a circle	metres per second ($m\,s^{-1}$)
Centripetal acceleration	a_c	Rate of change of velocity	metres per second per second ($m\,s^{-2}$)
Centripetal force	F_c	Net force responsible for causing object to undergo UCM	newton (N)

CHAPTER 1

Analysis of UCM utilises the relationships below:

- Period is the time per revolution, and frequency is the number of revolutions in a period of time (s), so these are reciprocals of each other: $T = \frac{1}{f}$
- The speed (magnitude of velocity) is just distance/time. If the time is chosen as one period, then the distance is the circumference: $v = \frac{2\pi r}{T}$ [Formulae sheet]
- The magnitude of centripetal acceleration can be derived from vector subtraction, but has been given to us: $a_c = \frac{v^2}{r}$ [Formulae sheet]
- From Newton's second law, $F = ma$ and so: $F_c = \frac{mv^2}{r}$ [Formulae sheet]

Now, angular quantities are measured in radians, where one radian is the angle subtended at the centre of a circle from the ends of an arc length that is equal to the radius.

The ratio of circumference to radius for a circle is $\frac{2\pi r}{r}$ or 2π, the number of radians in 360°.

So, the number of radians, θ, corresponding to an arc length l must be $\frac{l}{r}$, meaning that $l = r\theta$.

It follows that the angular velocity, ω, is the number of radians per second $\left(\frac{\Delta\theta}{t}\right)$.

That is, $\omega = \frac{\Delta\theta}{t}$ and so $\omega = \frac{2\pi}{T}$. [Formulae sheet]

Also note that $\omega = \frac{\Delta\theta}{t} = \frac{\Delta l}{rt} = \frac{v}{r}$, and so $v = r\omega$, which is handy to know!

The total energy of any object undergoing UCM will remain constant because the kinetic energy is a constant throughout (constant speed) and the gravitational potential energy value will be equal at the same position in each orbital cycle. Because the energy of an object only changes when work is done, it can be presumed that no work is done in UCM.

1.2.2 Special cases of uniform circular motion

There are several situations in which deeper analysis of UCM is warranted: motion in a **vertical circle**, in a **conical pendulum**, and on a **banked curve**. In each case the centripetal force is provided by combining the actions of two forces and, consequently, vector analysis is required to fully analyse the question.

Vertical circle

Only the conditions at the top and bottom of the circle are considered in this course. As can be seen in Figure 1.6, when an object is moved through a vertical circle, the centripetal force is provided by a combination (and therefore vector sum) of the weight force of the object, F_g, and the tension force, F_T, due to the string or rope pulling on the object. It can be seen that the weight force is constant and downwards and, because centripetal force is constant in magnitude but always directed towards the centre, the tension force will be greatest at the bottom of the circle and least (possibly even zero) at the top. In an alternative vertical circle situation, a vehicle moving through a vertical circle in a 'loop-de-loop' situation would be influenced by a normal force rather than a tension force, but the other aspects of the analysis are the same.

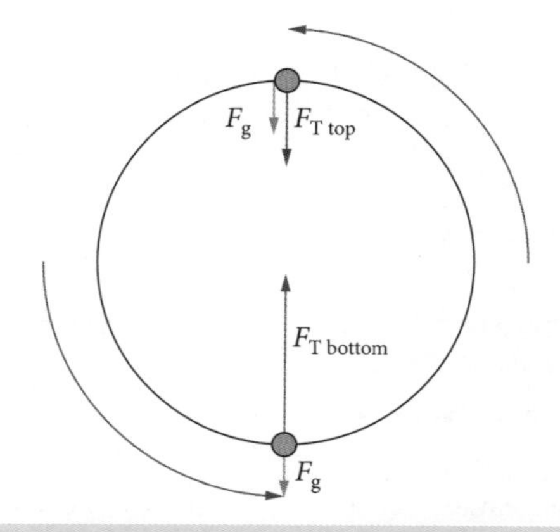

FIGURE 1.6 An object on a string or rope moves in a vertical circle. The centripetal force is constant in magnitude and always directed towards the centre. It is the vector sum of the gravitational force (F_g) and the tension force (F_T).

Note

In reality, it is almost impossible for an object executing motion in a vertical circle to be in uniform circular motion, but it is considered a reasonable approximation for the purposes of the HSC Physics course.

Conical pendulum

As can be seen in the Figure 1.7, when an object is moved through a horizontal circle the centripetal force is again provided by a combination (and therefore vector sum) of the weight force of the object, F_g, and the tension force, F_T. Quantitative analysis of a conical pendulum often involves trigonometry, using the length of the string and the radius of the circle to determine the angle that the string makes with the vertical. This angle is the same as the angle between F_g and F_T and, of course, $F_c = \frac{mv^2}{r}$ applies.

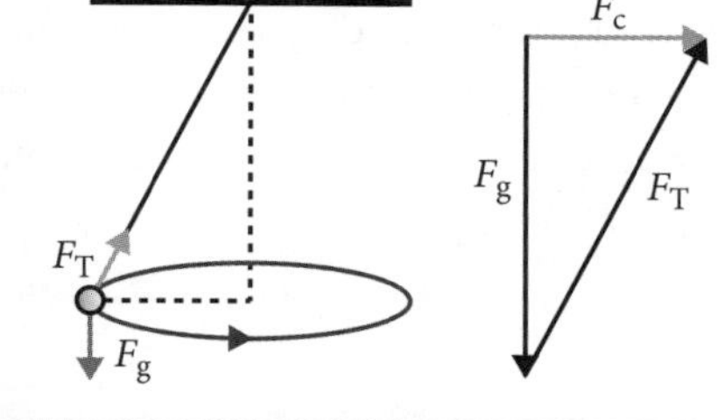

FIGURE 1.7 An object moves in a horizontal circle on the end of a string or rope. The centripetal force is constant in magnitude. The centripetal force (F_c) is the vector sum of the gravitational force (F_g) and the tension force (F_T).

Banked curve

A vehicle moving in a horizontal circle on a flat surface relies on the force of friction to provide the net force towards the centre of the circle. In reality, corners are often banked in order to provide an additional force towards the centre of the circle. This can enable a vehicle to corner with greater safety than if the corner was flat. Often, analysis of a banked curve considers a situation in which there is no friction between the vehicle and the surface and all the force towards the centre of the circle is generated as a consequence of the banking effect.

As can be seen in Figure 1.8, when a vehicle moves through a horizontal circle on a banked curve, the centripetal force is provided by a combination (and therefore vector sum) of the weight force of the object, F_g, and the normal force, F_N. Quantitative analysis of a banked curve usually involves trigonometry, using the relationship between the angle made to the horizontal by the bank and the relationship between F_g and F_N.

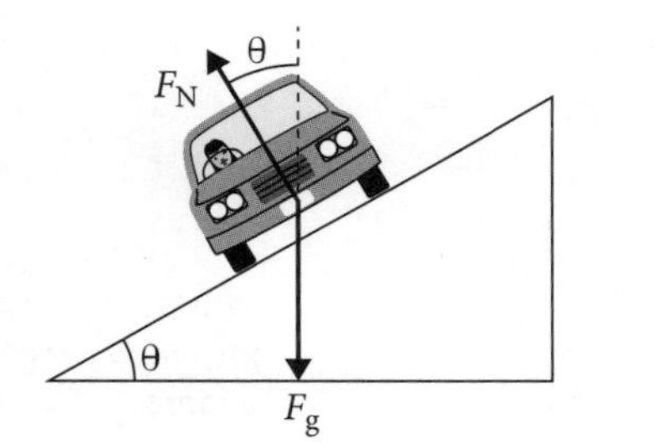

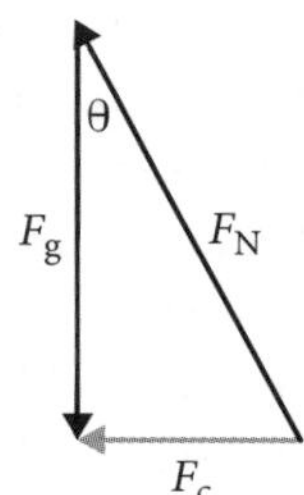

FIGURE 1.8 A car moves in a horizontal circle on a banked curve. The centripetal force (F_c) is constant in magnitude. It is the vector sum of the gravitational force (F_g) and the normal (F_N).

Note

Both banked curves and conical pendulums can also be analysed by resolving the normal force (on the banked curve) and the tension force (of the conical pendulum) vectors respectively to analyse the horizontal component (F_{Nx} and F_{Tx} in each case). This horizontal component is equal to the centripetal force and quantitative analysis will yield the same outcome as the other method. The two methods are equally valid, and it comes down to personal preference and proficiency. In both cases analysis will show that $\tan\theta = \frac{v^2}{rg}$.

1.2.3 Rotation and torque

Torque is a measure of the turning force applied to an object. Torque can result in rotation of an object, such as turning a tap or opening a door. Torque is defined as the product of the applied force and the perpendicular distance between the line of action of the force and the pivot. This is written as:

$$\tau = r_{\perp}F = rF\sin\theta$$

Torque can be increased by increasing any of the three factors in the equation: the distance from the pivot to the point at which the force is applied, the magnitude of the force applied, or the angle between the applied force and the line joining the pivot to the point of application of that force. The torque is greatest when this angle is 90°.

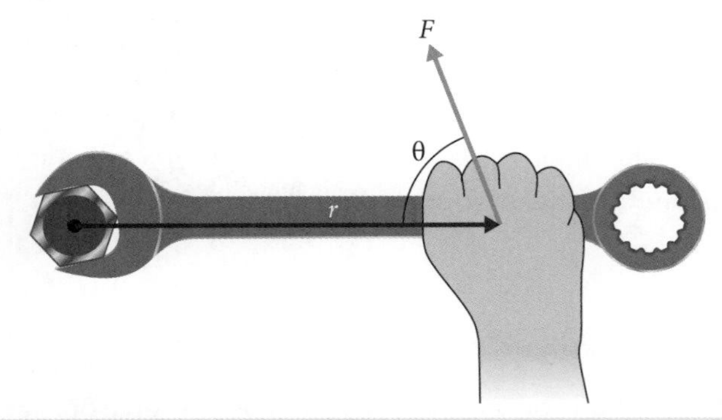

FIGURE 1.9 Torque is applied to the spanner by the hand.

1.3 Motion in gravitational fields

Motion in a **gravitational field** is distinguished from projectile motion effectively by scale. When considering motion in gravitational fields, it is considered that the motion being analysed is the motion of objects in a radial gravitational field (field lines look like spokes of a bicycle wheel), rather than a uniform gravitational field (field lines look like bars of a cage) used in projectile motion. (Clearly, if only a small range of altitude is considered, the 'radial spokes' look much like 'parallel bars'.) A simple rule of thumb is that if a diagram was drawn to represent the situation and in the diagram Earth looks like a circle, then the gravitational field is considered to be radial (Figure 1.10). This physics is used to analyse the motion of such objects as satellites, moons and planets.

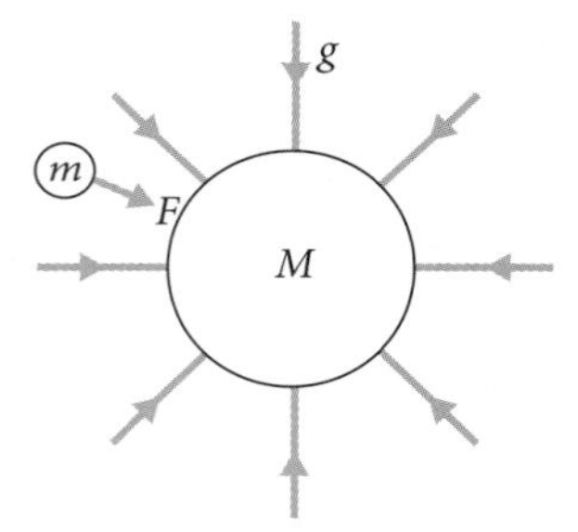

FIGURE 1.10 Mass *M* provides a radial gravitational field in which mass *m* experiences a force.

1.3.1 Gravitational forces and gravitational fields

All objects exert a gravitational force on all other objects according to the relationship described by Newton's law of universal gravitation: $F = \frac{GMm}{r^2}$, where r is the distance between the centres of the two masses M and m, and G is the universal gravitational constant. This gravitational force is always a force of mutual attraction between the masses. Figure 1.11 illustrates the force that M exerts on m.

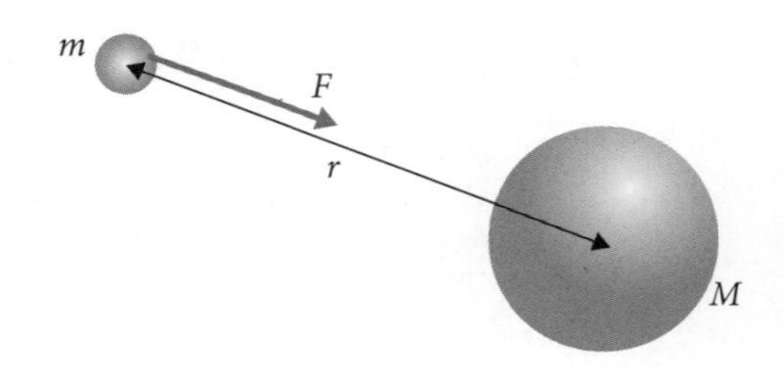

FIGURE 1.11 The mass *M* exerts an attractive gravitational force on mass *m*.

Newton's third law tells us that the force that M exerts on m must be equal in magnitude and opposite in direction to the force that m exerts on M. It can be seen in the equation for the gravitational force that the order of the mass values in the calculation will not affect the result.

As with electrostatic and magnetic forces acting at a distance via fields, we describe this gravitational force as being mediated by, or acting as a result of, a gravitational field. All objects are surrounded by a gravitational field.

The intensity or strength of any field is the ratio of the force exerted to the unit of whatever the field affects (e.g. charge for electric fields, mass for gravitational fields). The strength of a gravitational field is the gravitational force per mass, or $\frac{1}{m} \times \frac{GMm}{r^2} = \frac{GM}{r^2}$.

Since $F = ma$, then $\frac{F}{m} = a$, and so the **gravitational field strength** is the acceleration due to gravity:

$$g = \frac{GM}{r^2}$$

The gravitational field strength is defined as the force per unit mass acting on a test mass in the field and has units of N kg^{-1}. Quantitatively, the gravitational field strength is equivalent to the acceleration due to gravity, so on the surface of Earth $g = 9.8\,\text{N kg}^{-1}$ or $9.8\,\text{m s}^{-2}$. The units N kg^{-1} and m s^{-2} are equivalent. (You should be able to prove why!)

The gravitational field can be drawn as a series of field arrows, as seen in Figure 1.13. Arrows will always point towards the centre of the object. As with electric and magnetic fields, the density of the arrows indicates the relative strength of the field.

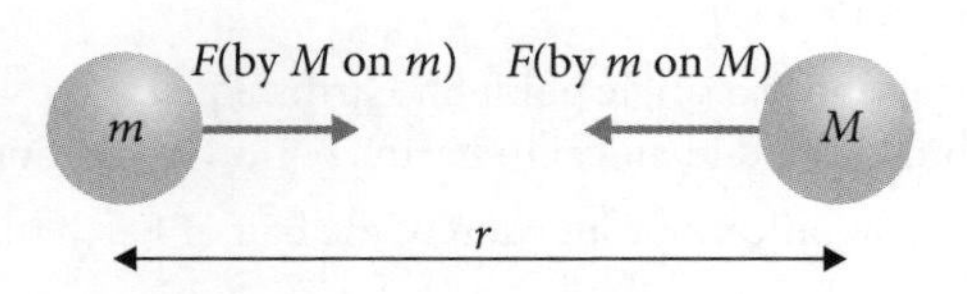

FIGURE 1.12 The gravitational force that mass M exerts on mass m is equal and opposite to the gravitational force that mass m exerts on mass M.

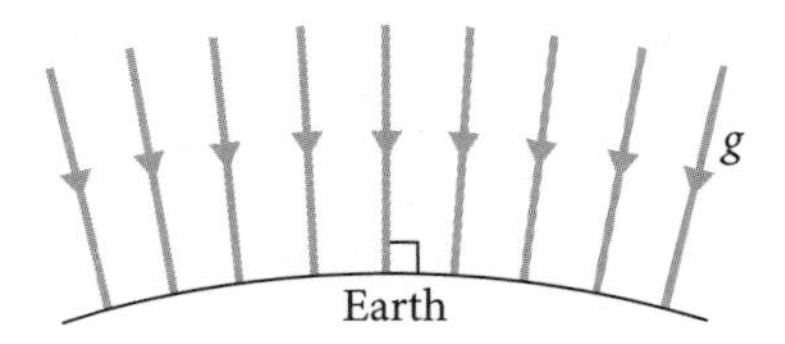

FIGURE 1.13 The gravitational field of Earth shows that field lines are approximately parallel in this close-up version, but arrows point towards Earth's centre.

1.3.2 Satellite motion

The gravitational force provides a centre-directed force (F_c) that can act on a much smaller mass moving perpendicular to that force at a constant speed, which will, therefore, undergo uniform circular motion. This UCM is called an **orbit** (there are non-circular orbits that are not considered in this section), and the object is called a **satellite**.

Equating the centripetal force required for UCM with the gravitational force that provides the centre-directed force, it can be shown that

$$\frac{mv^2}{r} = \frac{GMm}{r^2}$$

and by making velocity the subject, it is possible to derive an equation for the **orbital velocity** necessary for a satellite to sustain a stable orbit: $v = \sqrt{\frac{GM}{r}}$. Note that this shows that the orbital velocity is independent of the mass of the orbiting object.

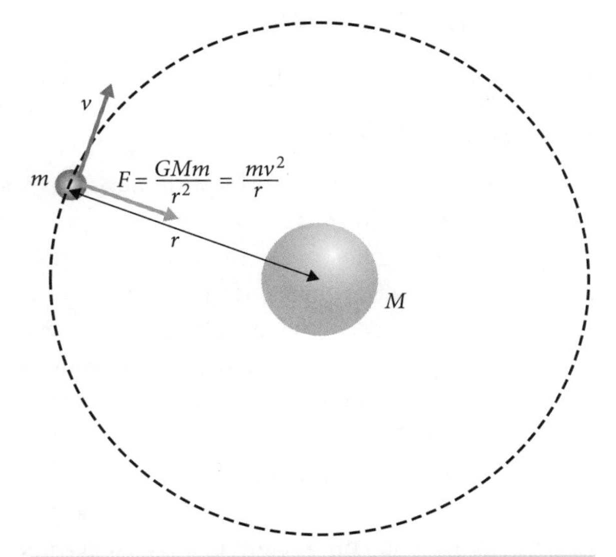

FIGURE 1.14 The gravitational force provides a centripetal force acting on an orbiting object.

Furthermore, the centripetal acceleration is given by $a_c = \frac{v^2}{r} = \frac{GM}{r^2} = g$ and g was shown to be equal to the acceleration due to gravity in section 1.3.1. Effectively, this means that orbiting objects are in freefall – constantly falling, but never getting any closer to the central mass because of their velocity.

Low-Earth orbits and geostationary orbits

There are many types of artificial satellites that are used for a range of purposes, but in this course the focus is on two groups – satellites in a **low-Earth orbit** and satellites in a **geostationary orbit** – and the link between their orbital properties and applications (seen in Table 1.2).

TABLE 1.2 The orbital properties and applications of low-Earth orbit satellites and geostationary satellites

Satellite type	Distance above surface (km)	Orbital period	Detail and applications
Low-Earth orbit	150–2000	90–160 minutes	• Has a range of orbital radii • Has detailed coverage of very small proportion of the part of Earth above which it is located at any moment; however, by doing multiple laps of Earth per day, data volume is high • At a low altitude, thus high-resolution data collection is viable, but significant atmospheric drag is experienced • Used in remote sensing, such as environmental monitoring and meteorological forecasting, and espionage
Geostationary orbit	35 786	24 hours	• Has a specific orbital radius • Designed to stay above a single point on Earth's equator by having an orbital period identical to Earth's period of rotation • Has coverage of significant proportion of the half of Earth that is facing its way • At a high altitude, so high-resolution data collection is difficult, but there is no atmospheric drag • Used in communications, meteorological studies and defence

1.3.3 Kepler's laws

Johannes Kepler proposed three powerful astronomical laws based on analysis of substantial collected data.

Kepler's first law

All planets move in elliptical orbit with the Sun as one focus.

An **ellipse** is an oval-like shape made when a point is moved in a plane such that the sum of the distances of that point from two points (the two foci) is constant. Circles are specialised ellipses in which the two foci are at the same location. The eccentricity of an ellipse is a measure of how stretched or distorted the circle is. Most planetary orbits are not very eccentric and, hence, we can treat them as circular at this level.

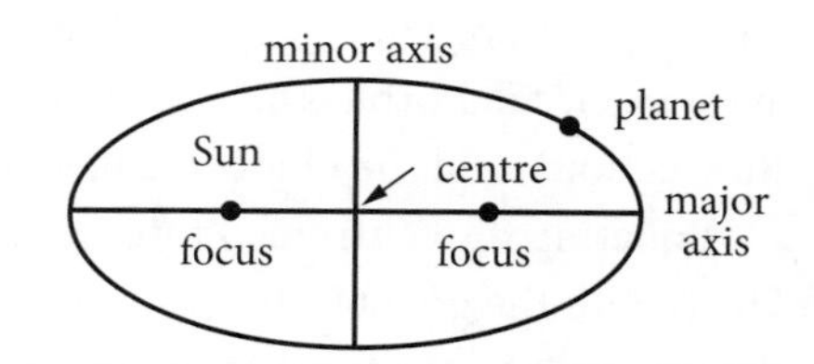

FIGURE 1.15 Kepler's first law, the law of ellipses, illustrates that the path of the planets about the Sun is elliptical in shape, the Sun being at one focus.

The total energy (sum of kinetic energy, K, and gravitational potential energy, U) of an object moving in a stable elliptical orbit is constant, although both K and U will change.

Kepler's second law

A line joining a planet to the Sun sweeps out equal areas in equal time periods.

The velocity of a planet changes throughout its orbit – it moves faster the closer it is to the Sun.

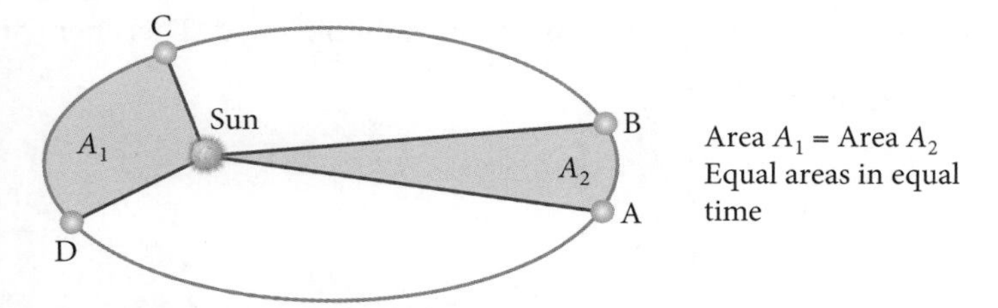

FIGURE 1.16 Kepler's second law states that segments of equal area AB and CD are swept out in equal time intervals.

Kepler's third law

Kepler found a relationship between the orbital radius of a planet and its period. Although the orbits of planets are elliptical, a circle is a special case of an ellipse, so it is reasonable to explore the relationship for satellite motion:

$$\frac{mv^2}{r} = \frac{GMm}{r^2}$$

Simplifying: $$v^2 = \frac{GM}{r}$$

$$\frac{4\pi^2 r^2}{T^2} = \frac{GM}{r}$$

$$\frac{r^3}{T^2} = \frac{GM}{4\pi^2} \quad \text{[Formulae sheet]}$$

> **Note**
> If a question involves two objects orbiting a common mass and the solution involves the ratio $\frac{r^3}{T^2}$, it is not necessary to convert all data into SI units. As long as the *T* and *r* data for both masses use the same units then the process is valid.

The implication of this law is that a collection of objects orbiting a common central mass (such as the planets orbiting the Sun) will all have the same value for the ratio of their orbital radius cubed to their orbital period squared: $\frac{r^3}{T^2}$.

1.3.4 Satellite energy

In Year 11, you learned that **gravitational potential energy** (U) was given by the equation $U = mgh$. This equation was adequate for objects near Earth's surface, but for objects further from the surface the value of g decreases and this approximation is no longer adequate.

A superior analysis of gravitational potential energy is made possible by defining a position at an infinite distance from a planet at which the gravitational potential energy must be zero. Because work must be done to move an object from within the gravitational field of the planet to this point, it can be said that the gravitational potential energy at any point must be negative. Using a mathematical technique called integration, which is beyond this course, an equation can be derived for the gravitational potential energy of an object in a gravitational field: $U = -\frac{GMm}{r}$.

If the object is moving, then it also has kinetic energy given by $K = \frac{1}{2}mv^2$.

If this object is in orbit, then its orbital velocity, as was seen in section 1.3.2, can be given by $v = \sqrt{\frac{GM}{r}}$.

Combining these two equations, we can derive an equation for kinetic energy in terms of m and M: $K = \frac{1}{2} \times \frac{GMm}{r}$. It can be seen then that the kinetic energy has half the magnitude of the gravitational potential energy for an orbiting object.

Total energy can be determined by adding the gravitational potential energy and the kinetic energy:

$$U + K = -\frac{GMm}{2r}$$

Representing the energy values of a satellite at a range of orbital radii graphically reveals three related curves for which energy is inversely proportional to orbital radius in each case. Note that K is proportional to $\frac{1}{r}$ and U is proportional to $-\frac{1}{r}$.

If a satellite's orbit changes, then energy must be supplied or liberated. This energy is calculated easily from the difference in total energy, $-\frac{GMm}{2r}$.

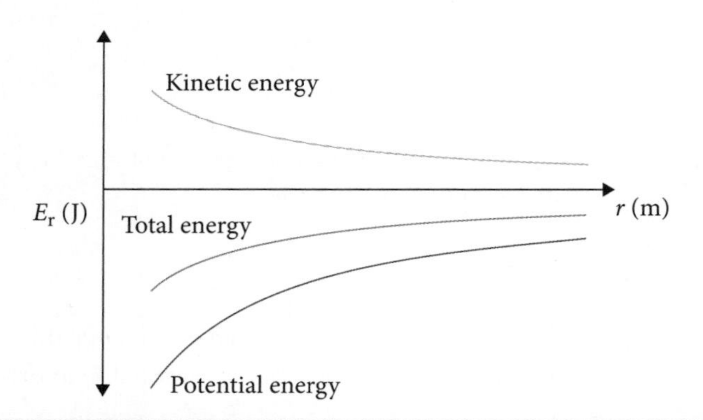

FIGURE 1.17 Graphs of gravitational potential energy (U), kinetic energy (K) and total energy (T) of a satellite at various orbital radii. Note that (i) all three curves represent the inversely proportional nature, (ii) the kinetic energy curve and total energy curve are equal in magnitude at all points and both have half the magnitude of the gravitational potential energy curve at all points, and (iii) both the gravitational potential energy and total energy curves are always negative.

Escape velocity

The magnitude of the gravitational potential energy of an object on the surface of a planet is equal to the amount of energy required to completely remove that object from the planet's gravitational field. Given enough initial kinetic energy, an object would be able to escape the gravitational field of the planet. The minimum speed required for an object to have sufficient kinetic energy to escape a planet's gravitational field is called the **escape velocity**, v_{esc}.

This can be derived by equating kinetic energy with the energy required to escape the gravitational field:

$$K = \frac{1}{2}mv^2 = \text{absolute value of } U = \frac{GMm}{r}$$

So $v^2 = \frac{2GM}{r}$

and so $v_{esc} = \sqrt{\frac{2GM}{r}}$.

Note

Escape velocity has no practical application but does provide a way of comparing the gravitational field strength of bodies from which an object may be launched.

Glossary

angular displacement The angle, in radians, through which an object has moved in its circular arc; symbol θ

angular velocity The rate, in radians per second, at which an object is moving through its circular arc; symbol ω

banked curve A circular surface for which the outside edge is higher than the inside edge

centripetal acceleration The rate of change of velocity of object undergoing uniform circular motion, in metres per second per second; symbol a_c

centripetal force The net force responsible for causing object to undergo uniform circular motion; symbol F_c

conical pendulum A mass moving in a horizontal circle at the end of a length of string

ellipse A shape created by tracing an arc such that the sum of the distances from two given points (foci) is constant

escape velocity The minimum initial velocity of an object sufficient to permanently escape a gravitational field

frequency The number of revolutions completed per second; symbol f

geostationary orbit An orbit in which the satellite maintains a constant position above a point on the equator

gravitational field The region around a massive object in which other masses experience a gravitational force

gravitational field strength The gravitational force per kilogram acting on a mass

gravitational potential energy The potential energy of a system due to the gravitational forces exerted on an object in a gravitational field

initial horizontal velocity The horizontal component of the velocity at which a projectile is launched; symbol u_x

initial velocity The velocity at which a projectile is launched; symbol u; a vector with components u_x and u_y

A+ DIGITAL FLASHCARDS
Revise this topic's key terms and concepts by scanning the QR code or typing the URL into your browser.

https://get.ga/aplus-hsc-physics-u34

initial vertical velocity The vertical component of the velocity at which a projectile is launched; symbol u_y

launch angle The angle at which a projectile is launched; the angle of the initial velocity, usually given as angle θ

low-Earth orbit An orbit in which a satellite is at an altitude of between approximately 160 km and 2000 km

maximum height The highest point in a projectile's trajectory at which the vertical component of the velocity is zero

orbit A regularly repeating, gravitationally curved trajectory of an object around a star, planet or moon

orbital velocity The velocity at which a satellite can sustain a stable orbit

period Time taken to complete one revolution in seconds; symbol T

projectile An object moving freely under the influence of gravity after launch

range The final horizontal displacement of a projectile

satellite An object that orbits a central body

tangential velocity The linear speed of any object moving along a circular path

torque A turning force; symbol τ

trajectory The path followed by a projectile; parabolic in nature

uniform circular motion Motion in a circle at a constant speed

vertical circle Motion in a circle for which the top and bottom of the arc are directly vertically aligned with the centre of motion

Exam practice

Multiple-choice questions

Solutions start on page 144.

Projectile motion

Question 1

A tennis ball is struck at 40 m s^{-1} at an angle of 20° to the horizontal. Which value is closest to the ball's maximum height above its launch?

A 72 m **B** 0.7 m **C** 1.9 m **D** 9.5 m

Question 2

A golf ball follows a path like the one shown below. Four points on the journey of the golf ball are marked.

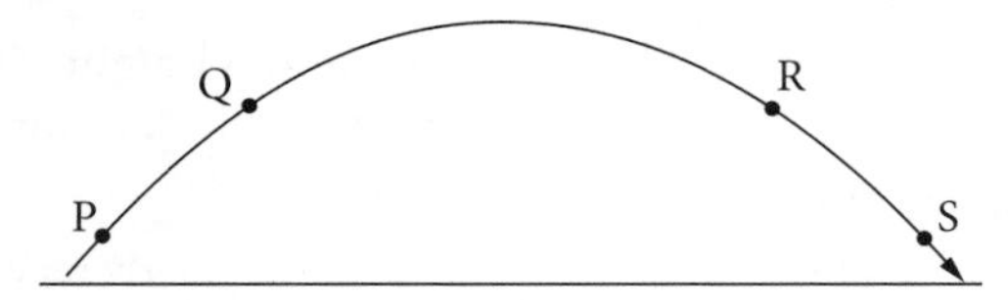

Which option best indicates the horizontal and vertical velocity vectors at each of the four positions?

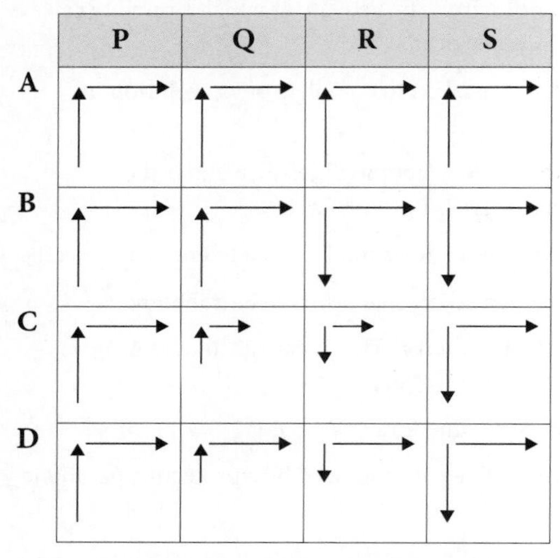

Question 3

Two balls (ball X and ball Y) are kicked at the same velocity from the ground on a flat surface. Ball X is kicked at 30° above horizontal and ball Y is kicked at 60° above horizontal. Which statement is true?

Hint
If you are struggling with this, choose a simple velocity (such as 10 m s^{-1}) and, by substituting, apply the equations of motion to determine values for range and maximum height and compare.

A Ball Y will have a greater range but a smaller maximum height than ball X.

B Ball X will have the same range but a smaller maximum height than ball Y.

C Balls X and Y will have the same range and maximum height as each other.

D Ball X will have a greater range but a smaller maximum height than ball Y.

Question 4

A projectile is launched at an angle of 45° above the horizontal on flat ground. Data about its range, maximum height and time of flight is collected. The projectile is then launched at the same initial velocity from the same location at an angle of 60° above the horizontal. How would the data that is collected have changed?

	Range	Maximum height	Time of flight
A	Increase	Decrease	Decrease
B	Decrease	Increase	Increase
C	Increase	Increase	Increase
D	Decrease	Decrease	Increase

Question 5

Which three factors will determine the time of flight of a projectile launched and landing on flat ground?

A Launch angle, launch velocity, acceleration due to gravity

B Initial horizontal velocity, range, final horizontal velocity

C Initial vertical velocity, acceleration due to gravity, maximum vertical displacement

D Acceleration due to gravity, range, initial vertical velocity

Question 6

A hot air balloon is moving horizontally at 10 m s^{-1} at an altitude of 300 m above a flat park. As the balloon passes over a picnic blanket in the park, a sandbag falls from the balloon. How far from the picnic blanket will the sandbag land?

A 612 m **B** 300 m **C** 3000 m **D** 78 m

Question 7

Consider the points marked P and Q on the path of the projectile in the diagram. Which of the statements about the motion of the projectile is correct?

A The projectile is moving up more slowly at P than it is moving down at Q.

B The total energy at P is less than the total energy at Q.

C The gravitational acceleration is greater at P than at Q.

D The horizontal component of the velocity is greater at P than at Q.

Question 8

Two heavy balls (P and Q) are rolled off the same cliff at the same time. The diagram shows their trajectories.

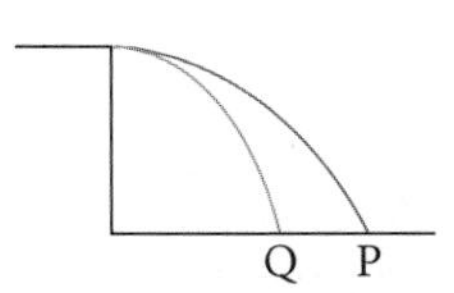

Which statement about their trajectories is true?

A P and Q land at the same time at the same speed.

B P lands before Q because it rolled off the cliff at greater speed.

C P and Q land at the same time but at different speeds.

D Q lands before P because it has travelled a shorter distance.

Question 9

An object is launched vertically on horizontal flat ground and lands at the same level from which it is launched. Which graph best represents the speed of the object during its time of flight?

A
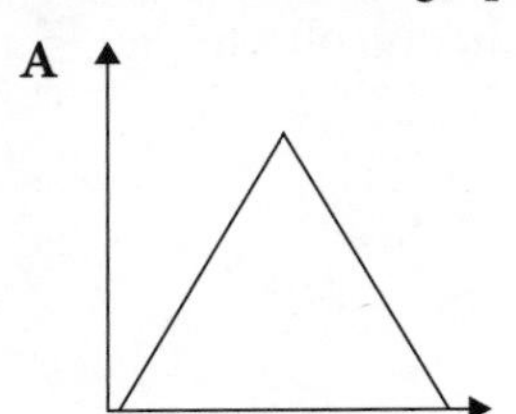

B
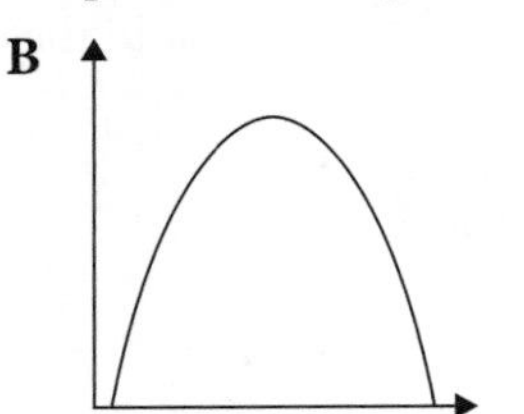

C
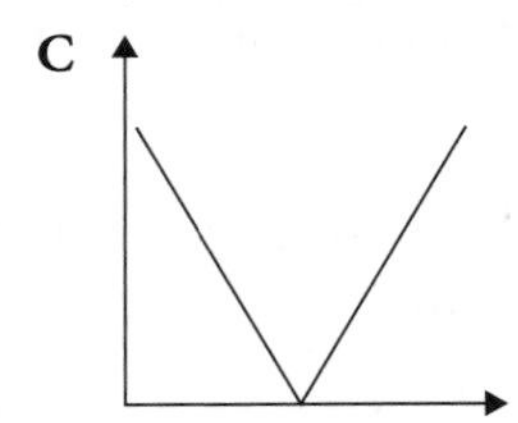

D
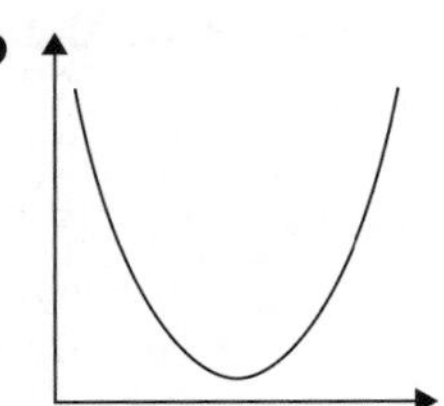

Question 10 ©NESA 2008 SI Q4

An investigation was performed to determine the acceleration due to gravity. A ball was dropped from various heights and the time it took to reach the ground from each height was measured. The results were graphed with the independent variable on the horizontal axis.

Which graph best represents the relationship between the variables?

A
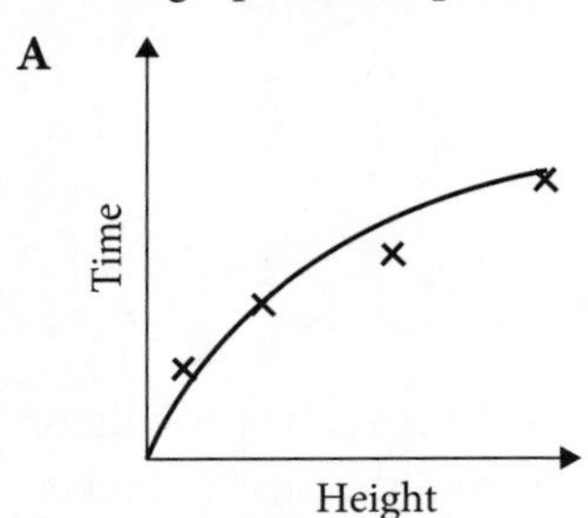

B
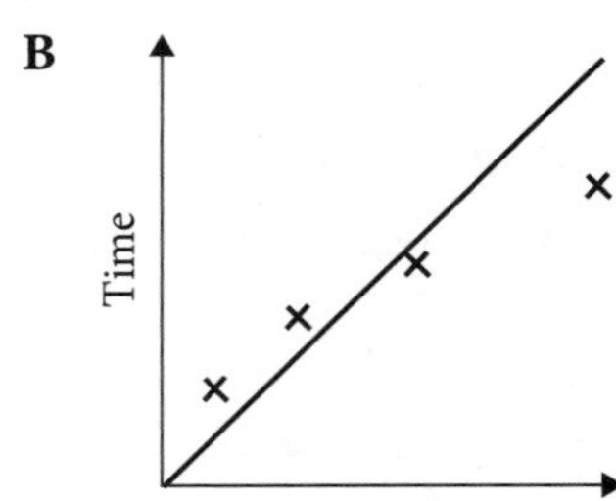

C
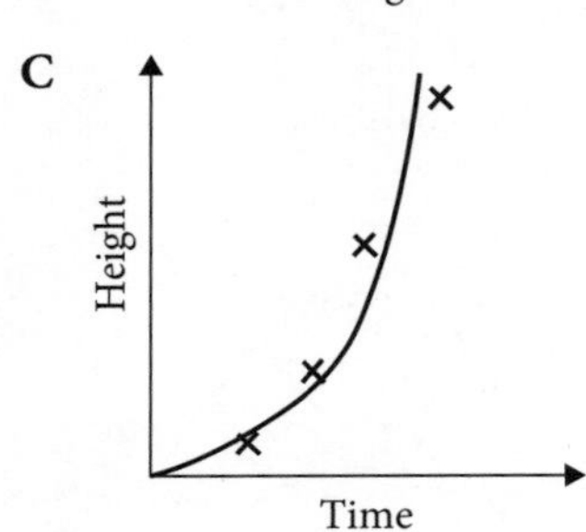

D
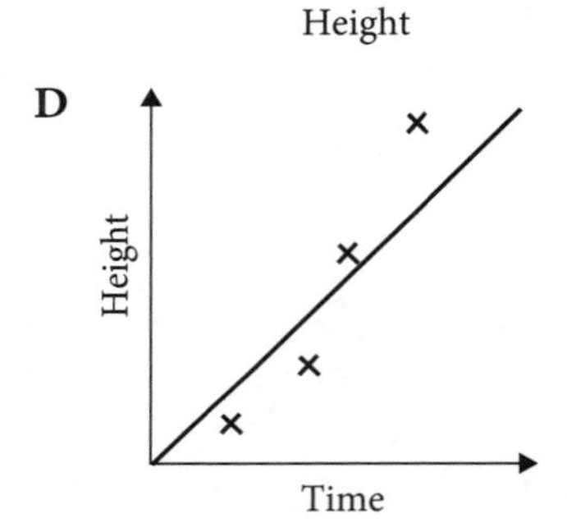

Question 11

A ball is dropped and bounces a few times. The velocity of the ball was plotted as a function of time. Which type of graph is the best representation of the motion?

> **Hint**
> Consider carefully what the gradient of the *v* vs *t* graph represents and what the regions above and below the *t* axis represent.

A
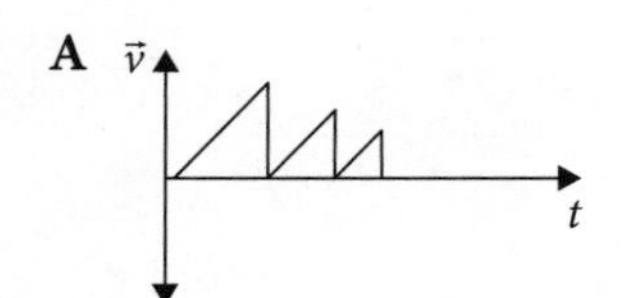

B
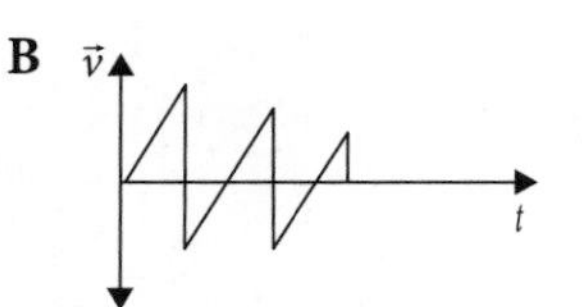

C
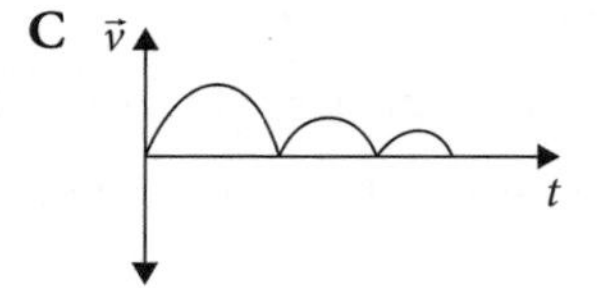

D
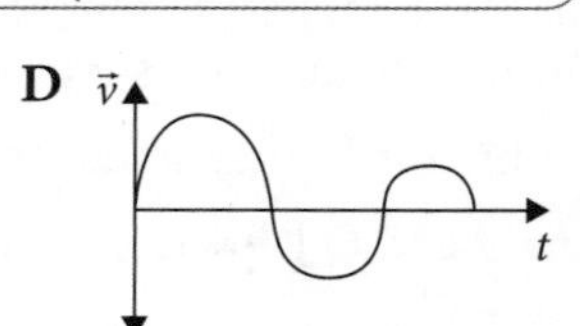

Question 12

An object is fired horizontally from a surface with an initial velocity of u. At the stage of its flight when it is moving at 45° to the horizontal, its velocity is now closest to

A $1.4u$ **B** $2u$ **C** u **D** u^2

Question 13 ©NESA 2016 SI Q17

A projectile was launched horizontally inside a lift in a building. The diagram shows the path of the projectile when the lift was stationary.

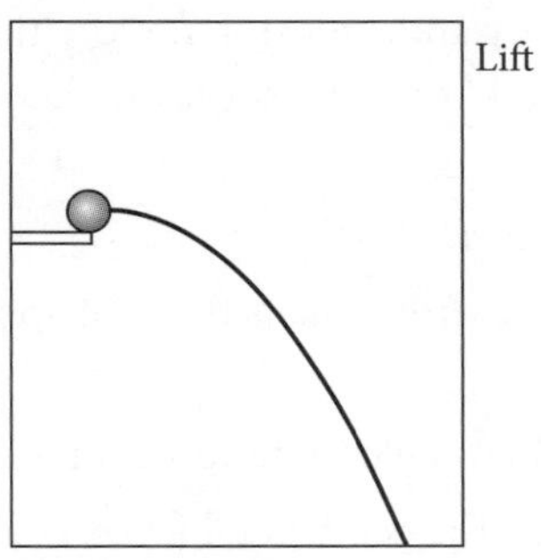

The projectile was launched again with the same velocity. At this time, the lift was slowing down as it approached the top floor of the building.

Which diagram correctly shows the new path of the projectile (dotted line) relative to the path created in the stationary lift (solid line)?

A

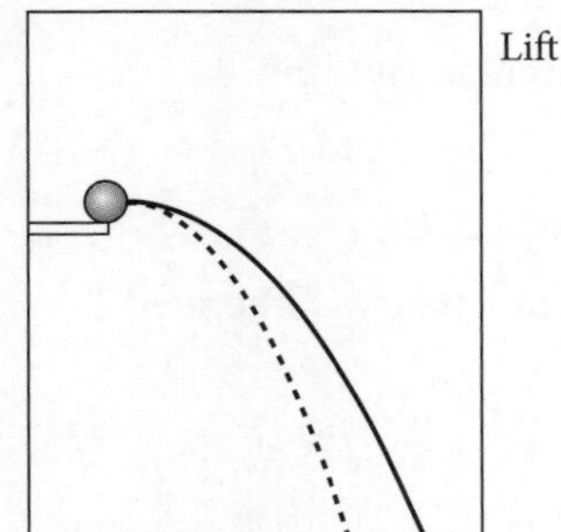

B

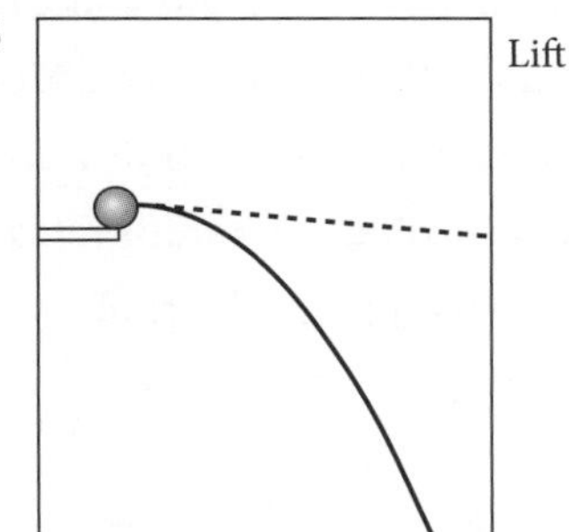

C

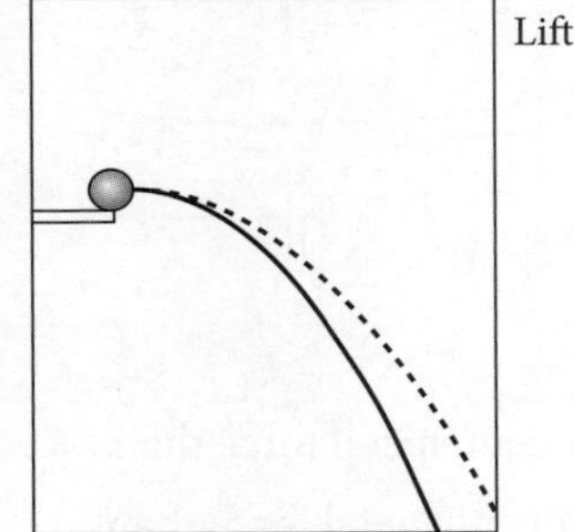

D

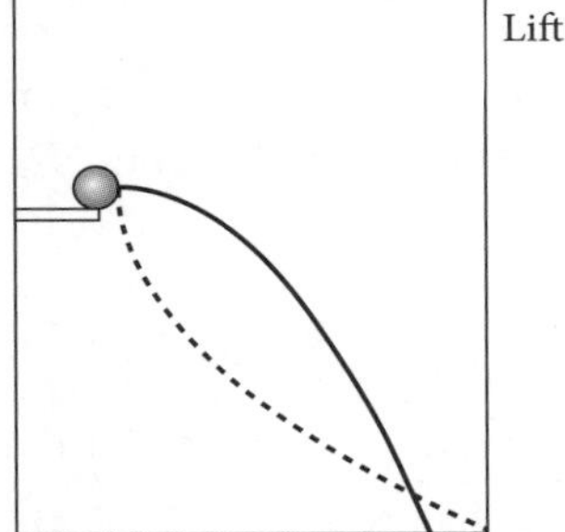

Circular motion

Question 14

Angela notes that two-thirds of a minute after her carriage was at the highest point of the Ferris wheel, it was at the lowest point. The best estimate of the frequency of the Ferris wheel is

A 80 s **B** 40 s **C** 0.0125 Hz **D** 0.025 Hz

Question 15

If the Ferris wheel in Question 14 has a diameter of 45 m, what is the speed of one of the carriages?

A 1.8 m s^{-1}

B 3.5 m s^{-1}

C 7.1 m s^{-1}

D It is not possible to calculate it from this data.

Question 16

Consider a mass swung in uniform circular motion in a vertical circle on the end of a rope, as shown in the diagram. The object will be under the influence of both a force due to gravity and a tension force acting through the rope. These two forces will add to a net force, which is a centripetal force.

Which statement is true of the situation described?

A The tension force is always in the same direction as the centripetal force.

B The tension force is always smaller than the gravitational force.

C The tension force is always larger than the centripetal force.

D The tension force is constant throughout the motion.

Top

Bottom

Question 17 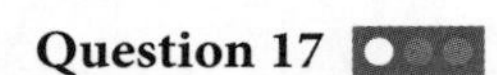

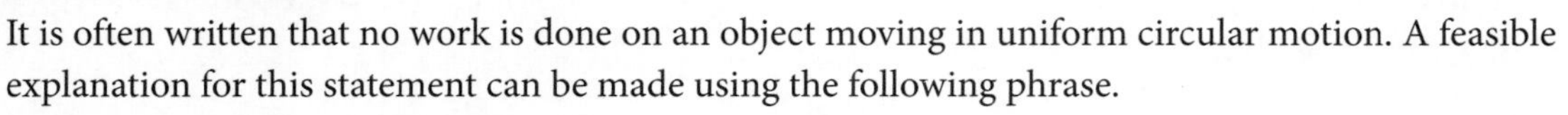

It is often written that no work is done on an object moving in uniform circular motion. A feasible explanation for this statement can be made using the following phrase.

A No work is done because there is no force acting on objects in uniform circular motion.

B No work is done because objects in uniform circular motion don't move.

C No work is done because objects in uniform circular motion don't have any energy.

D No energy is required to keep the object in uniform circular motion, and no energy is liberated by the system.

Question 18 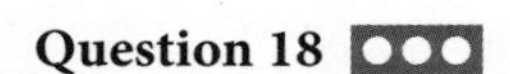

Two identical coins are placed on a rotating turntable as seen in the diagram.

Which option correctly identifies the relative period, velocity and centripetal force of coin M compared to coin N?

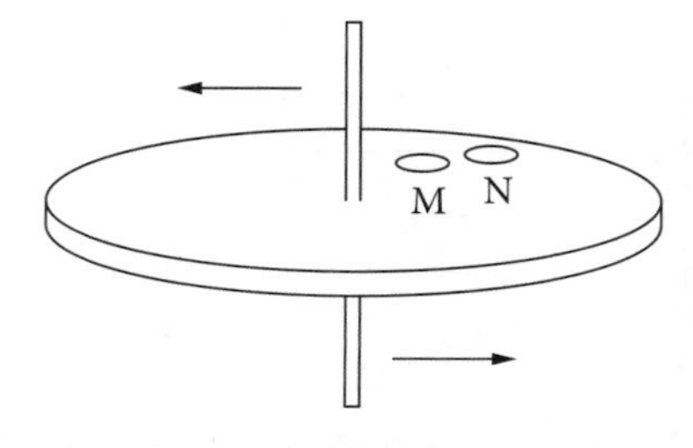

	Period	Velocity	Centripetal force
A	M has same period as N.	N is faster than M.	N requires a greater centripetal force than M.
B	M has longer period than N.	M is faster than N.	N requires a smaller centripetal force than M.
C	M has same period as N.	N is faster than M.	N requires a smaller centripetal force than M.
D	M has same period as N.	N is faster than M.	N requires the same centripetal force as M.

Question 19

In the diagram, two masses, m_1 and m_2, can be seen suspended from a frictionless pulley made from barrels of different radii, r_1 and r_2. If the pulley does not rotate when released, then which of the following statements must be true?

A m_1 must equal m_2.

B r_1 must equal r_2.

C $\frac{m_1}{m_2}$ must equal $\frac{r_1}{r_2}$.

D $m_1 \times r_1$ must equal $m_2 \times r_2$.

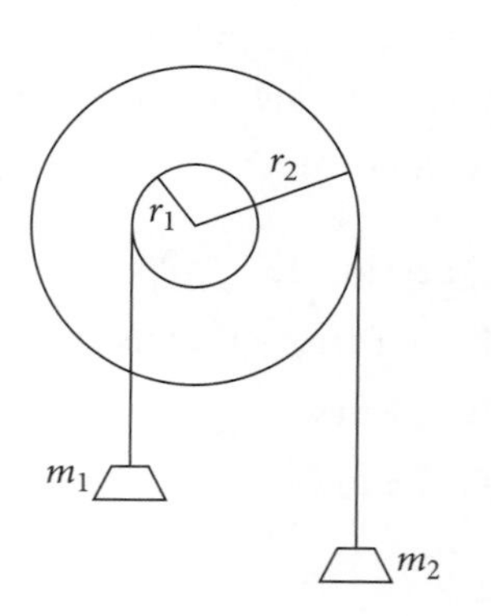

Question 20

At a tomato sauce factory, the empty bottles are gradually filled with sauce as they move in uniform circular motion on a large turntable. Which of the options below correctly identifies changes to the magnitude of the centripetal force acting on and the centripetal acceleration of the tomato sauce bottle during its journey?

	F_c	a_c
A	Increases	Increases
B	Stays same	Increases
C	Increases	Stays same
D	Stays same	Stays same

Question 21

An object of mass m makes n revolutions per second around a circle of radius r at a constant speed. What is the kinetic energy of the object?

A 0 **B** $\frac{1}{2}\pi^2 mn^2r^2$

C $2\pi^2 mn^2r^2$ **D** $4\pi^2 mn^2r^2$

Hint
Start with the equation for kinetic energy and substitute for v in the equation for the speed of an object moving in a circle. Use the information about n to work out the period of the uniform circular motion.

Question 22

An object hangs from a light string and moves in a horizontal circle of radius r.

Hint
Consider the horizontal and vertical forces acting on the mass and derive the equation for tan θ in terms of v, r and g and then substitute the expression for v in terms of ω.

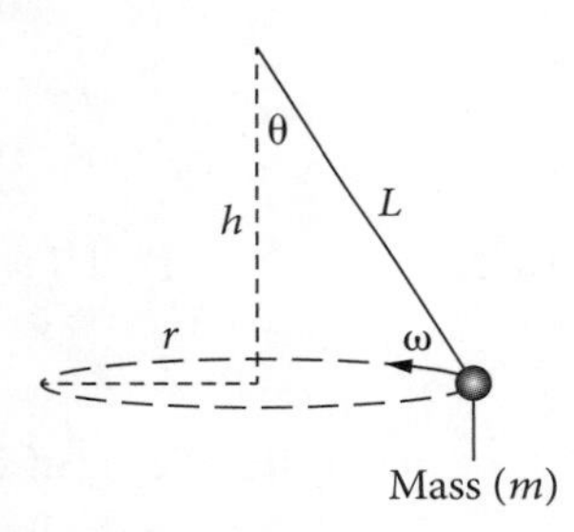

The string makes an angle θ with the vertical. The angular speed of the object is ω.

What is tan θ?

A $\frac{\omega^2 r}{g}$ **B** $\frac{g}{\omega^2 r}$ **C** $\frac{r^2\omega}{g}$ **D** $\frac{g}{r^2\omega}$

Question 23

Mass M is attached to one end of a string. The string is passed through a hollow tube and mass m is attached to the other end. Friction between the tube and the string is negligible.

Hint
Consider how the centripetal force is provided in this system.

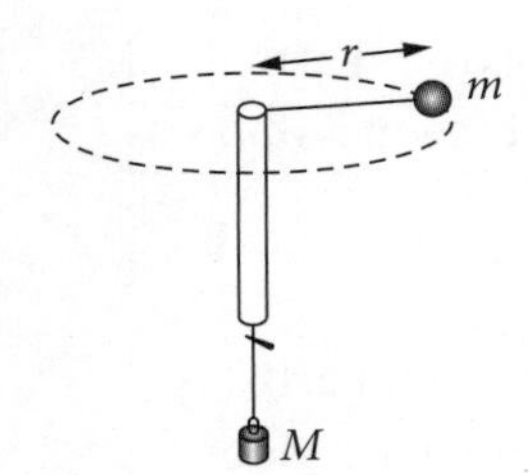

Mass m travels at constant speed v in a horizontal circle of radius r. What is mass M?

A $\frac{mv^2}{r}$ **B** mv^2rg **C** $\frac{mgv^2}{r}$ **D** $\frac{mv^2}{rg}$

Question 24 ©NESA 2016 SI Q18

A motorcycle travels around a vertical circular path of radius 3.6 m at a constant speed. The combined mass of the rider and the motorcycle is 200 kg.

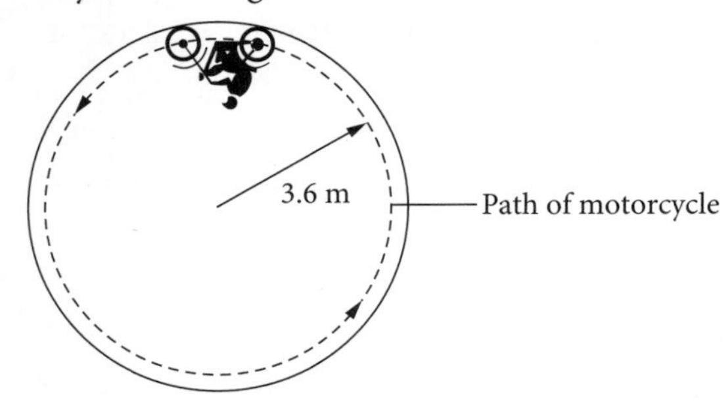

What is the minimum speed, in m s^{-1}, at which the motorcycle must travel to maintain the circular path?

A 0.42 **B** 1.9 **C** 5.9 **D** 35

Motion in gravitational fields

Use the data in the table below to answer Questions 25 and 26.

Mass of Mars	6.39×10^{23} kg
Mass of Deimos (a moon of Mars)	2.0×10^{15} kg
Orbital radius of Deimos	20 070 km
Orbital period of Deimos	30 hours

Question 25

What is the force due to gravity acting on Deimos due to Mars?

A 2.1×10^{14} N **B** 2.1×10^{20} N **C** 1.1×10^{7} N **D** 1.1×10^{10} N

Question 26

Given that the orbital radius of Mars' other moon, Phobos, is 30% of the orbital radius of Deimos, what would be its orbital period?

A 5 h **B** 9 h **C** 25 h **D** 180 h

Question 27

A spacecraft launched from Earth ejects its rocket 190 km above Earth's surface and continues the rest of the journey to the Moon without using any fuel. Which option below correctly indicates changes to the speed and total energy of the spacecraft as it moves away from Earth and towards the Moon?

	Spacecraft speed	Spacecraft total energy
A	Decreased moving away from Earth and increased approaching the Moon	Constant
B	Decreased moving away from Earth and increased approaching the Moon	Increased moving away from Earth and decreased approaching the Moon
C	Constant	Constant
D	Constant	Increased moving away from Earth and decreased approaching the Moon

Question 28

Planet X follows an elliptical orbit about a star and is known to have an average velocity of $22\,\text{km s}^{-1}$ when at its average distance from the star. Select the response that is most likely to be the planet's velocity (in km s^{-1}) when it is furthest from the star.

A 15 **B** 22 **C** 30 **D** 32

Question 29

Consider the point marked X situated between the planets P and Q, as shown on the diagram. P has a mass of $2M$ and Q has a mass of M. The distance from X to Q is R and the distance from X to P is $\frac{R}{2}$. R >>> the radius of either planet.

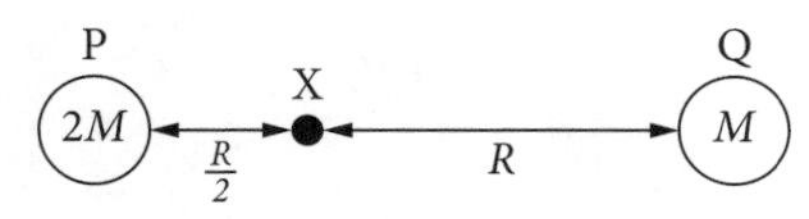

The gravitational field strength at X can be expressed as

A $\frac{3GM}{R^2}$ **B** $\frac{9GM}{R^2}$ **C** 0 **D** $\frac{7GM}{R^2}$

Question 30

Compared to a low-Earth orbit satellite, a geostationary satellite will

A experience less drag, 'see' more of Earth's surface at any instant, move slower.

B do more orbits of Earth each day, experience less drag, 'see' more of Earth's surface at any instant.

C move slower, do more orbits of Earth each day, experience less drag.

D 'see' more of Earth's surface at any instant, move slower, do more orbits of Earth each day.

Question 31 ©NESA 2012 SI Q18

The gravitational force due to Earth on a mass positioned at X is F_X and on the same mass positioned at Y is F_Y. The diagram is drawn to scale.

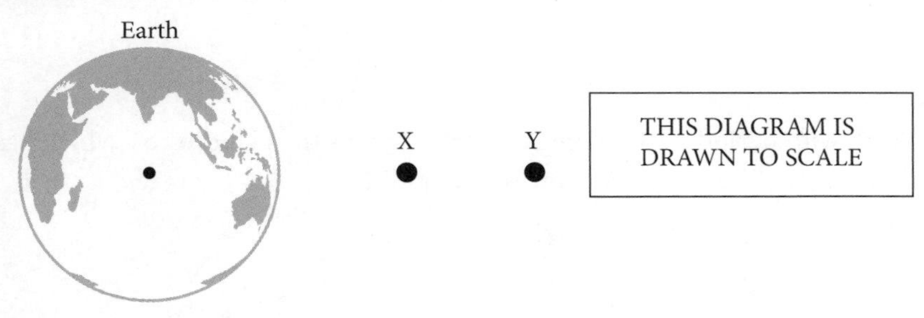

What is the value of $\frac{F_X}{F_Y}$?

A 1.5 **B** 2.0 **C** 2.25 **D** 4.0

9780170465304

Question 32 ©NESA 2015 SI Q11

Which of the following diagrams correctly represents the force(s) acting on a satellite in a stable circular orbit around Earth?

F_g = gravitational force	F_p = propulsive force
F_c = centripetal force	F_r = reaction force

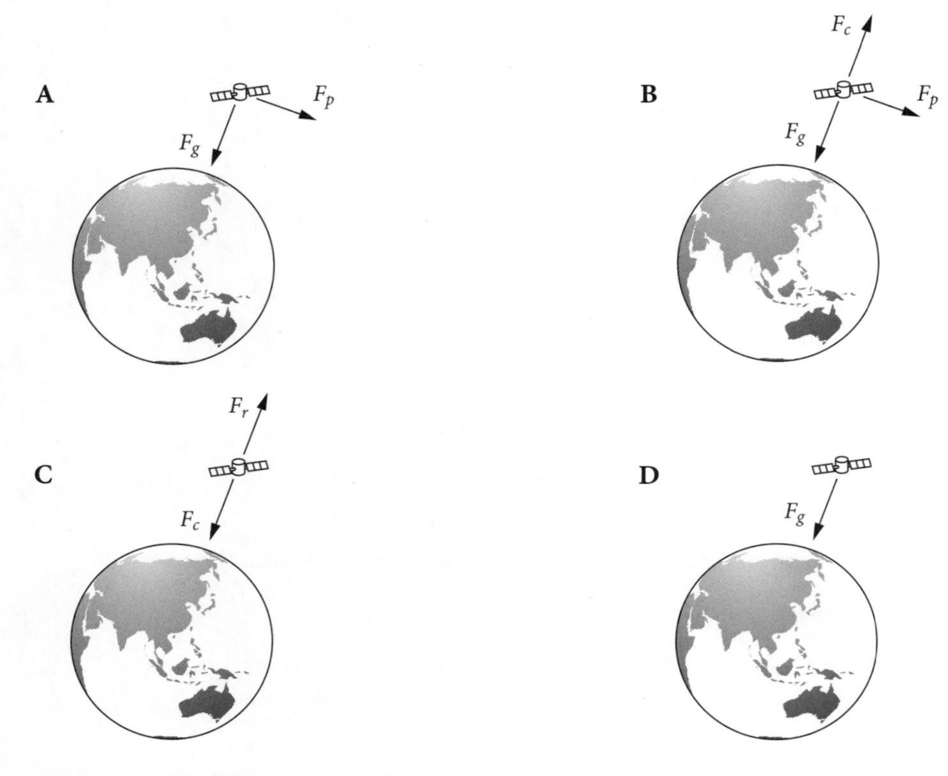

Question 33 ©NESA 2016 SIA Q14

A satellite orbits Earth with period T. An identical satellite orbits the plant Xerus, which has a mass four times that of Earth. Both satellites have the same orbital radius r.

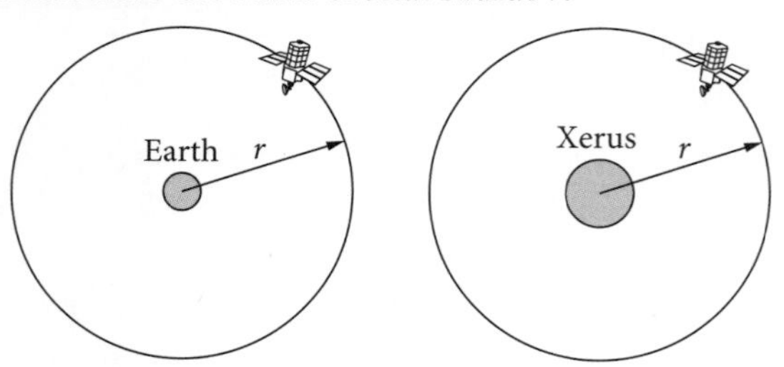

What is the period of the satellite orbiting Xerus?

A $\frac{T}{4}$ **B** $\frac{T}{2}$ **C** $2T$ **D** $4T$

Question 34

By what factor would the escape velocity from Earth change if the mass of Earth was doubled and its radius halved?

A Increase by a factor of 4

B Increase by a factor of 2

C Increase by a factor of $\sqrt{2}$

D Decrease by a factor of $\sqrt{2}$

9780170465304

Short-answer questions

Solutions start on page 150.

Projectile motion

Question 35 (4 marks)

An object is launched vertically upwards at 20 m s^{-1}. Sketch a velocity vs time graph for the motion of the ball until it returns to its launch position. Make upwards the positive direction and label the key quantitative values.

Question 36 (5 marks)

An arrow is fired at 75 m s^{-1} at an angle of 20° above horizontal from a height of 1.5 m above the ground.

a Determine the lowest speed of the arrow during its journey and describe at which point in the arrow's path that will occur. 2 marks

b Calculate the horizontal distance travelled by the arrow over its journey from the bow to the ground. 3 marks

Question 37 (4 marks)

A ball is rolled horizontally from the surface of a desk with a variety of initial velocities. The range of the ball is measured, and the data presented below.

Initial velocity of ball (m s^{-1})	Range of ball (cm)
0.7	40
1.2	65
1.8	100
2.3	125
2.7	145

Plot the data and use the graph to determine the height of the table.

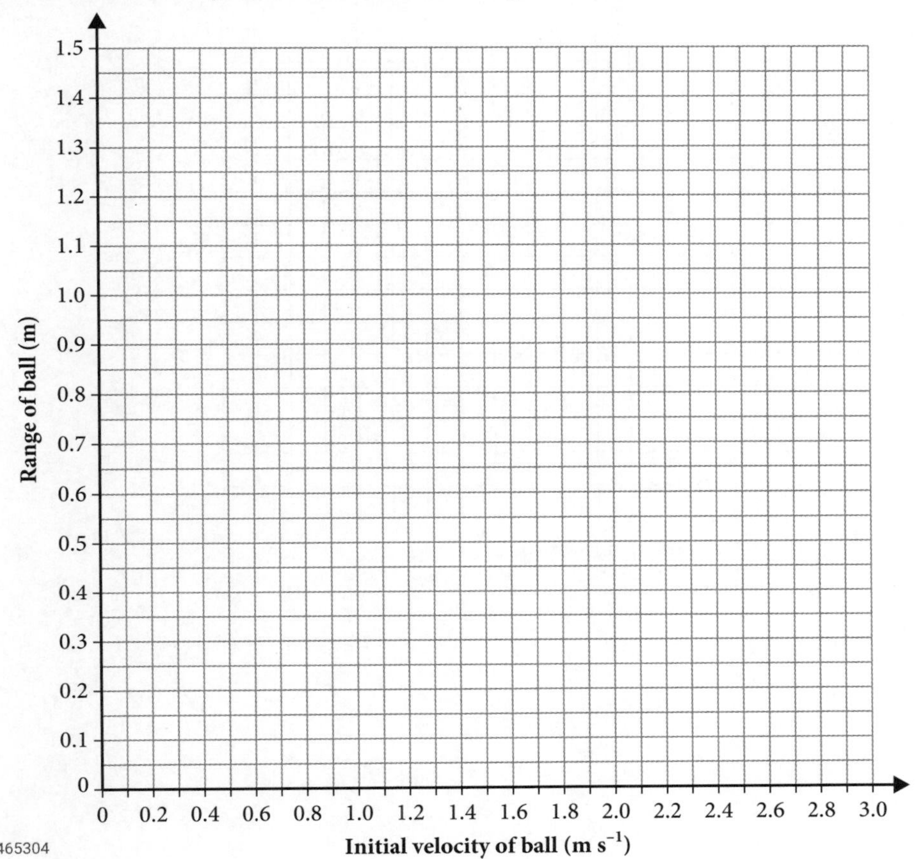

Question 38 (4 marks) ©NESA 2014 SIB Q30b

Cannonballs P and Q are fired so that they leave their barrels from the same height. Cannonball P is fired vertically upwards while cannonball Q is fired at an angle as shown.

Both cannonballs take 3 seconds to reach the same maximum height.

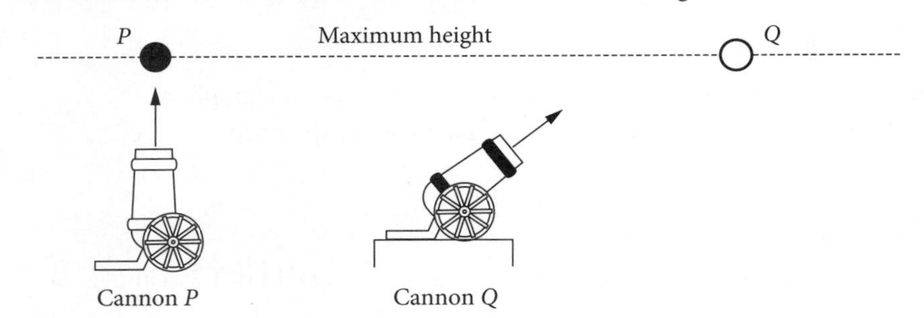

The position of cannonball P is plotted at the 3rd, 4th and 5th seconds of its flight.
The position of cannonball Q is plotted at the 3rd and 4th seconds of its flight.

Plot the positions of the balls at each second for the remainder of their flight. Show calculations.

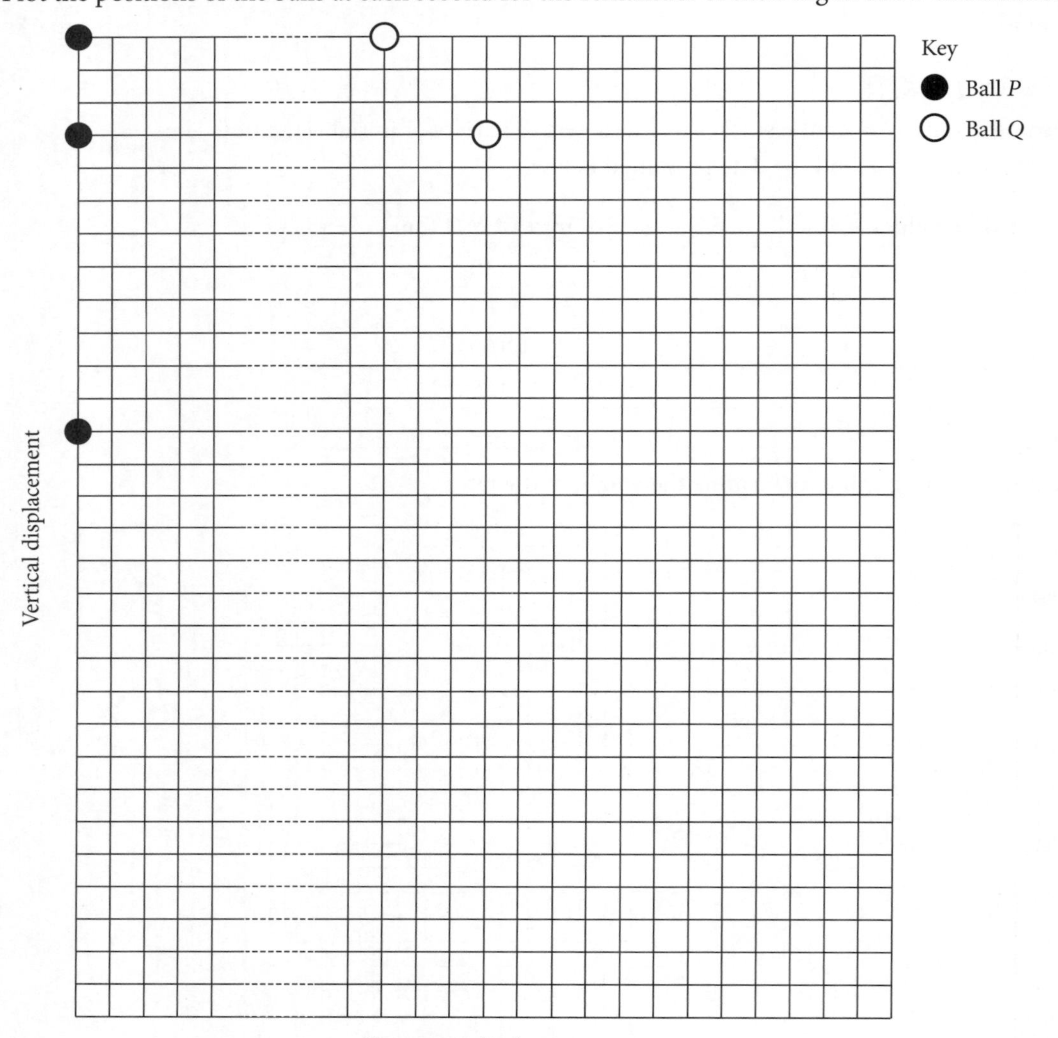

Question 39 (4 marks) ©NESA 2012 SIB Q27 ●●●

A toy bird is launched at 60° to the horizontal, from a point 45 m away from the base of a cliff.

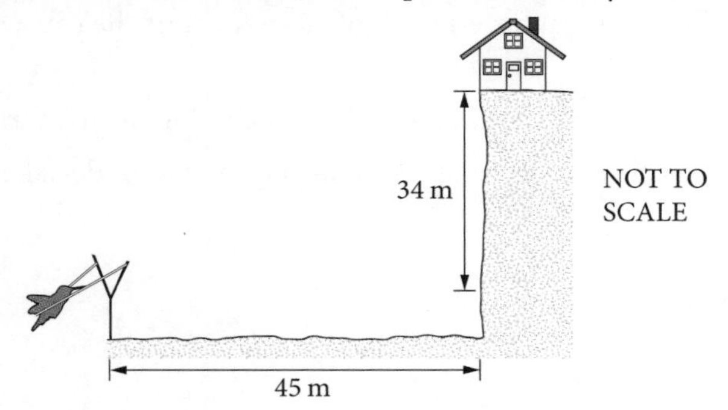

Calculate the magnitude of the required launch velocity such that the toy bird strikes the base of the wooden building at the top of the cliff, 34 m above the launch height.

Circular motion

Question 40 (2 marks) ●●●

Choose two forces from the list and describe how each force can be responsible for uniform circular motion by using an example:

gravitational force, friction, tension, electrostatic force, magnetic force, normal force

Question 41 (3 marks) ●●●

A student tries to undo a nut using a spanner by attaching a cable to the spanner and hanging a mass from the cable, as shown in the diagram.

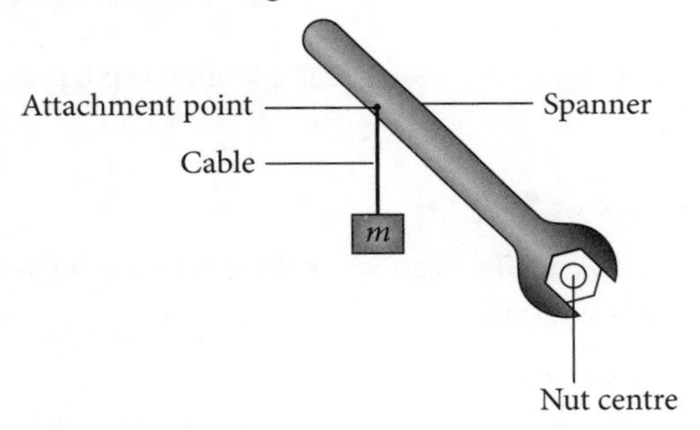

Describe and justify three separate changes to the set-up that the student could use to exert greater torque on the nut.

Question 42 (2 marks) ●●●

A skier approaches the top of a small hill that has a perfectly circular arc. If the radius of the arc of the hill is 10 m, what is the greatest speed with which the skier can travel over the hill to ensure that he does not lose contact with the slope?

Question 43 (6 marks) ●●●

A racing car of mass 800 kg is travelling around a flat corner of radius 200 m. The coefficient of friction between the tyres and the road is 0.9.

> **Hint**
> Use the coefficient of friction and the mass of the car to determine the frictional force that provides the centripetal force required.

a Calculate the maximum speed at which the car can travel around the corner. 3 marks

b The car is to be driven at a speed of $20\,m\,s^{-1}$ around a frictionless, banked corner of the same radius. Determine the angle, to the nearest degree, at which the corner must be banked to make the plan possible. 3 marks

Question 44 (7 marks)

A child at a school fair is swung in a horizontal circle on the end of a cable that makes an angle of 15° with the vertical. The child and seat have a mass of 50 kg (assume the cable has no mass) and move in a circle of radius 6 m.

a Draw a labelled diagram of the forces acting on the child, including the net force. 3 marks

b Calculate the tension in the cable and the time taken for the child to undertake one complete revolution. 4 marks

Motion in gravitational fields

Question 45 (7 marks)

The Hubble Space Telescope (HST) orbits Earth at an altitude of 540 km. The weight of the HST was found to be 1.2×10^5 N on Earth. Astronomers state that it has a weight of 1.0×10^5 N when in its stable orbit.

a Why do astronomers state the HST weighs less in orbit than on Earth? Justify your response with quantitative analysis related to the values of weight given. 2 marks

b What will be the orbital velocity of the HST? 2 marks

c Explain two changes to the motion of the HST if it is moved to a stable orbit at a lower altitude. 3 marks

Question 46 (3 marks)

It is estimated that the orbiting International Space Station has a kinetic energy of 1.23×10^7 MJ. Given that its mass is estimated to be 420 tonnes, determine its altitude.

Question 47 (2 marks)

Show that the escape velocity of an object from Earth will be less than from a planet of the same mass as Earth but a smaller radius.

Question 48 (6 marks) ©NESA 2015 SIB Q26

Consider the following two models used to calculate the work done when a 300 kg satellite is taken from Earth's surface to an altitude of 200 km.

You may assume that the calculations are correct.

Model *X*	Model *Y*
Data: $g = 9.8\ \text{m s}^{-2}$ $m = 300\ \text{kg}$ $\Delta h = 200\ \text{km}$ $W = Fs$ $= mg\Delta h$ $= 3 \times 10^2 \times 9.8 \times 2.0 \times 10^5$ $= 5.9 \times 10^8\ \text{J}$	Data: $G = 6.67 \times 10^{-11}\ \text{N m}^2\ \text{kg}^{-2}$ $r_{\text{Earth}} = 6.38 \times 10^6\ \text{m}$ $r_{\text{orbit}} = 6.58 \times 10^6\ \text{m}$ $M = 6.0 \times 10^{24}\ \text{kg}$ $m = 300\ \text{kg}$ $W = \Delta E_p$ $\Delta E_p = E_{p\ \text{final}} - E_{p\ \text{initial}}$ $= -\frac{GMm}{r_{\text{orbit}}} - \left(\frac{GMm}{r_{\text{Earth}}}\right)$ $= -1.824 \times 10^{10} - (-1.881 \times 10^{10})$ $= 5.7 \times 10^8\ \text{J}$

a What assumptions are made about Earth's gravitational field in models *X* and *Y* that lead to the different results shown? 2 marks

b Why do models *X* and *Y* produce results that, although different, are close in value? 1 mark

c Calculate the orbital velocity of the satellite in a circular orbit at the altitude of 200 km. 3 marks

Question 49 (4 marks) ©NESA 2012 SIB Q23

Consider the following thought experiment.

Two towers are built on Earth's surface. The height of each of the towers is equal to the altitude of a satellite in geostationary orbit about Earth. Tower A is built at Earth's North Pole and Tower B is built at the equator.

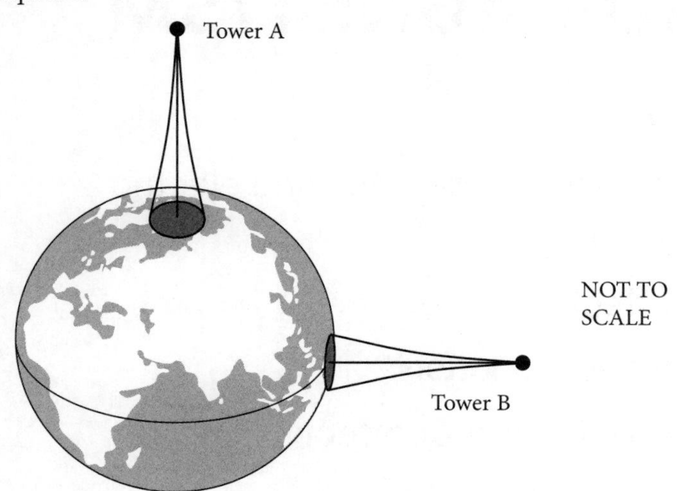

Identical masses are simultaneously released from rest from the top of each tower.

Explain the motion of each of the masses after their release.

Question 50 (3 marks)

Planet A and planet B orbit the distant star Regulus. The orbital radius of planet A is four times the orbital radius of planet B.

Determine the ratio of the orbital period of planet A to that of planet B.

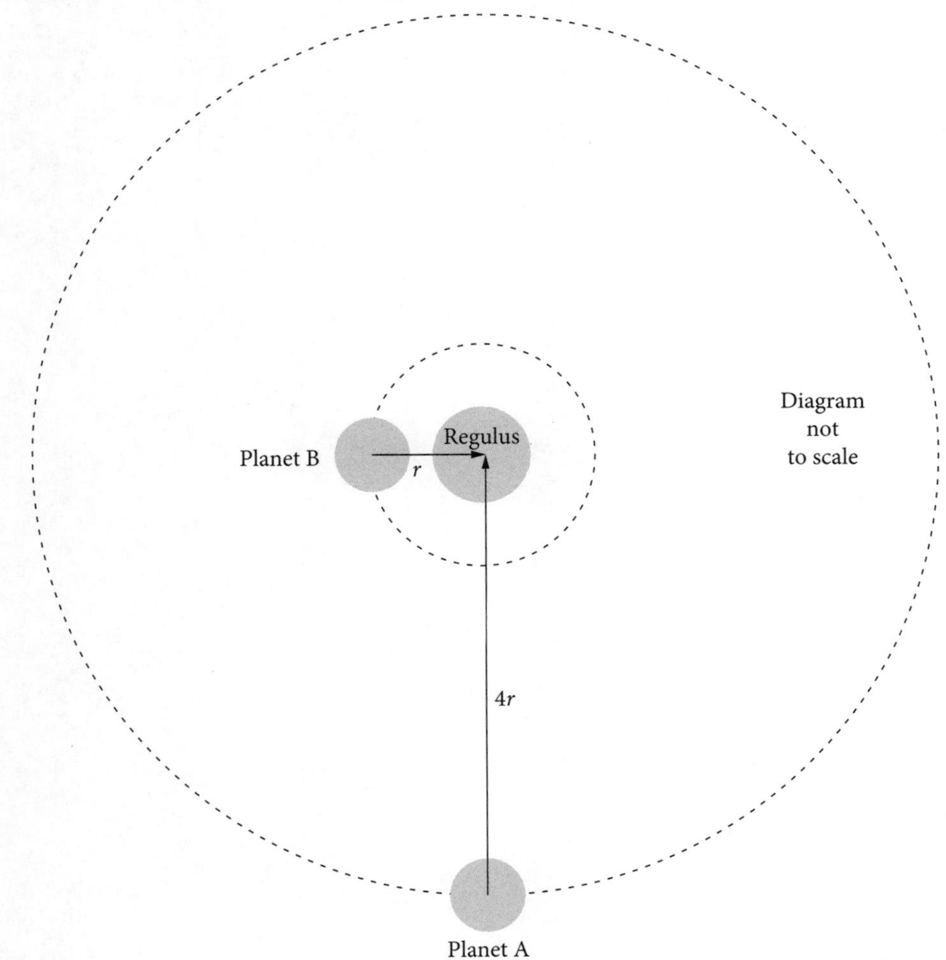

CHAPTER 2
MODULE 6: ELECTROMAGNETISM

Module summary 31

2.1 Charged particles, conductors, and electric and magnetic fields 33

2.2 The motor effect 38

2.3 Electromagnetic induction 40

2.4 Applications of the motor effect 44

Glossary 49

Exam practice 51

2

Chapter 2
Module 6: Electromagnetism

Module summary

Outcomes

On completing this module, you should be able to:

- develop and evaluate questions and hypotheses for scientific investigation
- design and evaluate investigations in order to obtain primary and secondary data and information
- conduct investigations to collect valid and reliable primary and secondary data and information
- select and process appropriate qualitative and quantitative data and information using a range of appropriate media
- analyse and evaluate primary and secondary data and information
- explain and analyse the electric and magnetic interactions due to charged particles and currents and evaluate their effect both qualitatively and quantitatively.

Working Scientifically skills

In this module, you are required to demonstrate the following Working Scientifically skills:

- develop and evaluate questions and hypotheses for scientific investigation
- design and evaluate investigations in order to obtain primary and secondary data and information
- conduct investigations to collect valid and reliable primary and secondary data and information
- select and process appropriate qualitative and quantitative data and information using a range of appropriate media
- analyse and evaluate primary and secondary data and information
- solve scientific problems using primary and secondary data, critical thinking skills and scientific processes
- communicate scientific understanding using suitable language and terminology for a specific audience or purpose.

9780170465304

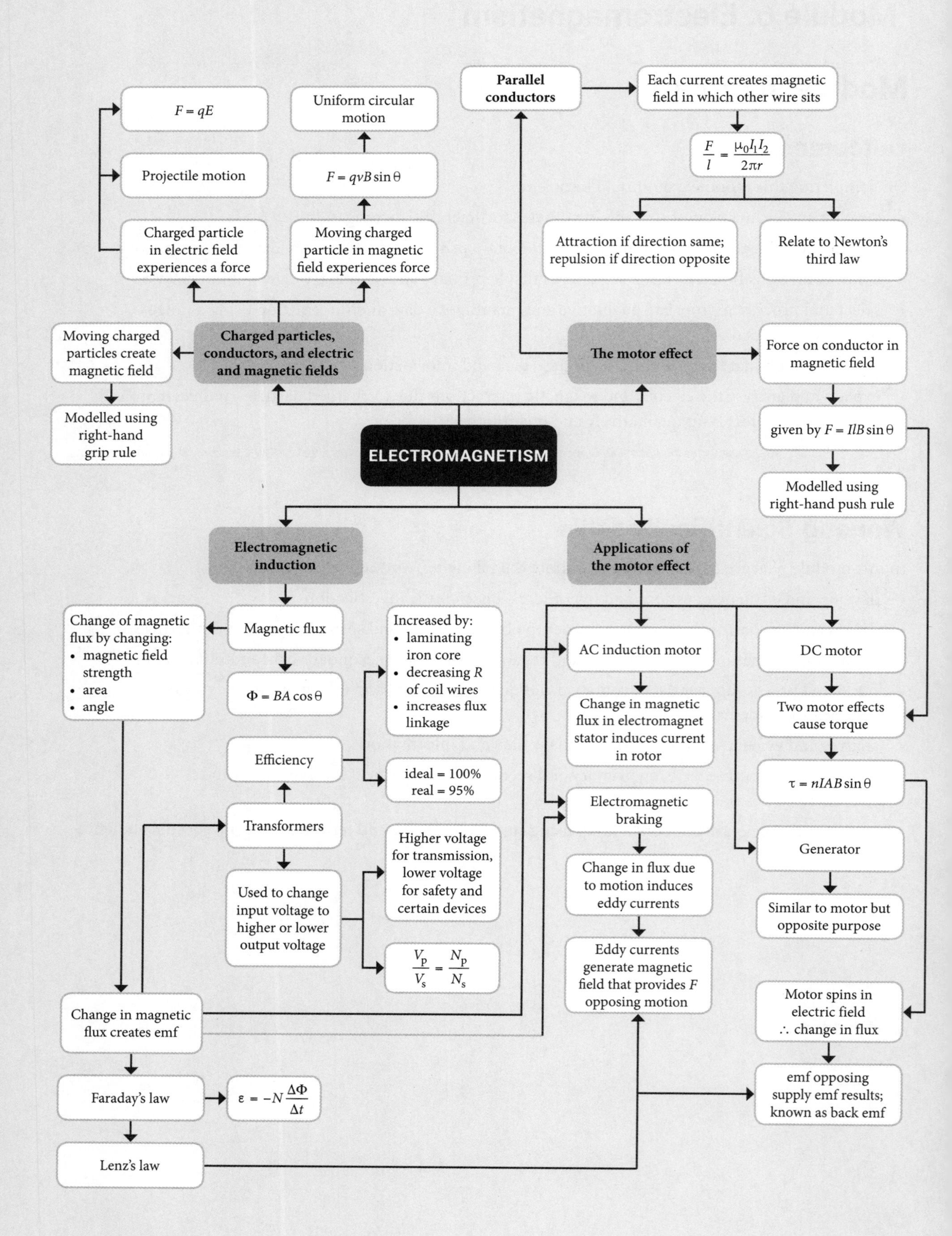

ELECTROMAGNETISM
Charged particles, conductors, and electric and magnetic fields
Moving charged particles create magnetic field
Modelled using right-hand grip rule
Charged particle in electric field experiences a force
$F = qE$
Projectile motion
Moving charged particle in magnetic field experiences force
$F = qvB\sin\theta$
Uniform circular motion
The motor effect
Parallel conductors
Each current creates magnetic field in which other wire sits
$\frac{F}{l} = \frac{\mu_0 I_1 I_2}{2\pi r}$
Attraction if direction same; repulsion if direction opposite
Relate to Newton's third law
Force on conductor in magnetic field
given by $F = IlB\sin\theta$
Modelled using right-hand push rule
Electromagnetic induction
Magnetic flux
Change of magnetic flux by changing:
• magnetic field strength
• area
• angle
$\Phi = BA\cos\theta$
Efficiency
Increased by:
• laminating iron core
• decreasing R of coil wires
• increases flux linkage
ideal = 100%
real = 95%
Transformers
Used to change input voltage to higher or lower output voltage
Higher voltage for transmission, lower voltage for safety and certain devices
$\frac{V_p}{V_s} = \frac{N_p}{N_s}$
Change in magnetic flux creates emf
Faraday's law
$\varepsilon = -N\frac{\Delta\Phi}{\Delta t}$
Lenz's law
Applications of the motor effect
AC induction motor
Change in magnetic flux in electromagnet stator induces current in rotor
Electromagnetic braking
Change in flux due to motion induces eddy currents
Eddy currents generate magnetic field that provides F opposing motion
DC motor
Two motor effects cause torque
$\tau = nIAB\sin\theta$
Generator
Similar to motor but opposite purpose
Motor spins in electric field ∴ change in flux
emf opposing supply emf results; known as back emf

2.1 Charged particles, conductors, and electric and magnetic fields

All objects that are charged will create an **electric field**, and any object that has a net **electric charge** will experience a force when placed in an electric field. A positively charged object will experience a force in the direction of the electric field lines, and a negatively charged object will experience a force in the opposite direction to the electric field lines.

Moving charged particles create a **magnetic field** as they move. This created magnetic field will interact with any external magnetic field, resulting in a pair of forces. The moving charged particle will experience one of the forces in this pair. The direction of this force can be predicted and will be perpendicular to both its velocity and the external magnetic field.

2.1.1 Charged particles in electric fields

Recall from Year 11 Physics that:

- Neutral objects can become electrically charged by losing electrons (and becoming positively charged) or gaining electrons (and becoming negatively charged). The symbol for charge is q or, sometimes, Q. The magnitude of that charge can be described in units of coulombs (C).
- Charged objects exert a force on other charged objects, the magnitude of which can be calculated using Coulomb's law: $F = \frac{1}{4\pi\varepsilon_0}\frac{q_1q_2}{r^2}$. Like charges repel and unlike charges attract.
- Therefore, a charged object is said to have generated an electric field – a surrounding region in which another charged particle will experience a force. Arrows (also known as lines of force) are used to represent the electric field. The direction of the arrow at any location indicates the direction of the force that would be experienced by a positive charge placed at that point. If the arrows are more closely packed in one region than another, it indicates that the field is stronger in that region. The shape of an electric field varies, depending on the arrangement and shape of the charged object or objects generating the field. The most important electric field to understand for Year 12 Physics is the uniform electric field, indicated by the parallel and evenly spaced lines, associated with parallel, oppositely charged plates.

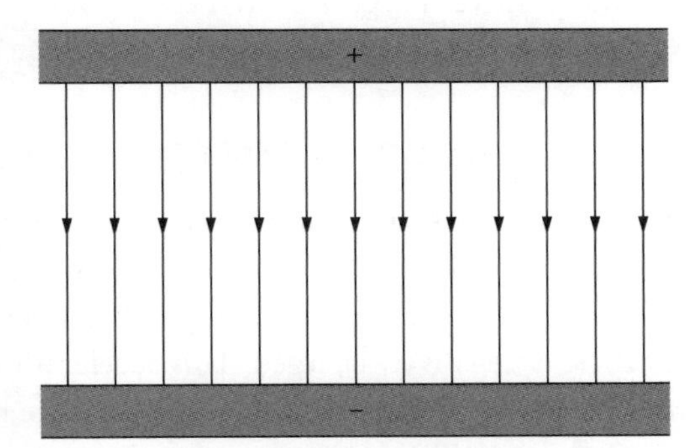

FIGURE 2.1 A uniform electric field between parallel, oppositely charged plates is represented by parallel and evenly spaced electric field lines.

The intensity or strength of the electric field is indicated by the symbol E, which, as should be expected, is measured by the force experienced by a unit charge of q (1 C) at that point in the field. So $E = \frac{F}{q}$. It can be seen that E must be a vector in the same direction as F with units of $\mathrm{N\,C^{-1}}$. The equation is usually written as $F = qE$.

- Much like the difference in the gravitational potential energy of an object at two positions in a gravitational field, an object in an electric field may have different electrical potential energies at two different positions in that field. A volt (V) is defined as the change in electrical potential energy per coulomb of charge: $1\,\mathrm{V} = 1\,\mathrm{J\,C^{-1}}$ or $V = \frac{W}{q}$. This equation is usually written as $W = qV$.

The difference in electrical potential energy per coulomb of charge is called **electrical potential difference** (usually abbreviated to potential difference) and has the symbol V and is measured in volts (it is often referred to as voltage). To increase the electrical potential energy of a charged object, work must be done on the object according to the equation $W = qV$. Since $W = Fs$, provided the force is uniform and thus the field is uniform ($F = qE$), then $\Delta U = W = Fs = qEd$, where d is the distance moved due to W (Figure 2.2).

Thus $W = qEd$ means that $E =$ and so $E = \frac{V}{d}$ (since $V = \frac{W}{q}$).

- Furthermore, we can describe the electric field strength in terms of electrical potential difference using the equation $E = \frac{V}{d}$. This is most significant when considering the uniform electric field associated with parallel, oppositely charged plates separated by a distance d and connected to a power source of potential difference, or voltage, V. By considering a situation in which the positive charge is moved all the way from the negative plate to the positive plate, you can see that d, therefore, is the distance between the plates.

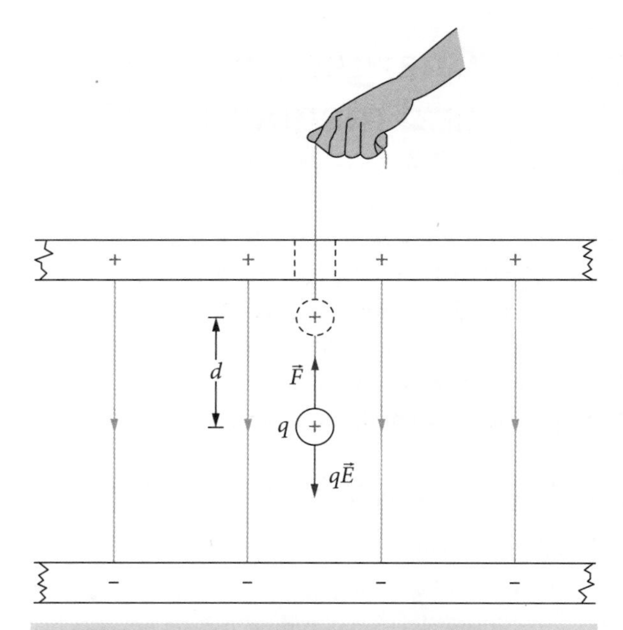

FIGURE 2.2 A positively charged object in a uniform electric field between parallel, oppositely charged plates will experience a force in the direction of the field lines. Work must be done to move the object in the opposite direction.

2.1.2 Motion of charged particles in uniform electric fields

Given that an electric field is a place where a charged particle will experience a force, it stands to reason that the charged particle will accelerate in that field if it is able. The acceleration will be in the direction of the field lines if the particle is positively charged, and in the direction opposite to the field lines if the particle is negatively charged. This acceleration can be determined by equating the force of the charged particle in the electric field, $\vec{F} = q\vec{E}$, with the net force causing acceleration, $\vec{F} = m\vec{a}$, to derive the equation $\vec{a} = \frac{q\vec{E}}{m}$.

For uniform acceleration in one direction only, the motion of a charged particle in a uniform electric field is analogous to the motion of an object in a uniform gravitational field. This was described as projectile motion in the preceding chapter. As such, quantitative analysis of the motion can be done using the same processes that were used for projectile motion.

> **Note**
> This projectile motion will *only* occur in the field. Before entering and after leaving the field, the motion will be linear.

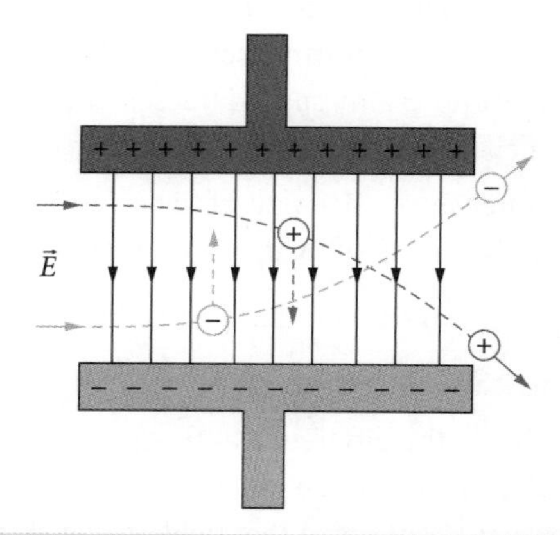

FIGURE 2.3 When a charged object enters a uniform electric field, it will experience a force parallel to the field lines. If the initial motion had a component perpendicular to the field, then the object will undergo a form of projectile motion.

Just as with projectile motion, the motion of the charged particle can be resolved into two perpendicular components. Since there will be a component of the motion perpendicular to the electric field (which can be considered the *x* component) that is constant ($a = 0$), the equations of motion can be rewritten for the perpendicular aspect of the motion. A subscript *x* can be used to indicate that these equations refer only to the perpendicular motion.

$v_x = u_x$ and
$s_x = u_x t$

For motion parallel to the electric field (the *y* component), where acceleration is constant and will be determined by the magnitude and sign of the charge, the three equations of motion can be rewritten using a *y* subscript.

$v_y = u_y + at$
$s_y = u_y t + \frac{1}{2}at^2$
$v_y^2 = u_y^2 + 2as_y$

A parabolic arc may be described by the motion of the charged particle fired into an electric field, as was possible for an object fired into a gravitational field.

Note
This analogy with projectile motion from Module 5 creates a good opportunity for cross-modular questions from examiners.

2.1.3 Charged particles in magnetic fields

The Year 11 Physics course introduced some of the key ideas in magnetism.

- **Magnetic flux** – the concept of magnetism 'flowing' though space, **magnetic flux density** or magnetic field strength – a vector quantity with symbol *B* and units of tesla (T).
- The magnetic field that is created by magnetic objects and within which another magnetic object will experience a force.
- The magnetic field is represented by arrows (lines of force) with the direction of the arrows indicating the direction of the force on a north pole at that point in the field. Consequently, magnetic field lines will always come out of a north pole and go into a south pole. Arrows closer together in a field diagram show that the region has a greater magnetic field strength or magnetic flux density than other regions in the diagram.
- The shape of a magnetic field varies, depending on the arrangement and shape of the source creating the field. The most important magnetic field to understand for Year 12 Physics is the uniform magnetic field, indicated by parallel and evenly spaced field lines.
- Moving charged particles in a current-carrying conductor create a magnetic field in the shape of concentric circles around the conductor, which can be modelled using the right-hand grip rule (Figure 2.4). The strength of the magnetic field at a distance *r* from a straight conductor carrying a current *I* is given by $B = \frac{\mu_0 I}{2\pi r}$.

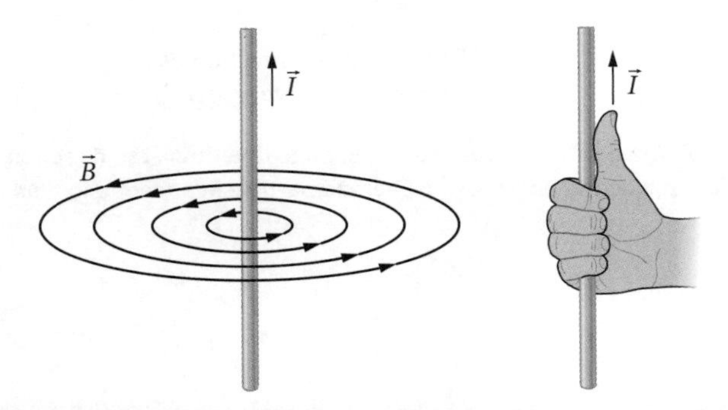

FIGURE 2.4 The magnetic field around a current-carrying conductor can be modelled using the right-hand grip rule.

- A current-carrying conductor made into the shape of a solenoid or coil will create a field similar to that created by a bar magnet, with distinct north and south poles (Figure 2.5).

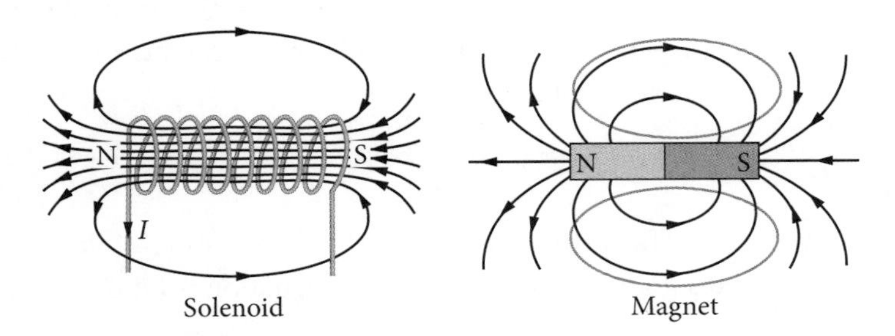

FIGURE 2.5 The magnetic field around a current-carrying solenoid or coil has similarities to the magnetic field created by a bar magnet.

- Dots and crosses can be used to illustrate a field that goes out of and into a surface respectively (Figure 2.6).

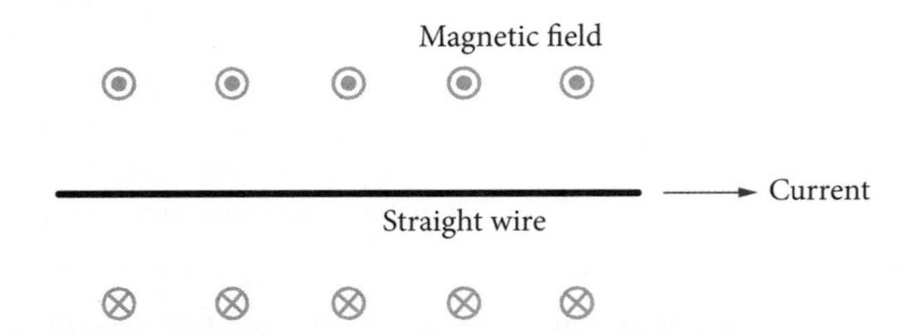

FIGURE 2.6 Dots represent field lines coming 'out of the page' and crosses represent field lines going into the page, seen here illustrating the field associated with the current in the conductor.

Moving charged particles create magnetic fields. If the moving charged particles are placed in an external magnetic field, then this created magnetic field may interact with an external magnetic field and, consequently, the moving charged particles may experience a force in the external magnetic field. In fact, as long as there is a component of motion of the charged particle perpendicular to the external field, then the moving charged particle will experience a force. The magnitude of this force is given by the equation $F = qvB\sin\theta$, where θ is the angle between the velocity vector and the magnetic field vector. The direction of the force on a positive charge is determined by the right-hand push rule (Figure 2.7).

> **Note**
> If the charge is negative, then the force will be indicated by the direction of the back of the hand or by using the left hand. The choice of which method to use is an individual one.

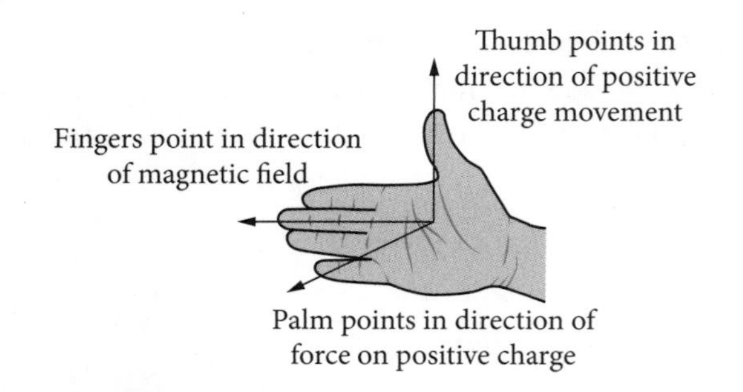

FIGURE 2.7 The right-hand push rule is used to determine the direction of the magnetic force acting on a moving charged particle in a magnetic field.

As the magnetic force is perpendicular to the velocity of the charged particle, it will act as a centripetal force, and the charged particle will undergo uniform circular motion and execute part or all of a circular arc (Figure 2.8).

Note
This analogy with uniform circular motion from Module 5 creates a good opportunity for cross-modular questions from examiners.

The radius of the circular arc for a particle entering the magnetic field at right angles to the field lines (hence $\theta = 90°$) can be determined by equating the centripetal force with the magnetic force:

$$F_c = F_B$$

$$\frac{mv^2}{r} = qvB$$

$$r = \frac{mv}{qB}$$

The orbital period for the circular arc can be derived by substituting $v = \frac{2\pi r}{T}$ to yield $T = \frac{2\pi m}{qB}$.

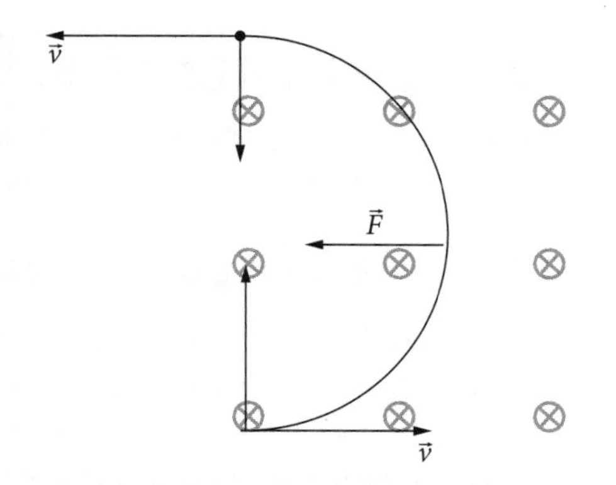

FIGURE 2.8 The right-hand push rule indicates that a centripetal force is experienced by a charged particle moving in that field. The particle will move in a circular arc.

Note
This uniform circular motion will only occur in the field. Before entering and after leaving the field the motion will be linear.

2.1.4 Moving charged particles in electric and magnetic fields

The motion of objects in various fields can be compared (Table 2.1).

TABLE 2.1 Comparing the three field forces

Field type	Created by	Nature of force	Exerts a force upon
Gravitational	Any object with mass	Always attractive, i.e. always in same direction as field lines	Any object with mass
Electric	Any object with net charge	Parallel to field lines and dependent on type of charge; positive in direction of field lines and negative in direction opposite to field lines	Any object with net charge
Magnetic	Moving charges (including spinning electrons in permanent or temporary magnets)	Perpendicular to field lines and dependent on type of charge and direction of motion of the charge; determined by right-hand push rule for positive charges (opposite force direction for negative charges)	Only *moving* charges with a component of motion perpendicular to the field (and magnetised objects)

The motion of objects that results from the force exerted by this field is compared in Table 2.2.

TABLE 2.2 Comparing the motion of relevant objects within the three fields

Field type	Resulting motion in uniform field
Gravitational	Acceleration parallel to field lines if initially stationary or with initial velocity parallel to field lines; otherwise parabolic trajectory similar to projectile motion
Electric	Acceleration parallel to field lines if initially stationary or with initial velocity parallel to field lines; otherwise parabolic trajectory similar to projectile motion
Magnetic	Circular motion if moving perpendicular to field lines; unaffected by field if moving parallel to field lines

2.2 The motor effect

A magnetic field will surround a current-carrying conductor and the conductor will, consequently, experience a force when it carries a current in a region where there is an external magnetic field.

2.2.1 Force on a current-carrying conductor in a magnetic field

Section 2.1.3 explored the force acting on charged particles that have a component of their motion perpendicular to an external magnetic field. Charges moving in a current-carrying conductor will experience this force, which will manifest itself as a force exerted on the conductor (as long as some component of the length is perpendicular to the field).

It can be shown by simple experimentation that the magnitude of the force will depend upon the current, I, the length of wire within the field, l, the strength of the magnetic field, B, and the angle the conductor makes with the field lines, θ.

The term used to describe this force acting on the current-carrying conductor in the magnetic field is the **motor effect**.

By substituting $\frac{l}{t}$ for v in the equation $F = qvB\sin\theta$ and using the knowledge that $I = \frac{q}{t}$ (by definition), the equation for the magnitude of this motor effect force can be derived:

$$F = lIB\sin\theta$$

The direction of the force can be determined by applying the same right-hand push rule as used in Figure 2.7, with the thumb still representing the motion of positive charge, this time inside a conductor and known as conventional current (Figure 2.9).

The origin of this force can be considered to be the result of the vector sum of the uniform external magnetic field and the circular magnetic field created by the current. The force on the wire is in the direction from the region of highest magnetic field strength towards the region of lowest magnetic field strength, as seen in Figure 2.10.

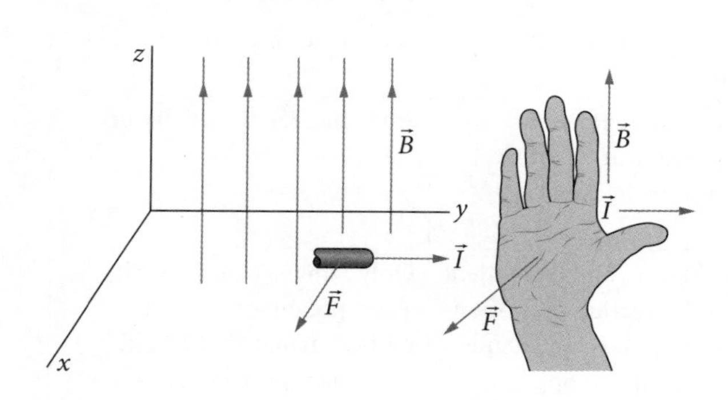

FIGURE 2.9 The right-hand push rule is applied to a conventional current flowing in a conductor in an external magnetic field.

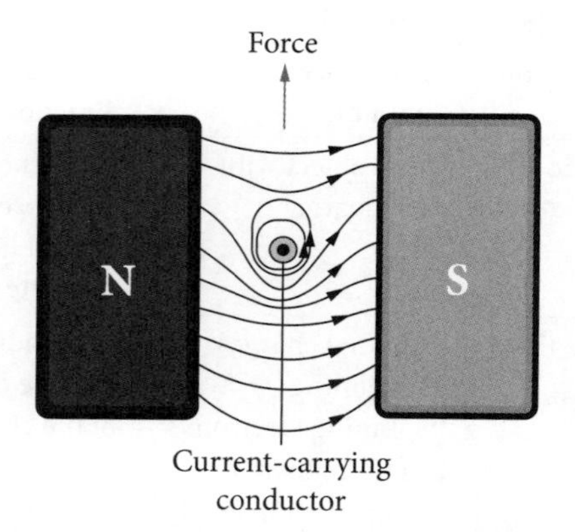

FIGURE 2.10 The magnetic field created by the current flowing in the conductor is added to the external magnetic field. Above the conductor, the fields are in opposite directions and the total field is weaker than below it, where the fields are in the same direction and add to make a stronger field.

2.2.2 Forces between parallel current-carrying wires

As seen in section 2.1.3, moving charged particles in a current-carrying conductor create a magnetic field in the shape of concentric circles around the conductor, which can be modelled using the right-hand grip rule. The strength of the magnetic field is given by $B = \frac{\mu_0 I}{2\pi r}$.

If a second current-carrying conductor is placed at a distance r parallel to the first, then each wire lies in the field of the other!

If the first wire carries a current I_1 and the second wire carries a current I_2, then the second conductor will experience a force given by $F = lI_2B\sin\theta$.

If a substitution is made for B, the magnetic field created by I_1, such that $B = \frac{\mu_0 I_1}{2\pi r}$, then the equation becomes:

$$\frac{F}{l} = \frac{\mu_0 I_1 I_2}{2\pi r}$$

This expression describes the force exerted by the conductor carrying I_1 on the conductor carrying I_2. It can be seen by applying Newton's third law that an equal force in the opposite direction will be exerted by the conductor carrying I_2 on the conductor carrying I_1. This can also be shown algebraically by performing the derivation with I_2 and I_1 swapped from the start.

Using the right-hand push rule, it can be shown that the force causes the two wires to be attracted if the currents flow in the same direction and to be repelled if the currents flow in the opposite direction.

Note

This is often remembered as 'like currents attract, unlike currents repel', which is the opposite of the observation for charged particles.

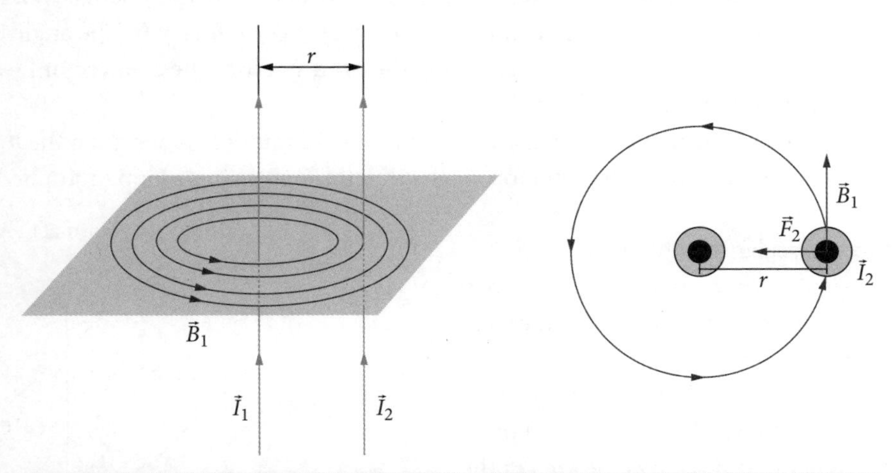

FIGURE 2.11 The magnetic field created by the current flowing in the left conductor, I_1, results in a force to the left on the right conductor. The same process could be used to show that the conductor on the left experiences a force to the right due to the conductor on the right. The magnetic field created by I_2 has been omitted to simplify the diagram.

A magnetic field diagram can be used to show that, if the two wires carry currents in the same direction, the field between the wires will be weaker than the field around the outside and that the converse is true if the currents flow in opposite directions (Figure 2.12). As seen in section 2.2.2, this analysis can be used to show the origin of the repulsive force acting on oppositely directed currents and the attractive force acting on similarly directed currents.

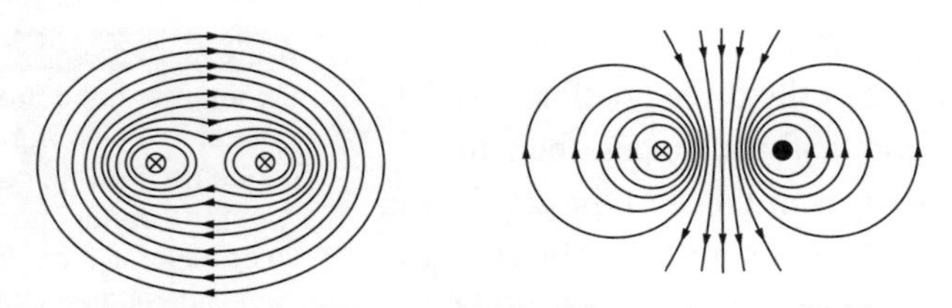

FIGURE 2.12 The sum of the magnetic fields created by the current flowing in the adjacent conductors can be seen to create a weaker field between them if the currents flow in the same direction and a stronger field between them if the currents flow in opposite directions.

The relationship between force and current in parallel wires is used to define the **ampere** – the SI unit of current. One ampere is required to flow in each of two infinitely long, straight parallel conductors separated by a metre to produce a force of 2×10^{-7} N.

2.3 Electromagnetic induction

The interconnection between electric and magnetic fields has been evident in the preceding sections. Electric current occurs when charge is transferred through a conductor. This is achieved by applying a voltage (i.e. an electric field) along the wire. The moving charges radiate a magnetic field. The changing electric field produces a changing magnetic field.

It will be seen in this section that a changing magnetic field will result in a changing electric field. In essence, if a magnetic field changes near a conductor or the conductor moves in a magnetic field, a current may flow in the conductor as a consequence. This process is known as **electromagnetic induction**.

2.3.1 Magnetic flux

Recall that the magnetic field strength or magnetic flux density can be described and is given the symbol B and measured in tesla (T). In understanding electromagnetic induction, it is useful to consider a related quantity – magnetic flux (Φ).

Magnetic flux density is a measure of the magnetic flux passing through a specific surface of specified area. Magnetic flux is defined as the product of the magnetic flux density perpendicular to a surface and the area of that surface. This can be determined by the equation $\Phi = BA\cos\theta$. The angle θ is measured between the field lines and a line representing the **area vector**. The area vector has a direction perpendicular (normal) to the plane of the surface.

In a situation in which a magnetic field passes through a coil of multiple loops, then the total magnetic flux will be the sum of the flux through each loop, or the flux through a single loop multiplied by the number of loops.

Magnetic flux is a scalar quantity with the unit weber (Wb) or tesla metres squared (T m^2). Magnetic flux density is a vector quantity with units of tesla (T) or webers per square metre (Wb m^{-2}).

It is important to understand that a change in magnetic flux can be achieved by changing (1) the strength of the magnetic field (B), (2) the area of the surface (A) or (3) the angle (θ) between the area vector and the field lines.

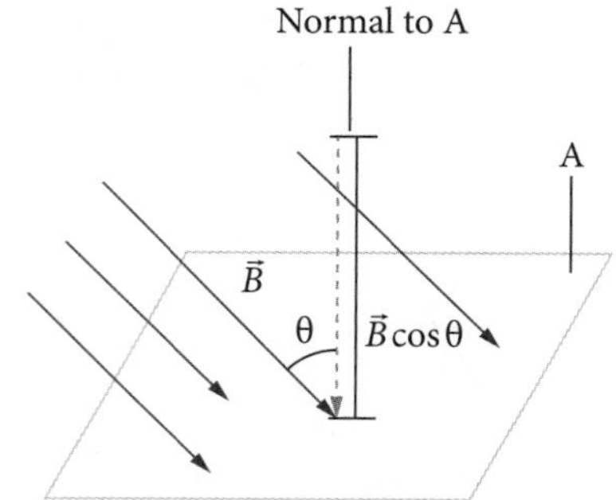

FIGURE 2.13 The magnitude of the magnetic flux is determined by the magnetic field strength, the area, and the angle between the magnetic field lines and the normal to the area or area vector.

> **Note**
> 'Flux' comes from the word for 'flow' and can help to give you the picture of the magnetic field flowing through space or permeating a space.

2.3.2 Processes of electromagnetic induction

A conductor may experience a change in magnetic flux as a consequence of one or more of the three possible reasons (changing B, A or θ). If magnetic flux changes, then an **electromotive force (emf)**, ε, is induced in the conductor. An electromotive force is the force experienced by charges in the conductor when they experience an electric field. It is a term used to describe an electrical potential difference and is similar to a battery voltage. The process is called electromagnetic induction.

> **Note**
> When magnetic flux changes, it can be visualised as magnetic field lines separating or being drawn closer together or moving past the conductor. When this occurs near a conductor, it could be visualised as the field lines passing through (cutting) the conductor. This concept of flux cutting can be invaluable in determining whether or not an emf is generated.

Faraday's law of electromagnetic induction states that the size of the emf is proportional to the negative rate of change of magnetic flux experienced by the conductor. This is written as $\varepsilon = -N\dfrac{\Delta\Phi}{\Delta t}$, where N is the number of turns of a coil.

The negative sign indicates the nature of the **induced emf** and is explained by Lenz's law (Figure 2.14).

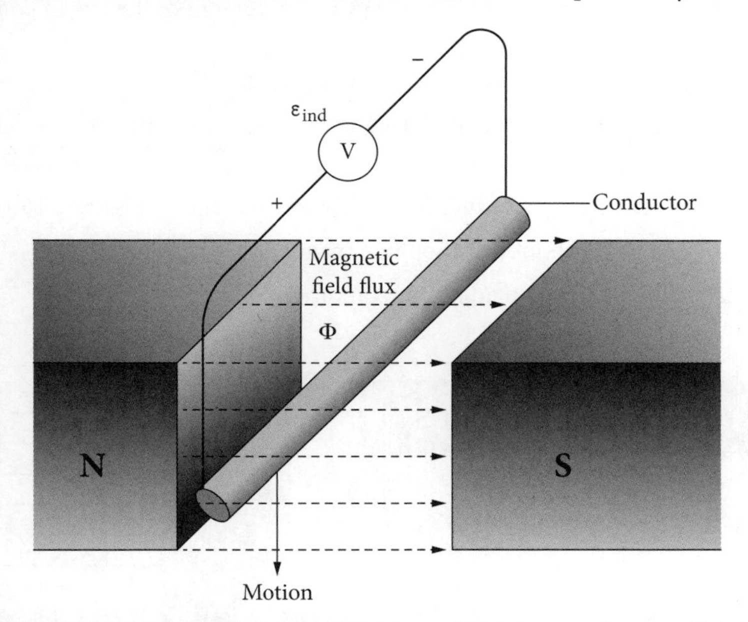

FIGURE 2.14 As the conductor is moved downwards through the magnetic field, it will experience a change of the magnetic flux (due to the area enclosed by the loop increasing) and so an emf will be induced, which causes positive charge to try to move to one end of the conductor. That end will tend to become positive and leaves the other negative.

Lenz's law describes a situation in which the induced emf causes a current to flow and states that the direction of the **induced current** creates a magnetic field which results in a force that will oppose the process that caused the change of magnetic flux.

> **Note**
> Lenz's law can be used to work out the direction of an induced current. Simply label a force opposite to the motion and then use the right-hand push rule to determine the direction in which the current must flow to create that force.

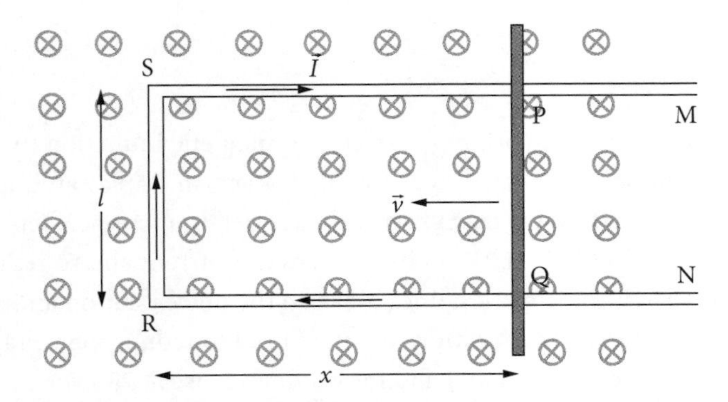

FIGURE 2.15 As the conductor is moved to the left through the magnetic field, it will experience a change of magnetic flux (due to the area enclosed by the loop decreasing) and so an emf will be induced from P to Q. Since there is a complete conducting loop, positive charge will move through the circuit in a clockwise direction. When charge flows from P to Q, that section of the circuit will experience a motor effect force to the right – in the direction opposite to the motion that is causing the change of flux.

It can be seen that changing the magnetic flux produces an electrical energy output. Thus, an energy input (work) must be made to produce this output. Similarly, it can be argued that in order to continue changing the magnetic flux, work must be done against an opposing force.

In each approach, the work done will be transformed into electrical energy by the process of electromagnetic induction and, consequently, Lenz's law is a statement of the law of conservation of energy.

Situations in which electromagnetic induction occurs include:

- relative translational motion between a conducting wire, conducting surface or solenoid, and a magnet (Figure 2.16a)
- rotational motion between a coil or conducting surface and a magnet (Figure 2.16b)
- a change in area enclosed by a conducting loop in a magnetic field (Figure 2.16c)
- changes in electrical supply to one coil adjacent to another coil.

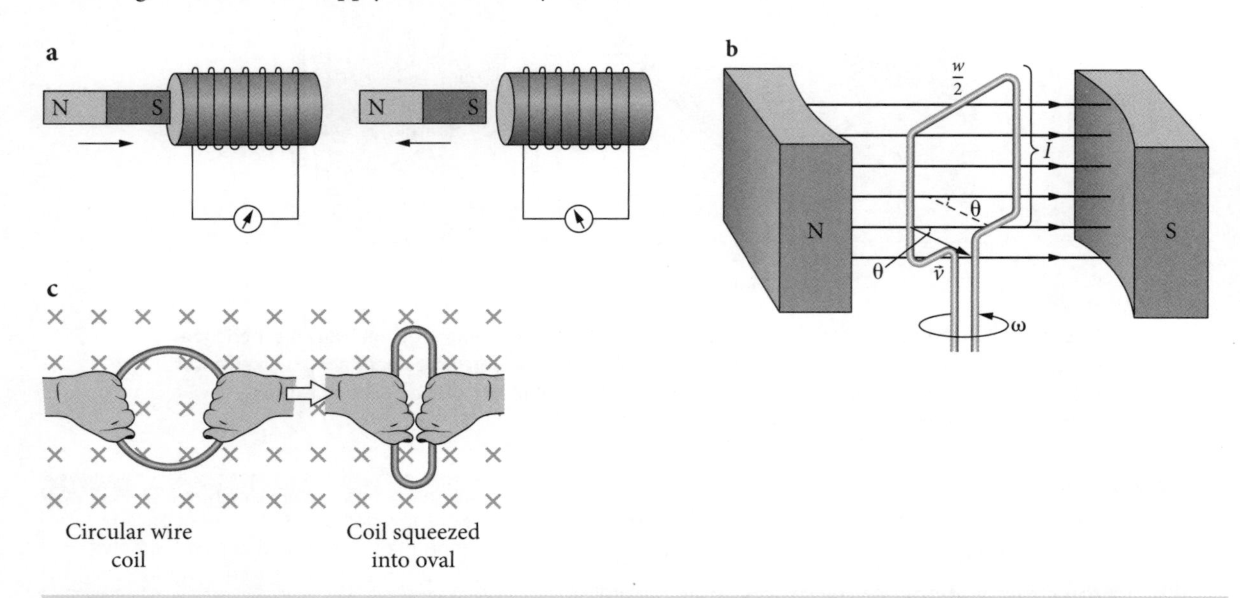

FIGURE 2.16 In each example, it can be seen that a change of flux results in electromagnetic induction: **a** relative motion between a solenoid and a magnet, **b** rotational motion between a conducting loop and a magnet and **c** a change in the area of a conducting loop in a magnetic field.

2.3.3 Transformers

A **transformer** is a device that uses the principles of electromagnetic induction to either increase AC voltage (step-up) or decrease AC voltage (step-down). Transformers play an important role in the electricity distribution network. **Step-up transformers** are used to increase (step up) voltages from the power station from about 25 kV to 330–500 kV. This reduces the current and so reduces **transmission power loss** (since $P_{loss} = I^2R$) in transmission lines between the power station and residential or industrial consumers. **Step-down transformers** are then used to reduce the voltage for industrial applications (sometimes in the realm of 11 kV) and for residential use at 230 V. Devices such as laptops feature further step-down transformers for reduction to lower voltages still.

Transformers are constructed of two conducting coils in proximity with one another and linked with a common **soft iron core**. The coil with the input AC voltage is called the **primary coil** and the other coil, from which the altered AC output will come, is called the **secondary coil**.

Note

Plug packs, commonly used to connect to your laptop or phone or the base of your desk lamp, will usually house a transformer. These transformers will be reasonably heavy (because of the iron core and copper wiring) and will get warm with use (unwanted energy transformation).

The supply AC voltage of the primary coil results in a changing magnetic flux being experienced by the secondary coil. As a consequence, an emf is induced in the secondary coil. The iron core through the coils acts to maximise the amount of magnetic flux that is 'channelled' from the primary coil to the secondary coil – called the **flux linkage**.

Since the magnetic flux in the iron core is proportional to the number of turns in the primary coil and the induced voltage is proportional to the number of turns in the secondary coil, then the relationship between the primary voltage and the secondary voltage can be stated:

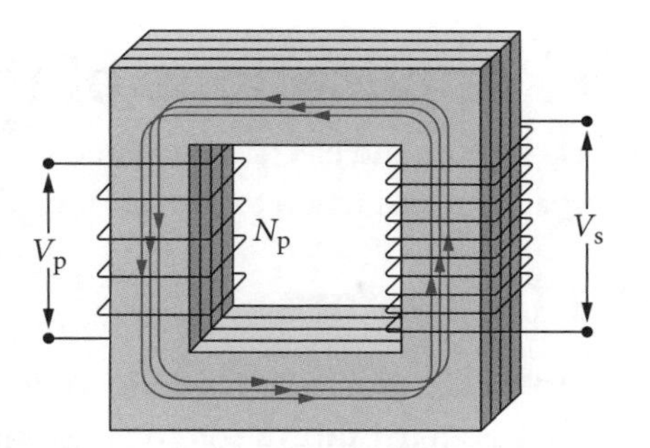

FIGURE 2.17 An AC supply to the primary coil results in a changing magnetic field to be 'channelled' through the soft iron core to the secondary coil. Here the change in flux induces an emf and a current, which will be AC in nature.

$$\frac{V_p}{V_s} = \frac{N_p}{N_s}$$

Controlling the ratio of number of turns in the primary coil to number of turns in the secondary coil can be used to effectively dictate the voltage of the secondary coil and, thus, whether it is a step-up or step-down transformer.

If the transformer was able to convert all input power into output power, it would be described as an **ideal transformer** and would be said to have an **efficiency** of 100%. As electrical power is described by the equation $P = VI$, for the transformer this would mean that $V_pI_p = V_sI_s$.

Note that if the transformer 'steps up' the voltage by a given factor, then the current will effectively be reduced (stepped down) by that factor in the process.

In reality no transformers are ideal. Large-scale industrial and commercial transformers can have efficiencies of better than 95%. The unwanted transformation of energy that occurs in real transformers is a consequence of three main processes.

1 The soft iron core is subjected to a change in magnetic flux and will, therefore, have **eddy currents** induced within it. Eddy currents are currents that circulate in a block or sheet of conductive material. The eddy currents that can circulate in the soft iron core will result in energy being converted to heat in the core (called resistive heating). This unwanted energy transformation can be reduced by minimising the size of the eddy currents by making the iron core out of many thin layers with an insulating material between them (laminating). **Laminated iron cores** are more efficient (Figure 2.18).

a

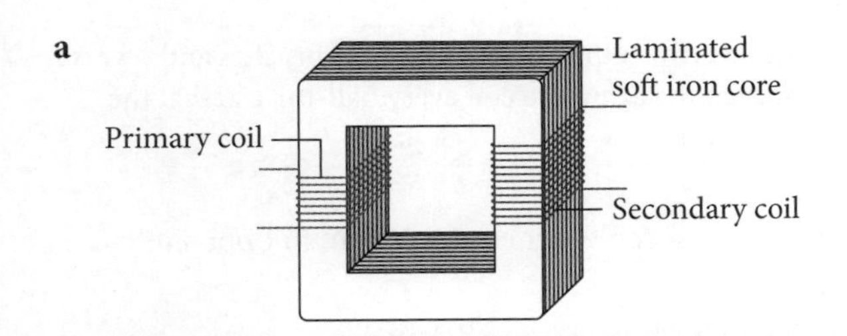

b

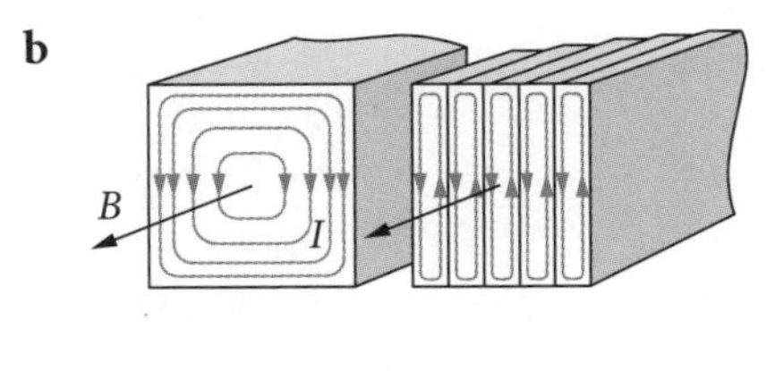

FIGURE 2.18 a The arrangement of laminations in the soft iron core. b The narrow 'slices' enable much smaller eddy currents to be induced.

2 The wires of the primary and secondary coils, although usually made from copper, will have some resistance. Power will be dissipated as heat by these wires, which can be considered using the equation $P_{loss} = I^2R$. This loss can be minimised by reducing the resistance of the wires by using thicker wires and/or cooling the transformer.

3 Incomplete flux linkage is a process in which magnetic flux generated by the primary coil is not contained in the core and thereby incompletely induces secondary emf. This reduces the efficiency of the transformer and is minimised by constructing an effective iron core.

2.4 Applications of the motor effect

Electrical motors and electrical generators rely on the simple principles of the motor effect and electromagnetic induction to provide a vast array of functions.

2.4.1 DC motors

A **DC motor** is a simple device designed to transform electrical energy into kinetic energy and is manifested as an **armature** coil rotating under the influence of a torque. The components of a DC motor are described in Table 2.3.

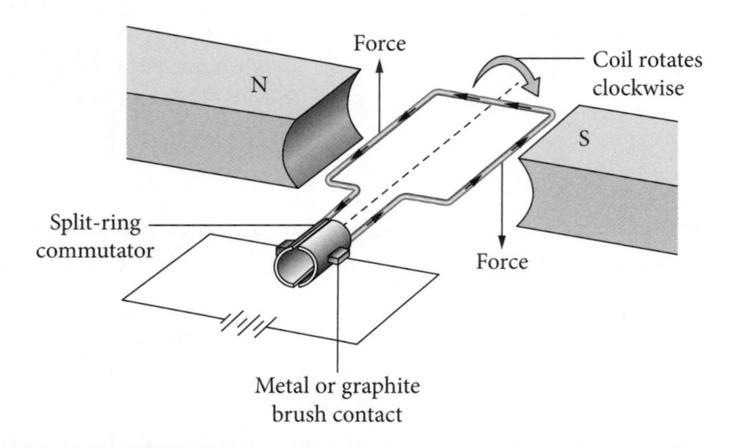

FIGURE 2.19 Electrical energy from the power source results in two motor effect forces when the current flows through the conducting loop. These two forces will generate a torque that causes the conducting loop or coil to rotate.

TABLE 2.3 Components of a DC motor

Component	Function
Armature coil	Built around an axis and thus able to rotate, it carries current at 90° to the magnetic field. The resulting motor effect forces create a torque that can cause rotation.
DC power source	Provides input **direct current (DC)** (I).
Brushes	Enable an electrical connection between the stationary conducting wires from the power source and the **split-ring commutator** that is rotating with the armature coil.
External magnetic field	Provides a (generally uniform) magnetic field of strength (B).
Split-ring commutator	Provides an electrical connection between the armature and the brushes and reverses the direction of the current through the armature coil every half-turn so that the torque is consistently in one direction.

Each side of the armature coil experiences a motor effect force. As these forces are in opposite directions at a distance from the pivot, a torque is generated.

The magnitude of the force on each side of the coil will be given by $F = lIB\sin\theta$.

As the side of the coil will be perpendicular to the uniform magnetic field throughout each rotation (and therefore $\theta = 90°$ and $\sin\theta = 1$), the equation can be simplified to $F = lIB$. On each side of the coil this force will be acting at a distance, r, equal to half the width, w, of the coil.

Recall from Module 5 that:

$\tau = r_{\perp}F = rF\sin\theta$

The total torque, therefore, will be given by adding the torques contributed by each side, so $\tau = 2 \times \frac{w}{2} \times lIB \times$ sine of the angle made between the line of the radius and the motor effect force. Since $2l \times \frac{w}{2}$ is the area (A) of the coil, then the torque formula for a motor with n turns of wire in the coil can be written as:

$\tau = nIAB\sin\theta$

where θ is the angle between the area vector (from section 2.3.1) and the magnetic field lines. Note that this θ is distinct from the one in the Module 5 torque equation and has been redefined. This means that torque will be a maximum when the plane of the coil is parallel to the field lines.

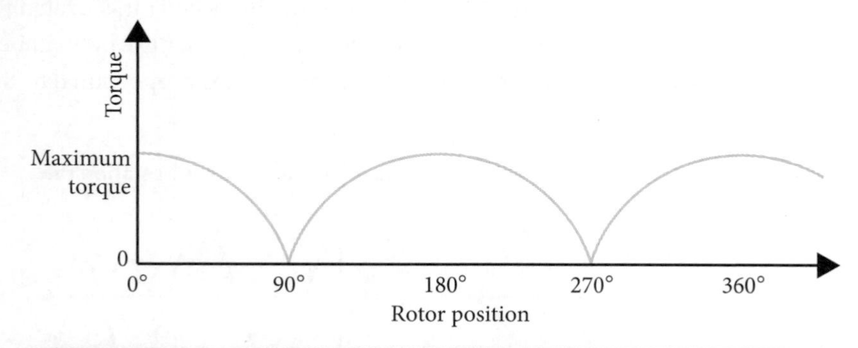

FIGURE 2.20 The action of the split-ring commutator reversing the direction of the current in the coil every half-turn results in a torque that changes over time but is always in one direction.

> **Note**
> In a real motor, the stages at which there is zero torque would rely on momentum for the motor to keep running. Starting the motor from one of these stages by switching the current on would be impossible. Several design features are used to minimise or overcome this issue. Radial magnets can be used to create more consistent torque throughout the rotation (although there will still be zero points). Multiple coils offset at an angle to one another ensure that there will not be a stage during the rotation at which the torque will be zero.

When in operation, a DC motor will have a conducting coil in motion in an external magnetic field. The coil will experience a change in magnetic flux and, consequently, act as a generator in which an emf will be induced. The direction of this emf will be such that it opposes the **supply emf** and, consequently, it is called a **back emf**. This will have the effect of reducing the total emf of the coil according to the relationship total emf = supply emf – back emf. The current through the coil will be determined by the total emf and the coil resistance, as in the version of Ohm's law expressed as $I = \frac{\varepsilon_{total}}{R}$.

The faster the motor spins, the greater the rate of change of magnetic flux and thus the greater the back emf. So the total emf experienced by a motor decreases as a motor spins faster. Because the total emf of the motor will determine the current through the coil, which will, in turn, determine the torque that enables the motor to spin faster, the back emf acts as a speed limiter for the motor. The speed limit will depend on motor design and factors such as load and friction. In an imaginary, ideal motor with no load, the maximum speed will occur when the back emf exactly equals the supply emf because this is the point at which the torque applied becomes zero.

When a motor is first connected to power or is prevented from spinning, there is no back emf and, consequently, the current is a maximum. Since the resistance of the copper coil is quite small, this current is usually quite large and can cause overheating damage associated with melting the insulation. (The wire of the coils is insulated to prevent a short circuit across coil loops.) This is called burning out the motor.

> **Note**
> Variable resistors are used to ensure burn-out does not occur when a motor is first switched on and is yet to start rotating at a sufficient rate to decrease current to safe operating levels.

2.4.2 Generators

A **generator** is a device that converts kinetic energy into electrical energy; that is, it is a device that performs a process in reverse to that performed by a motor. Electrical energy not produced by batteries or solar panels is produced by generators.

An external source of energy is required to cause the rotation of the armature coil, and the induced current will flow from the coil through an external circuit via the brushes.

A **DC generator** features the same components as a DC motor, but now the split-ring commutator acts to ensure the induced current in the external circuit flows in one direction only.

Similarly, in an **AC generator** the **slip rings** enable an **alternating current** to flow into an external circuit.

When a simple generator (see Figure 2.16b) is rotated uniformly, the magnetic flux changes in a sinusoidal fashion with time. Since, according to Faraday's law, the induced emf is related to the negative of the rate of change of magnetic flux, it too will have a sinusoidal nature but will be out of phase, as seen in Figure 2.21. It can be shown that the emf vs time graph is proportional to the negative of the gradient (rate of change) of the flux vs time graph using Faraday's law.

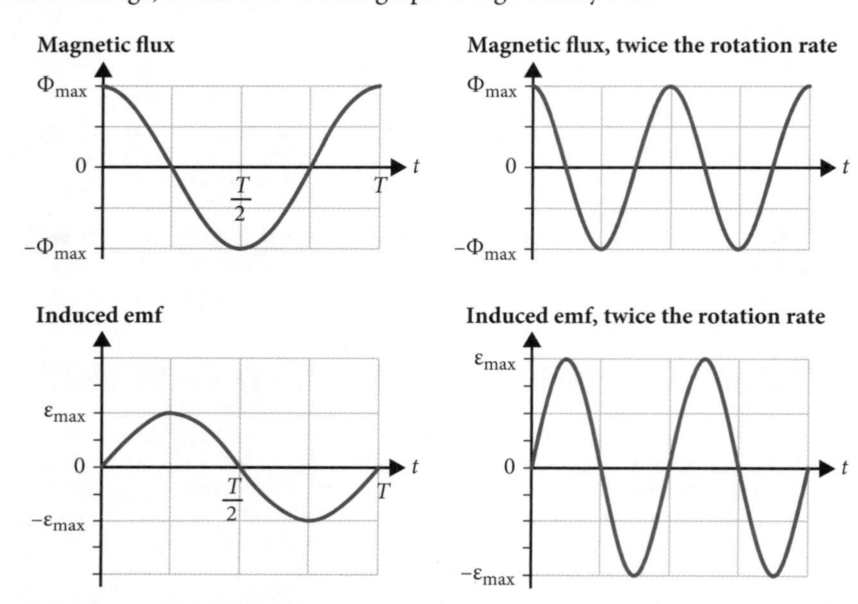

FIGURE 2.21 The emf generated can be seen to be proportional to the negative of the gradient of the magnetic flux–time graph at that time. Doubling the rate of change of flux results in twice the emf.

Note

Viewed perpendicular to the axis, the coil appears to 'open' and 'close' as the assembly rotates (see Figure 2.16b and c).

The 'magnetic flux' referred to above is really the magnetic flux 'threading' the coil.

When the coil is 'open', the flux threading the coil is maximum but the *rate* of flux change is virtually zero. When the coil is 'closed', the flux threading the coil is virtually zero but the *rate* of flux change is maximum.

Faraday's law refers to the production of emf (an available voltage), not current, and, therefore, not energy. We are only working against internal friction to turn the generator! The electrical energy only becomes available when the generator is connected to an external circuit and a current is enabled.

If an **electrical load** is placed in the external circuit, then electrical energy will be transformed by the load. As a consequence, through conservation of energy, the generator will be harder to turn than if the external circuit was incomplete. This can also be explained by the motor effect – once a current flows in the armature coils, within the external magnetic field, the coils will experience a force in the direction that opposes the direction in which the generator is turning.

2.4.3 AC induction motors

As the name implies, an **AC induction motor** uses electromagnetic induction to induce the current in the **rotor** and no electrical connection is required between the stationary part of the motor and the rotating part. This high-efficiency, low-wear design makes AC induction motors popular for many applications from circular saws to blenders.

The AC induction motor consists of several pairs of opposing electromagnets arranged in a circle and connected to phased AC power (see Figure 2.22a). This phased AC power ensures that the pairs of magnets produce magnetic fields that change over time in a way that the fields are sequentially out of step with each other (see Figure 2.22b). In this way, the **stator** electromagnets produce a north magnetic pole that progressively moves around the circle of electromagnets. Opposite this 'moving' north pole will be a moving south pole created by the other member of the pair.

Within the arrangement of electromagnets is a **'squirrel cage' rotor** consisting of parallel copper bars connected by conducting end plates (see Figure 2.22c). This rotor will be subjected to a change in magnetic flux as the magnetic field 'rotates' around the stator electromagnets. The change in magnetic flux induces an emf in the squirrel cage rotor and, consequently, currents circulate through the bars and end plates of the rotor. These currents within the magnetic field of the electromagnet result in a torque that causes the squirrel cage to rotate.

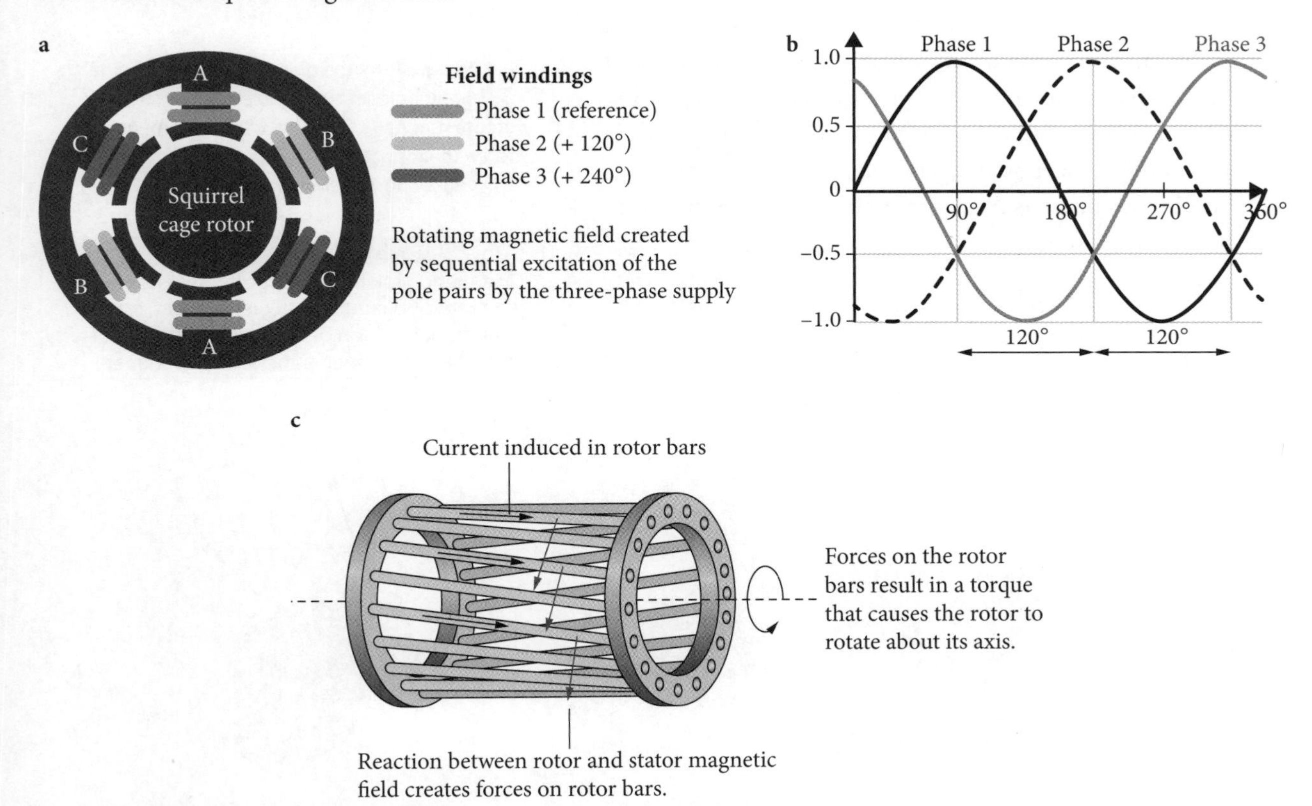

FIGURE 2.22 a The three pairs of stator electromagnets that produce a 'rotating' magnetic field in which the squirrel cage rotor will be made to turn. b The graph of the phased power supply to the stator magnets. c The squirrel cage rotor with induced currents flowing through bars and resulting in a motor effect force that creates the torque that turns the motor.

The rotation will be in a direction that minimises the change in magnetic flux (Lenz's law) – that is, in a direction that 'chases' the changing magnetic field.

It should be noted that the torque of an AC induction motor will be greater when the motor operates under load because it will fall behind the changing magnetic field and experience a greater change in magnetic flux.

Note

The substantial advantages of AC induction motors mean that they are the dominant AC motor used in devices, appliances and industry.

2.4.4 Lenz's law – DC motors and electromagnetic braking

Recall that Lenz's law states that an induced emf will be such as to produce a current that will create magnetic fields that oppose the change in flux that induced the emf. Essentially, this is a statement of the conservation of energy – if the induced current enhanced the change in flux, then this process would cycle upwards, creating greater currents and greater magnetic flux indefinitely.

In a DC motor, we can see Lenz's law in action through the back emf (discussed earlier) that is induced as a consequence of the coil of the motor moving in the motor's external magnetic field and thus acting as a generator.

When there is relative movement between a piece of conducting material and a magnetic field (such as when a magnet is dropped through a hollow copper tube or a metal disc is rotated past a magnet), a change of magnetic flux occurs. This change of magnetic flux will result in the induction of eddy currents in the conducting material. These eddy currents will create magnetic fields that will, as per Lenz's law, result in a force that opposes the relative motion. That is, there will be a resulting magnetic force that opposes the motion. This process is called **magnetic braking** and has many industrial applications, such as for train wheels and amusement parks, because of its advantages over traditional friction brakes.

Energy is seen to be conserved in magnetic braking, as kinetic energy is transformed into heat energy within the conductive material.

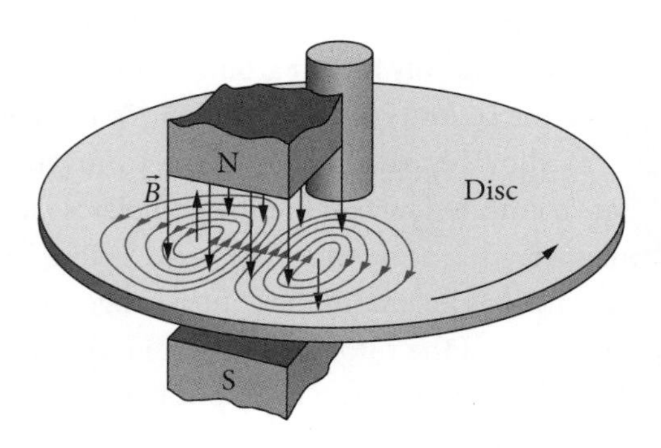

FIGURE 2.23 As the disc turns in the external magnetic field, the induced eddy currents will create magnetic fields resulting in a magnetic force that opposes the change in flux. This force will cause the wheel to slow down.

Note

Magnetic braking is most effective when the rate of change of flux is greatest, so traditional friction brakes are often used after the speed has been substantially reduced.

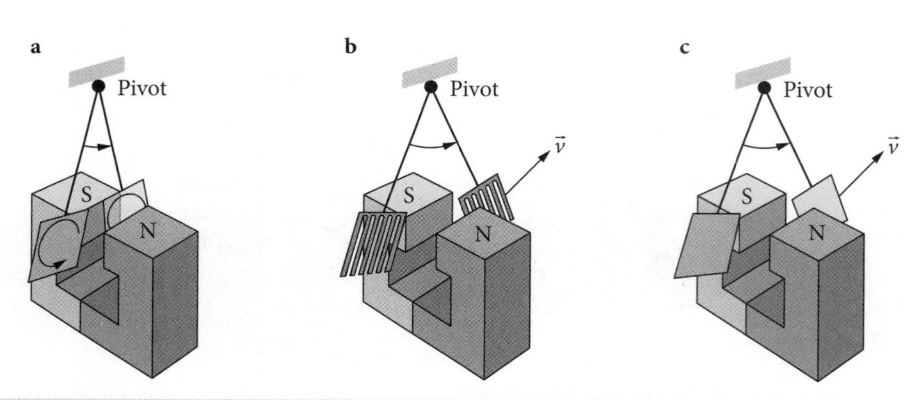

FIGURE 2.24 a Eddy currents induced in the conductor as it enters and leaves the magnetic field create magnetic fields that will result in forces that oppose the change in flux; that is, the motion. As a result, the pendulum slows rapidly. b Eddy currents induced in the conductor as it enters and leaves the magnetic field will be reduced in size because of the slits cut in the conductor. Hence, the magnetic fields created will be much weaker and will result in minimal forces that oppose the change in flux. As a result, the pendulum slows gradually. c Since the pendulum material is not conductive, there will be no eddy currents induced in the pendulum.

Glossary

AC generator A generator that produces an alternating current output

AC induction motor A motor using electromagnetic induction to create a current in the squirrel cage rotor from a changing magnetic field provided by stator electromagnets with an AC supply

alternating current (AC) A current that varies sinusoidally over time, changing in magnitude and direction over its period

ampere The SI unit of current; defined by the relationship between force and current in parallel conductors

area vector The vector associated with a plane surface, the direction of which is perpendicular to the plane surface

armature The rotating structure of a motor or generator, including wire coils

back emf The induced emf that occurs in a motor coil as a consequence of the change in flux that the coil experiences as the motor turns in a magnetic field

brushes A part of the stator of a motor or generator that acts to provide an electrical connection between the moving part of the device and the stationary part

DC generator A generator that produces a direct current output

DC motor A motor that uses direct current input to convert electrical energy into kinetic energy

direct current (DC) Current that always flows in one direction

eddy currents Circulating currents created by a change in flux experienced by a block or sheet of conductive material

efficiency The ratio of useful power output to input power; usually expressed as a percentage

electric charge A fundamental physical property of matter that causes it to experience a force when placed in an electric field; measured in coulombs

electric field The field created by an electrically charged object and in which another charged object will experience a force

electrical load An electrical component that transforms electrical energy

electrical potential difference A measure of the electrical energy difference between two points in an electric field and defined by the work done moving a unit of positive charge from one point in the field to another

electromagnetic induction The production of an electromotive force in a conductor as a result of a change in flux

https://get.ga/aplus-hsc-physics-u34

A+ DIGITAL FLASHCARDS

Revise this topic's key terms and concepts by scanning the QR code or typing the URL into your browser.

electromotive force (emf) The potential difference, produced by a battery or generator device, that may cause a current to flow in a conductor

Faraday's law A law that states that the induced emf caused by a change of flux is equal to the negative of the rate of change of flux

flux linkage The magnetic connection between coils of a transformer, or the amount of flux created by one coil that passes through the other coil

generator A device that uses electromagnetic induction to convert kinetic energy to electrical energy

ideal transformer A transformer that converts all input power into useful output power

induced current A current that results from an induced emf

induced emf An emf created by a changing magnetic flux

laminated iron core An iron core constructed of thin, insulated slices of iron to minimise eddy current production in the iron core

Lenz's law A law that states that the direction of an induced current is such that it will create a magnetic field, which will result in a force that opposes the change in magnetic flux that induced the current

magnetic braking The braking or slowing of relative motion of a magnet (or electromagnet) and a conductor as a result of the formation of eddy currents that generate a magnetic force opposing the motion

magnetic field A field created by moving charged particles and currents, which exerts a force on moving charged particles and currents

magnetic flux A measure of the amount of magnetic field permeating a region of space

magnetic flux density A measure of the strength and direction of a magnetic field at a point

motor effect The force experienced by a current-carrying conductor in a magnetic field

primary coil A coil connected to an AC supply that generates a changing magnetic field in a transformer

rotor The rotating part of a motor or generator

secondary coil A coil within which an alternating current is induced as a consequence of a changing magnetic field in a transformer

slip rings Components of an AC motor or generator that provide an electrical connection between the armature and the external circuit

soft iron core The component of a transformer that intensifies and contains the magnetic field of the primary coil so it is optimally linked (flux linkage) to the secondary coil

split-ring commutator The component of a DC motor or generator that provides an electrical connection between the armature and the external circuit

squirrel cage (rotor) The rotating part of an AC induction motor consisting of parallel conducting rods joined to circular copper end plates

stator The stationary part of a motor or generator

step-down transformer A transformer with an output voltage less than the input voltage as a result of having fewer turns on the secondary coil than on the primary coil

step-up transformer A transformer with an output voltage greater than the input voltage as a result of having more turns on the secondary coil than on the primary coil

supply emf The potential difference provided to a motor by a battery or other external source

transformer A pair of coils used to increase or decrease output alternating (AC) voltage by the process of electromagnetic induction

transmission power loss The power lost, in the form of heat, because of the resistance of conducting wires during the process of transmitting electrical energy from the power station to the consumer

Exam practice

Multiple-choice questions

Solutions start on page 158.

Charged particles, conductors, and electric and magnetic fields

Question 1

The diagram represents an electric field between a metal rod on a building and a charged cloud.

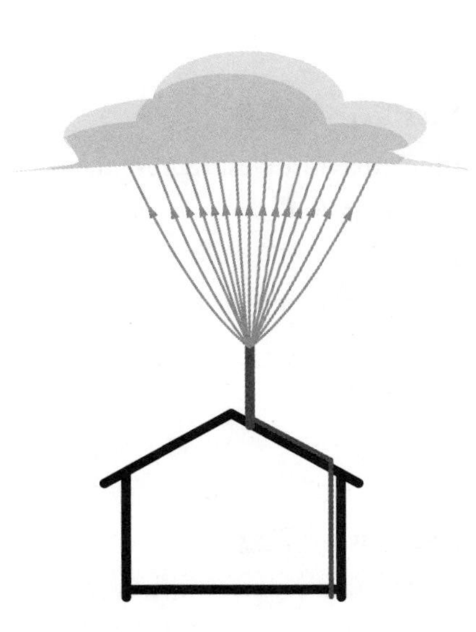

An electron placed in the electric field will move in the direction

A of the field with increasing acceleration.

B against the field with increasing acceleration.

C of the field with decreasing acceleration.

D against the field with decreasing acceleration.

Question 2

A negatively charged particle in a uniform gravitational field is positioned midway between two charged conducting plates.

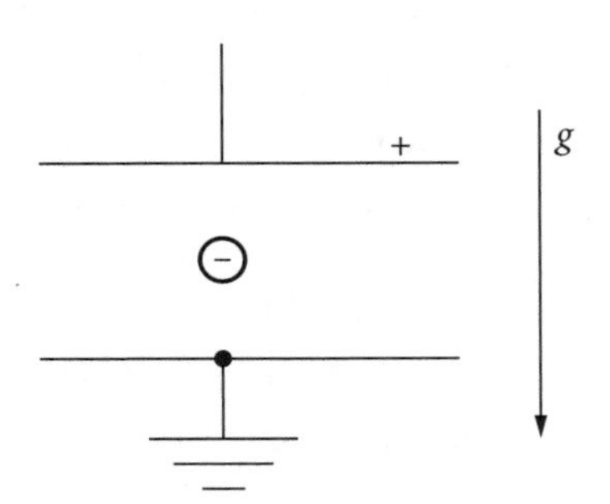

The potential difference between the plates is adjusted until the particle is at rest relative to the plates.

What change will cause the particle to accelerate downwards relative to the plates?

A Decreasing the charge on the particle

B Decreasing the separation between the plates

C Increasing the length of the plates

D Increasing the potential difference between the plates

Question 3

Two point charges, $+q$ and $-q$, are placed as shown.

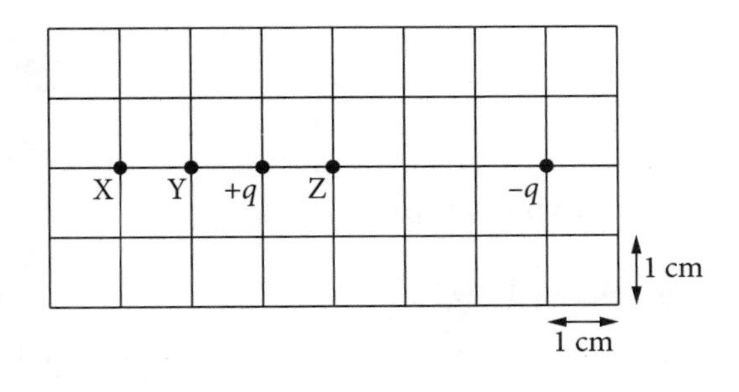

What is the order (from largest to smallest) of the magnitude of the electric field strength at the three points X, Y and Z?

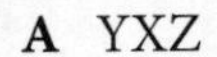

A YXZ **B** ZXY

C ZYX **D** YZX

Question 4

A proton travelling at velocity v enters a magnetic field at right angles to the field and follows a circular path of radius r.

If the strength of the magnetic field was doubled and the velocity halved, then the radius of the path of the proton would be

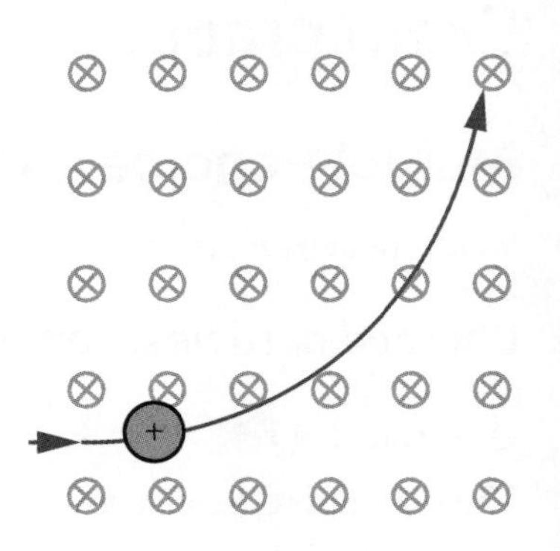

A $4r$

B $2r$

C $\frac{r}{2}$

D $\frac{r}{4}$

Question 5

Which diagram best represents the magnetic field created by two parallel wires carrying currents in the same direction?

A

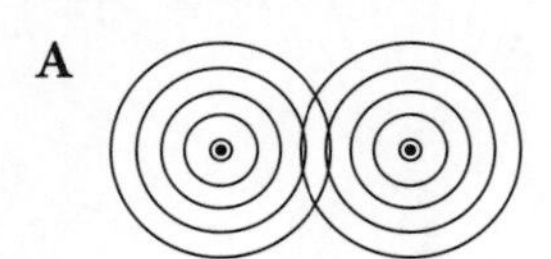

B

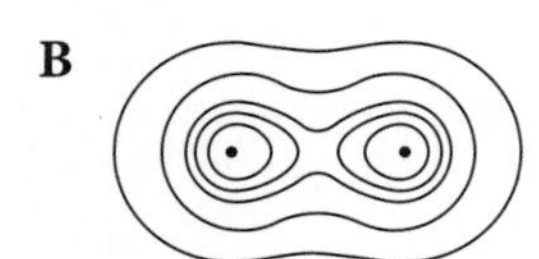

C

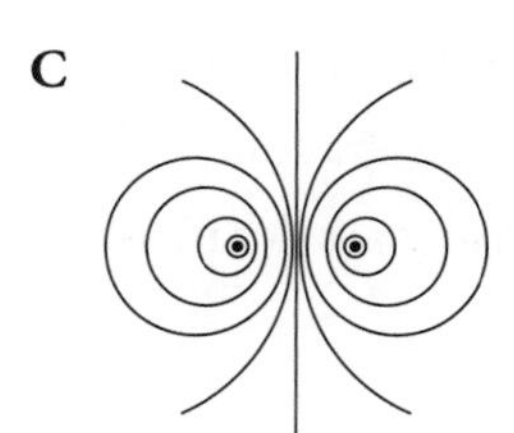

D

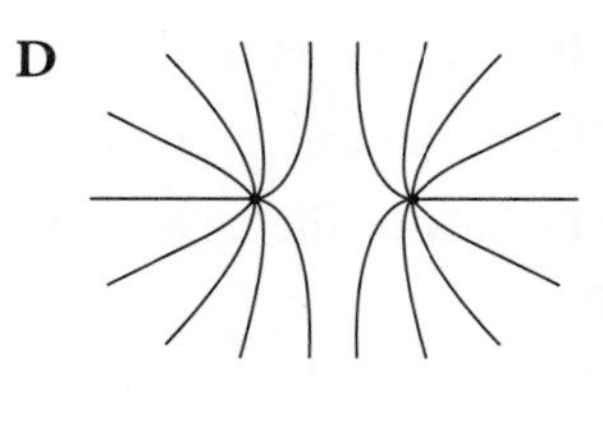

Question 6

The diagram represents the moment a photon of energy becomes an electron and an antielectron in a magnetic field. From the diagram, you can infer that the magnetic field is directed

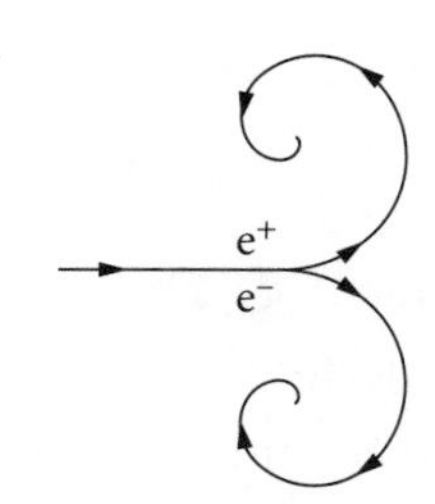

A out of the page and the particles are slowing down.

B out of the page and the particles are speeding up.

C into the page and the particles are slowing down.

D into the page and the particles are speeding up.

Question 7

The diagram shows four charged particles moving in directions as indicated adjacent to a current-carrying wire.

Which particle that will experience an initial force directly towards the wire?

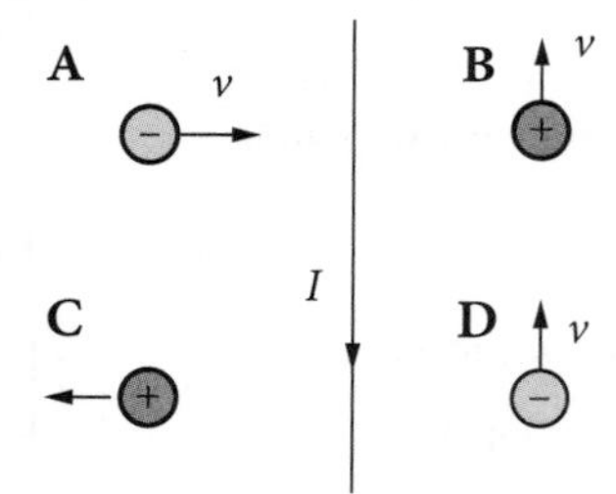

The motor effect

Question 8

When a wire with an electric current I is placed in a magnetic field of strength B, it experiences a force F. The direction of F is

A determined only by B.

B determined only by I.

C in the plane containing I and B.

D perpendicular to the plane containing I and B.

Question 9

A metal bar is placed across frictionless, parallel conductive rails. A uniform magnetic field is arranged as shown in the diagram. The bar is initially at rest when a current begins to flow from the power source.

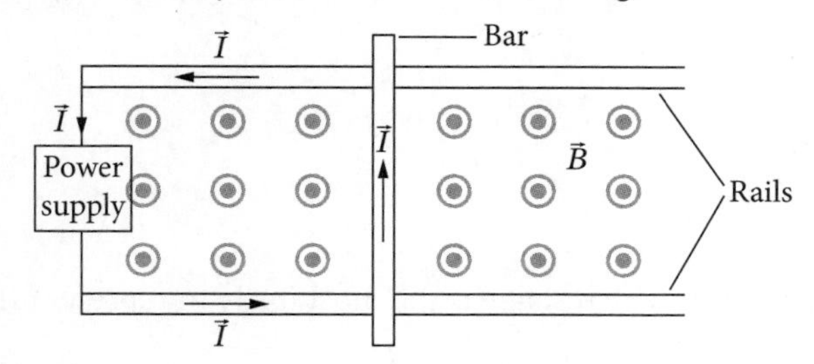

Which statement best describes the motion of the bar when the current begins to flow?

A The bar moves to the right.

B The bar remains at rest.

C The bar moves to the left.

D The bar twists, with the top moving left and the bottom moving right.

Question 10

A conductor placed between the poles of a magnet is seen to move vertically upwards when a current passes through it. Which diagram illustrates the orientation of the conductor and the direction of current flow that would enable this?

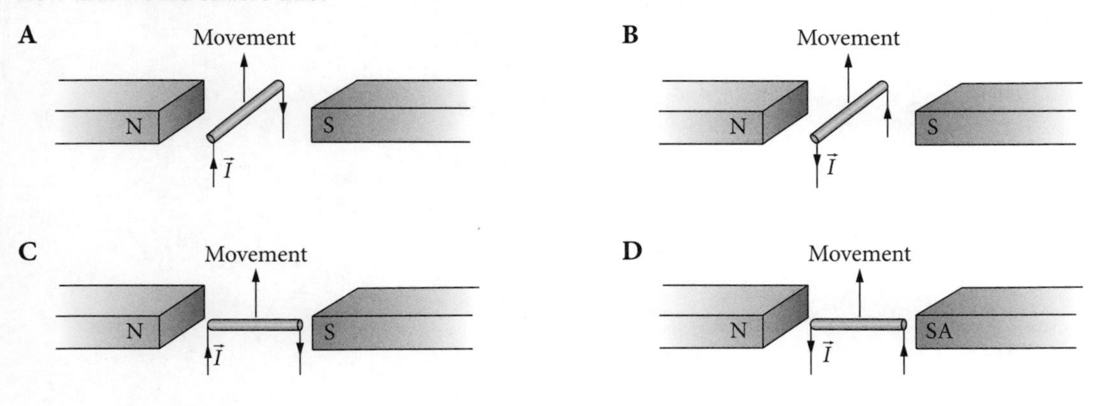

Question 11

Two parallel conducting wires carry current of equal magnitude and both experience a repulsive force of F N. Which changes would result in an attractive force of $8F$ N?

A Changing the direction of the current in both wires, doubling the current in both wires and halving the distance between the wires

B Changing the direction of the current in one wire, doubling the current in one wire and halving the distance between the wires

C Changing the direction of the current in one wire, doubling the current in both wires and halving the length of the wires

D Changing the direction of the current in one wire, doubling the current in both wires and halving the distance between the wires

9780170465304

Question 12

A flexible conducting wire is shaped as shown.

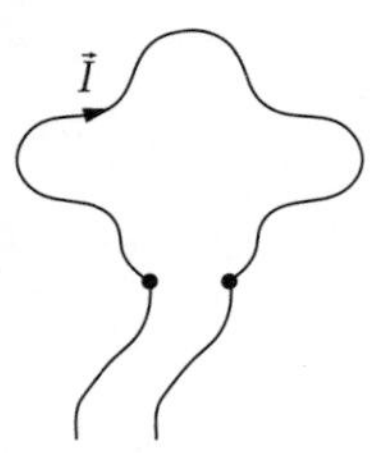

If the loop is connected to a battery, which shape is the most likely to represent the final shape?

A

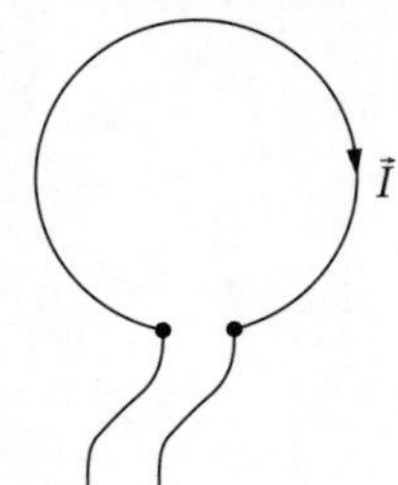

B

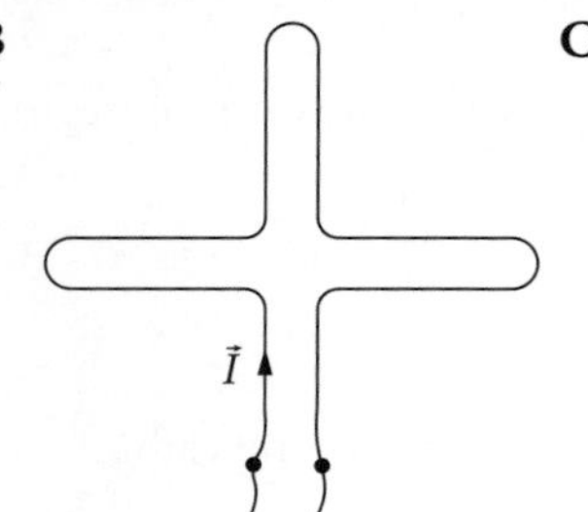

C

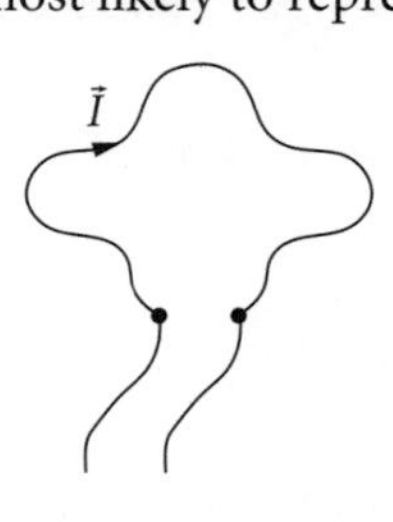

D

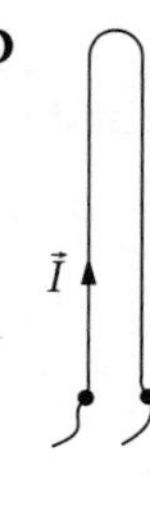

Question 13

A conducting wire XY is moved through a field as shown in the diagram.

Magnetic field 5.2 mT

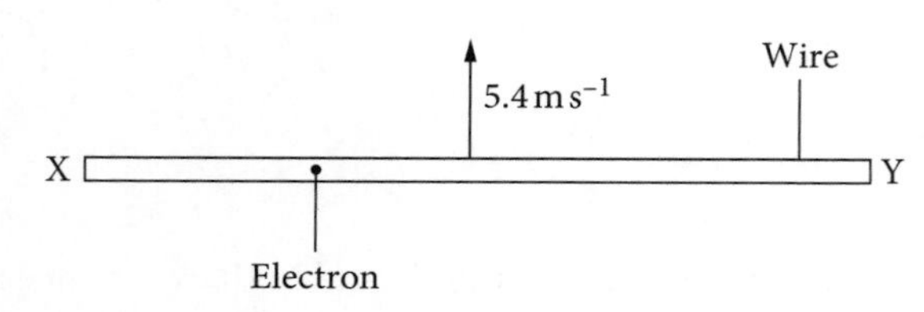

What will be the force on the electron in the conductor?

A 4.5×10^{-21} N towards X

B 4.5×10^{-18} N towards X

C 4.5×10^{-18} N towards Y

D 4.5×10^{-21} N towards Y

Question 14

A 10 cm long wire PQ lies horizontal to a uniform magnetic field of strength 200 mT. A current flows through the wire. This results in a 40 mN force acting upwards on the wire.

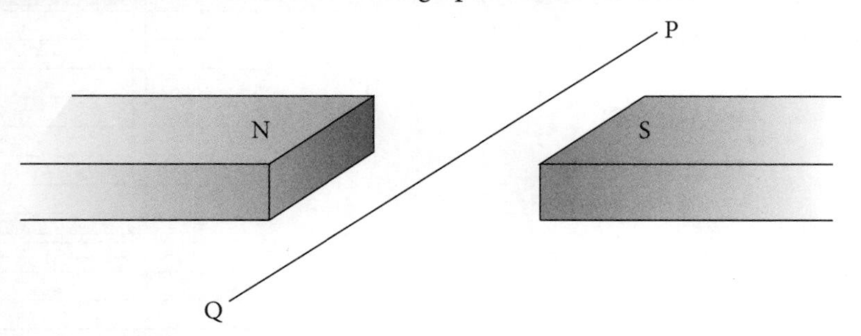

Which option lists the direction and magnitude of the current in the wire?

	Current direction	Current magnitude (A)
A	P to Q	2.0
B	P to Q	0.02
C	Q to P	2.0
D	Q to P	0.02

Electromagnetic induction

Question 15

A coil of wire is rotated about its axis within a uniform magnetic field created between the poles of two permanent magnets as shown.

Which diagram represents the position that shows the greatest magnetic flux through the coil?

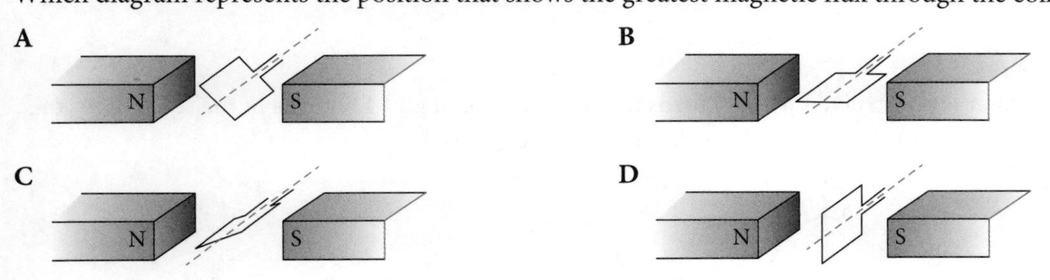

Question 16

A coil is made from three square loops of wire. The sides of each loop are 4 cm. The coil is positioned so that its plane makes a 60° angle with a magnetic field of 0.015 T, as shown in the side view below.

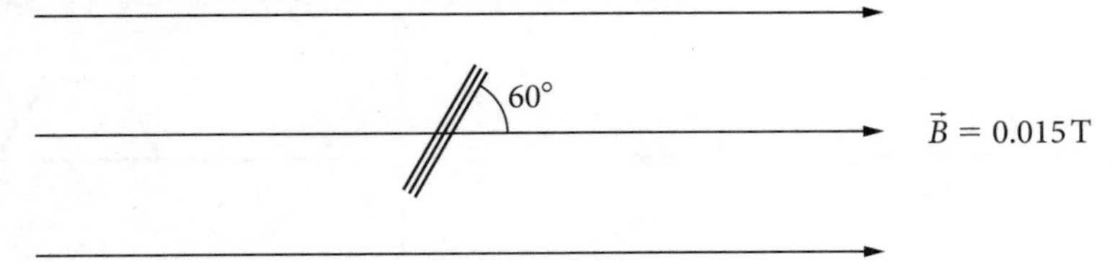

What is the magnitude of the magnetic flux through the coil?

A 6.2×10^{-5} Wb **B** 2.1×10^{-5} Wb **C** 3.6×10^{-5} Wb **D** 1.6×10^{-3} Wb

Question 17

The diagram shows a loop of wire in a magnetic field.

What would be the change in flux experienced by the loop of wire if it is rotated about its axis through 90° from the position shown?

A 7.2 Wb **B** 0.072 Wb

C 25.92 Wb **D** 0.025 92 Wb

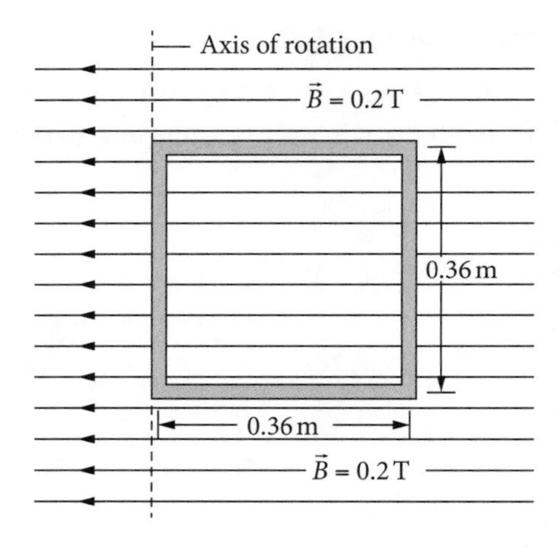

Question 18

The diagram below shows a

A step-up transformer with an output voltage that will be approximately double the input voltage.

B step-down transformer with an output voltage that will be approximately double the input voltage.

C step-up transformer with an output voltage that will be approximately half the input voltage.

D step-down transformer with an output voltage that will be approximately half the input voltage.

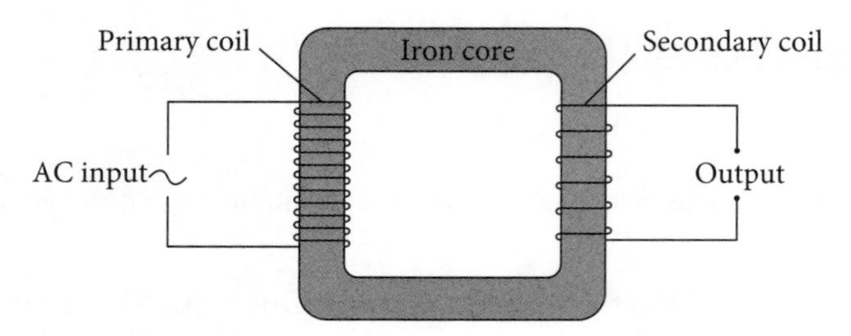

Question 19

An ideal transformer is constructed with 80 turns in the primary coil and 400 turns in the secondary coil. The input voltage to the primary coil changes as shown in the graph.

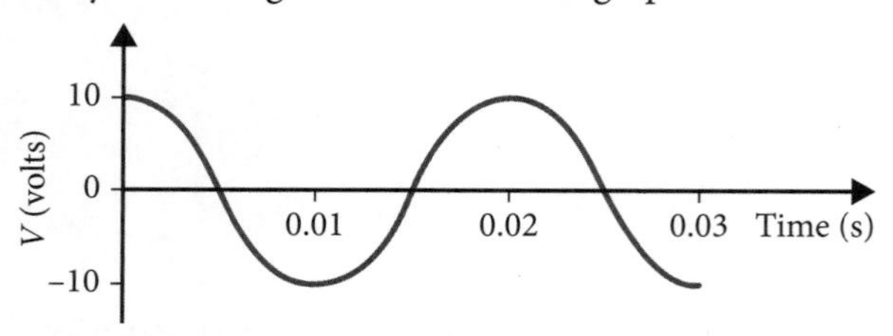

Which graph best represents the change in the output voltage of the transformer?

A

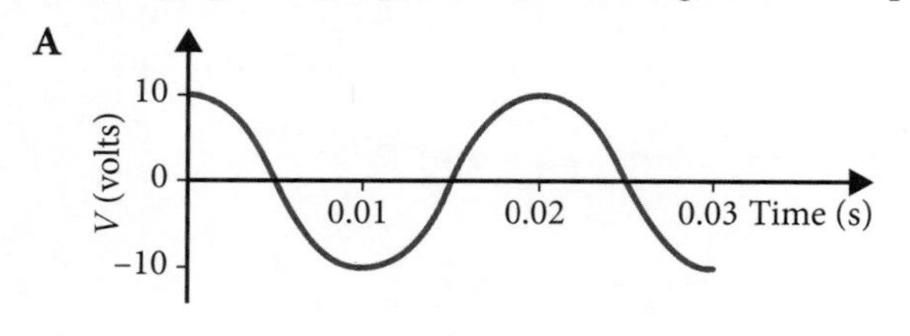

B

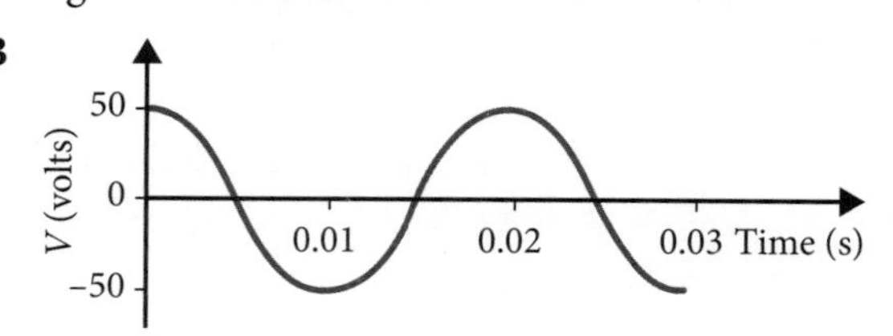

C

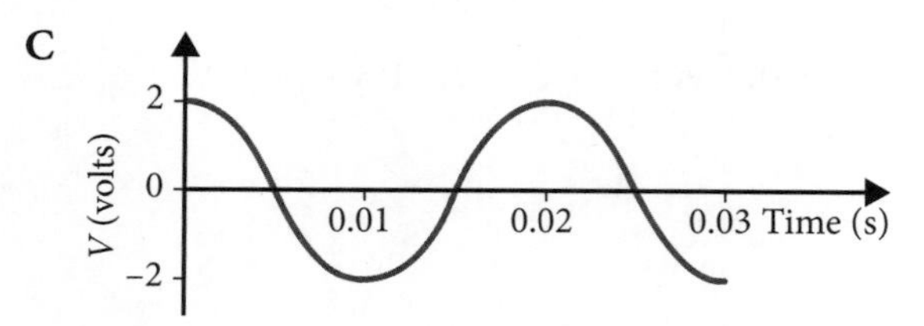

D

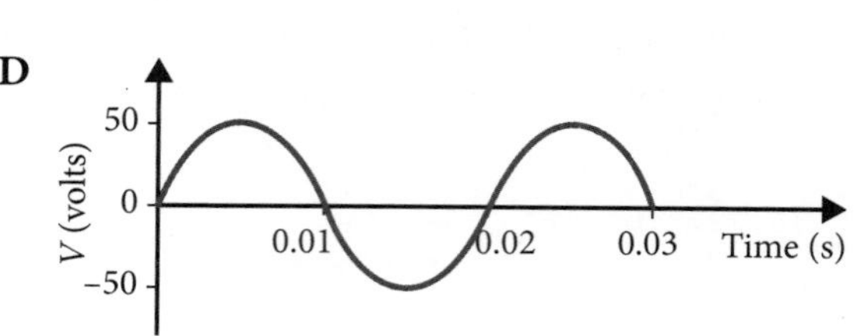

Question 20

A metal conducting rod is moved through a magnetic field at a velocity v as shown in the diagram.

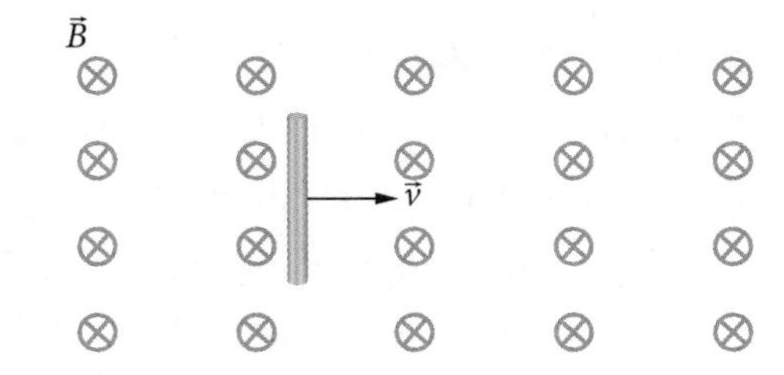

Which statement best describes what would be observed?

A The top end of the rod will become more positive.

B The bottom end of the rod will become more positive.

C The top end of the rod will become more positive and a sustained force opposing the motion will be felt.

D The bottom end of the rod will become more positive and a sustained force opposing the motion will be felt.

Question 21

A metal ring is dropped through a magnetic field as shown.

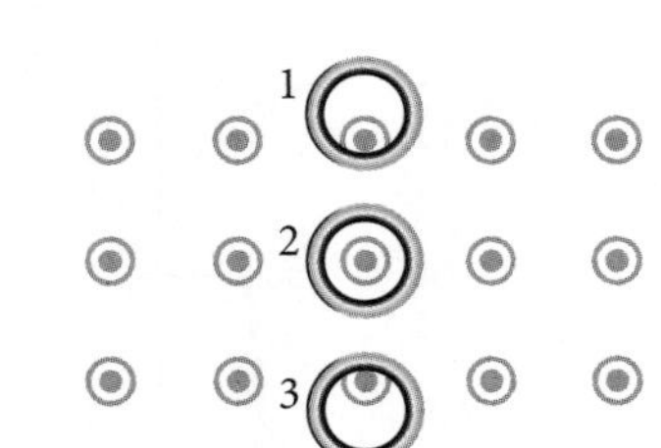

The current in the ring would be

A clockwise continuously.

B anticlockwise continuously.

C clockwise in position 1 and then anticlockwise in position 3.

D anticlockwise in position 1 and then clockwise in position 3.

Question 22

A bar magnet is dropped down the axis of a circular conducting ring as shown in the diagram.

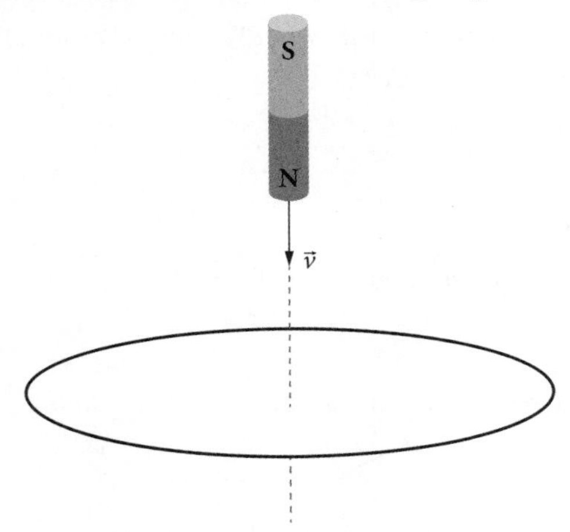

Which graph best represents the current through the ring as a function of time?

A

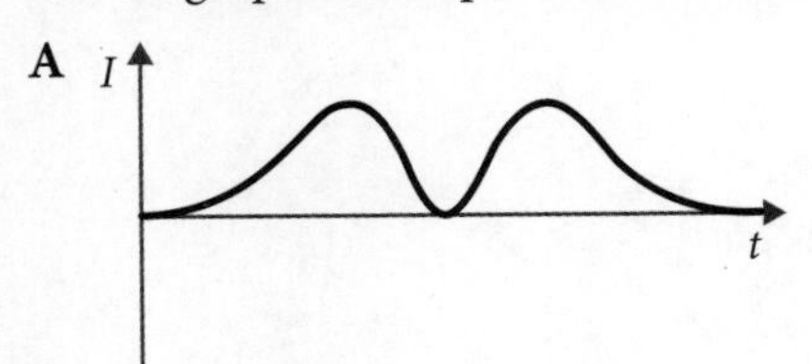

B

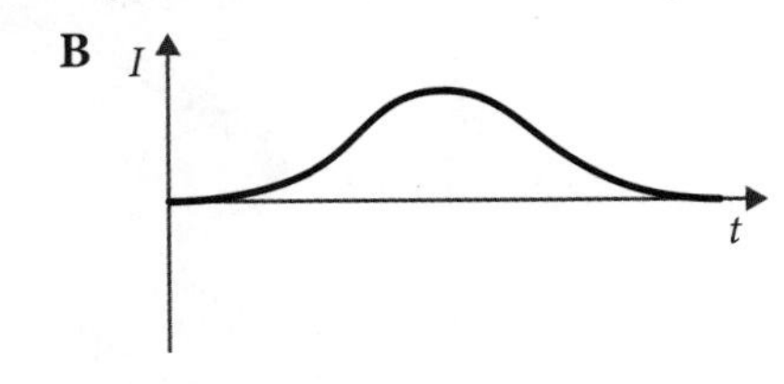

C

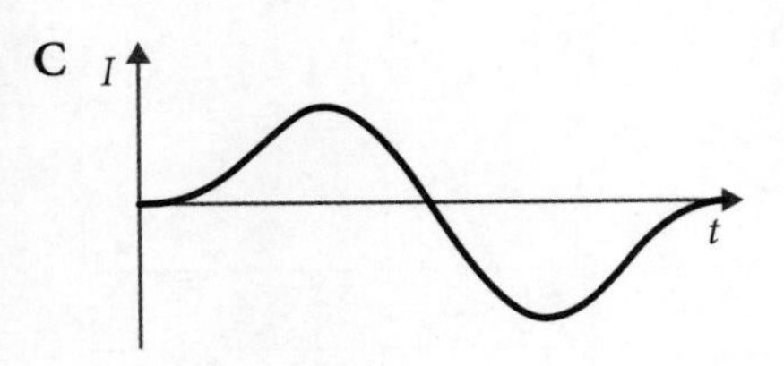

D

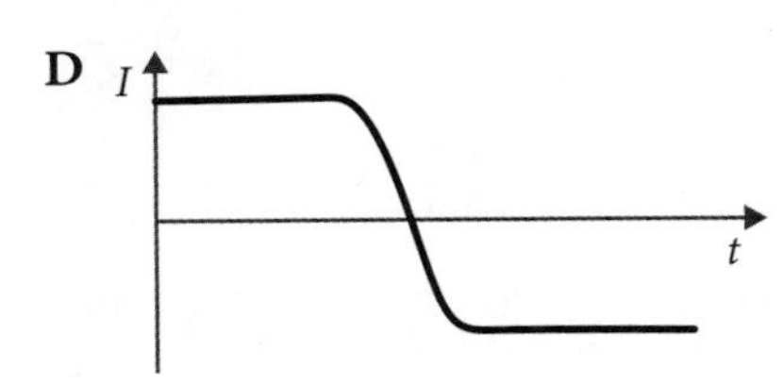

Applications of the motor effect

Question 23

The torque generated by a DC motor is greatest when the

A coil is rotating at its maximum speed.

B forces acting on opposite sides of the coil are in the same direction.

C plane of the coil is parallel to the field.

D plane of the coil is perpendicular to the field.

Question 24

A simple AC generator is connected to a cathode ray oscilloscope so that the generator output can be represented on the screen as seen below.

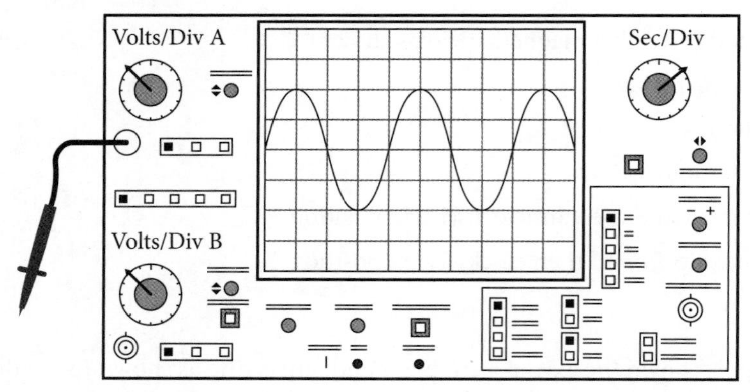

What changes could be made to the generator that would result in the amplitude of the curve doubling without any other change occurring to the curve?

A Rotating the generator at double the speed

B Double the number of coils

C Including an additional set of slip rings

D Including an additional set of magnets at 90° to the first set

Question 25

Which option correctly labels the parts of the device below?

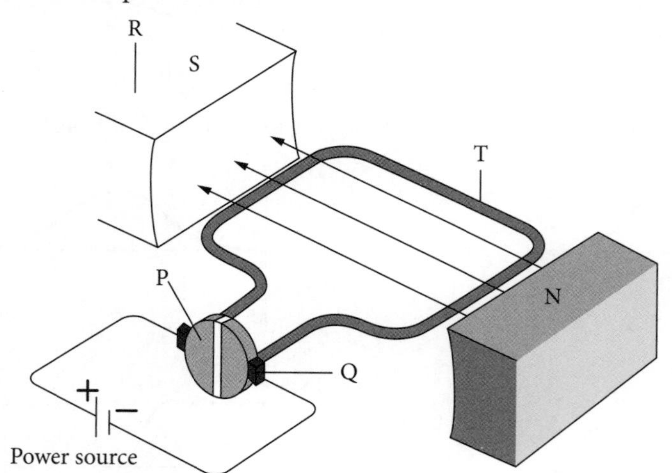

	P	Q	R	T
A	Armature	Split-ring commutator	Stator	Brush
B	Split-ring commutator	Brush	Stator	Armature
C	Armature	Split-ring commutator	Brush	Stator
D	Split-ring commutator	Brush	Armature	Stator

Question 26

The diagram is a depiction of an

A AC generator.

B AC motor.

C DC generator.

D DC motor.

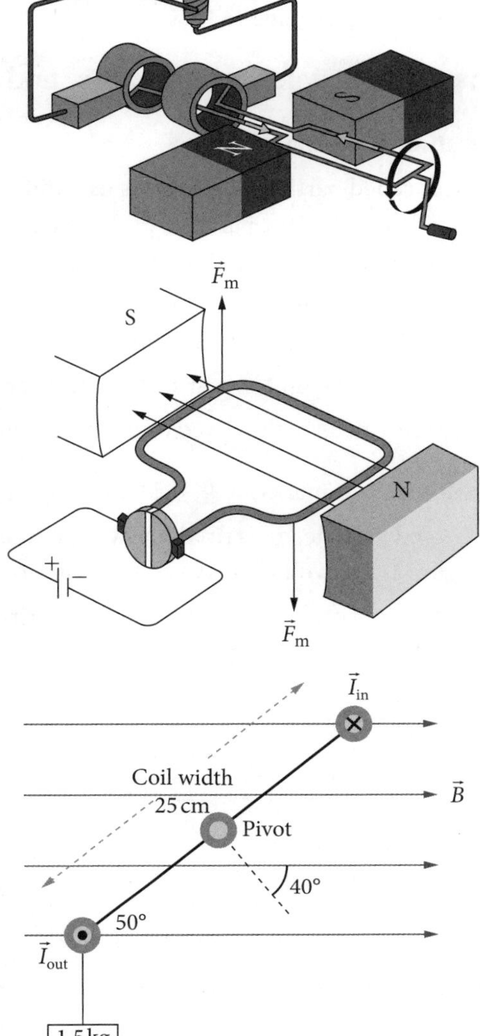

Question 27

Consider the diagram shown. If the coil was moved to a position 30° anticlockwise of its current position, what would happen to the magnitude of the force F_m?

A It will be unchanged.

B It will increase by a factor of sin 30°.

C It will decrease by a factor of sin 30°.

D It will decrease by a factor of sin 60°.

Question 28

A 1.5 kg mass is attached to one side of a motor coil, the end view of which is shown. The coil stalls in the position shown in the diagram. The width of the motor coil is 25 cm.

What torque is the motor coil experiencing as a result of the magnetic field?

A 2.36 N m **B** 1.18 N m

C 1.41 N m **D** 0.121 N m

Question 29

The design of a speaker features a coil and magnet, which enable the motor effect. The arrangement of stationary magnetic poles and the shape of the magnetic field can be seen in the diagram.

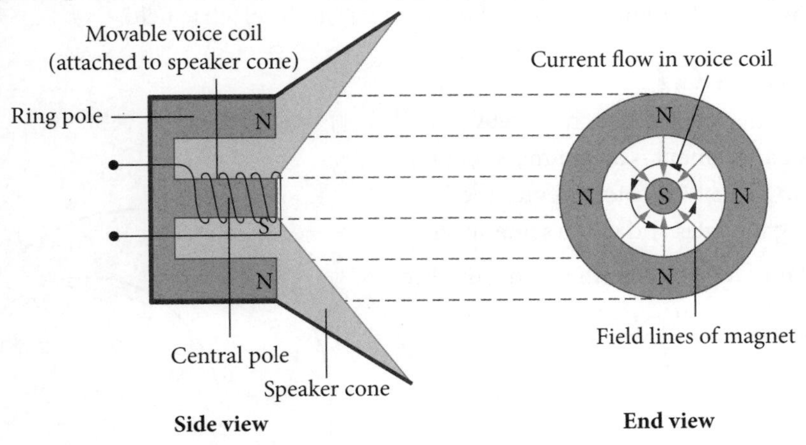

If the current flowed in the direction indicated in the end view of the speaker, then the motion of the movable 'voice coil' indicated in the side view would be

A to the left. **B** to the right.

C in a circle around the south pole. **D** impossible to predict using the information given.

Short-answer questions

Solutions start on page 163.

Charged particles, conductors, and electric and magnetic fields

Question 30 (6 marks)

Devices called particle accelerators use uniform electric fields to bring charged particles to very high speeds. One part of a particle accelerator features two adjacent charge-conducting sections with a potential difference of 1.25×10^4 V. A proton is released midway between the conducting sections.

a Describe how the electric field can result in the proton having energy. 2 marks

b Show that the proton gains 1.00×10^{-15} J of energy from the process. 2 marks

c Hence, calculate the speed of the proton as it reaches the negative conducting section. 2 marks

Question 31 (6 marks) ©NESA 2020 SII Q34

A charged particle, q_1, is fired midway between oppositely charged plates X and Y, as shown in Figure 1. The voltage between the plates is V volts.

The particle strikes plate Y at point P, a horizontal distance s from the edge of the plate. Ignore the effect of gravity.

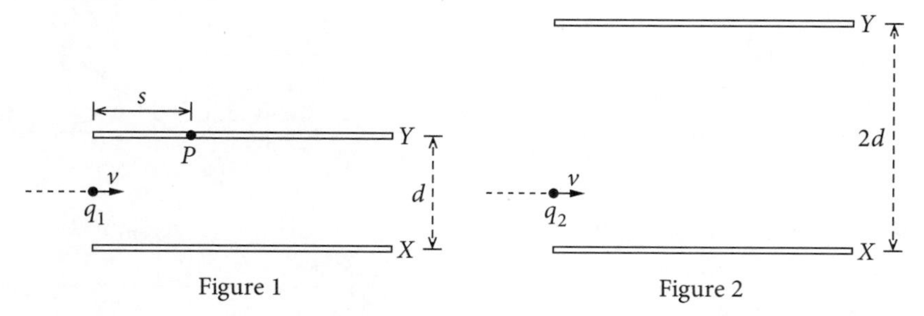

Plate Y is then moved to the position shown in Figure 2, with the voltage between the plates remaining the same.

An identical particle, q_2, is fired into the electric field at the same velocity, entering the field at the same distance from plate X as q_1.

a Compare the work done on q_1 and q_2. 3 marks

b Compare the horizontal distances travelled by q_1 and q_2 in the electric field. 3 marks

Question 32 (3 marks)

The diagram shows the paths of two identically charged isotopes of an element – one of mass m_1 and the other of mass m_2. The two isotopes enter the magnetic field at right angles and at the same speed v.

Show that the distance (d) between the points marked 1 and 2 is given by $d = 2\dfrac{v(m_1 - m_2)}{qB}$.

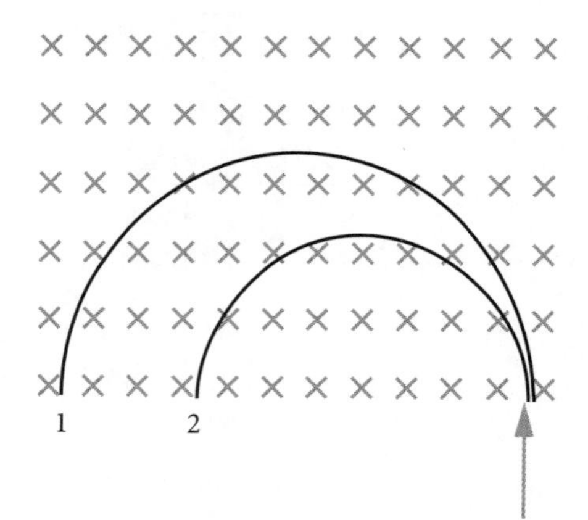

Question 33 (4 marks)

An electron enters a uniform magnetic field, of magnetic field strength 1.45×10^{-3} T, at right angles to the magnetic field. The electron completes a semicircle within the magnetic field. Show that the time the electron takes to complete a semicircle in the magnetic field is independent of the velocity of the electron and, thus, calculate the time taken to complete the semicircle.

The motor effect

Question 34 (5 marks)

A light conducting wire is suspended between the poles of a horseshoe magnet. When a current passes through the wire, it can be seen to swing outwards and to come to rest at the position shown.

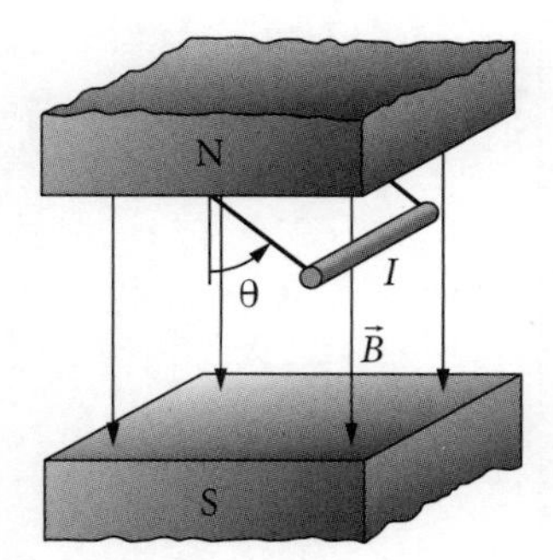

a In which direction is the current flowing in the wire? 1 mark

b Explain why the rod comes to the rest in the position illustrated. 2 marks

c Explain what would be observed if the current through the wire was reduced slowly to zero and then increased in the reverse direction. 2 marks

Question 35 (4 marks)

Two parallel, adjacent loops of wire, from which an electromagnet is constructed, carry a DC current, as shown in the diagram.

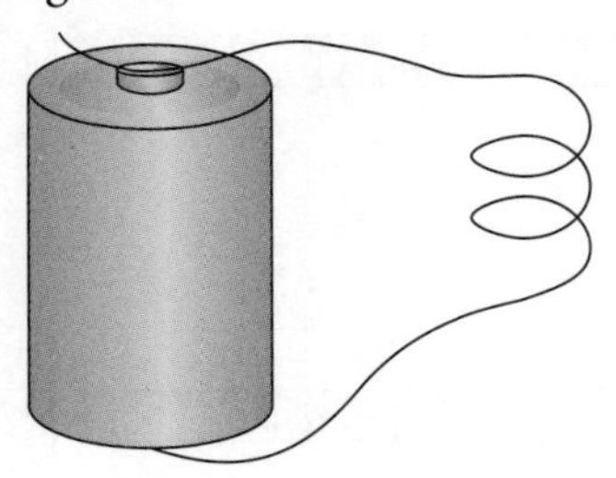

a Describe the force that each loop would exert on the other. 2 marks

b If the DC was replaced with AC, in what ways would the force change? 2 marks

Question 36 (6 marks)

Three long parallel conductors are arranged as shown, each separated by 10 cm vertically. $I_1 = 40.0$ A, $I_2 = 30.0$ A and $I_3 = 50.0$ A.

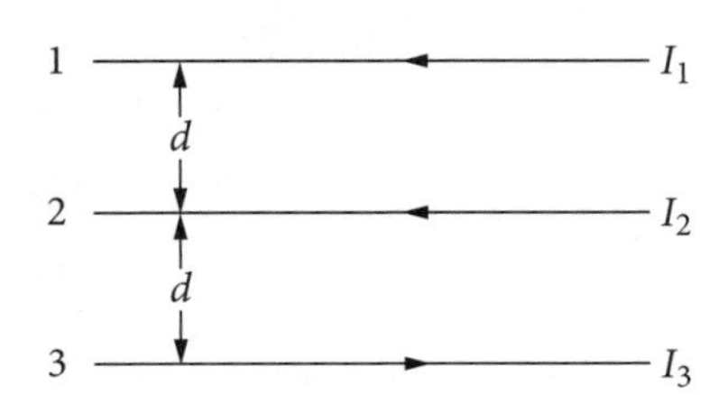

a What is the magnitude and direction of the total force on each metre of wire 1 due to currents in the other two wires? 3 marks

b If wire 3 is removed so only wires 1 and 2 remain, what would be the mass per metre of wire 2 if it is suspended at its location? 3 marks

Question 37 (4 marks)

The apparatus in the diagram can be used to measure magnetic flux density. The rectangle in the image consists of a conducting coil of 20 loops with a width of 40 cm and a length of 90 cm. An anticlockwise current of 0.50 A flows through the coil. With this current and no mass on the left side of the balance, the balance is in equilibrium.

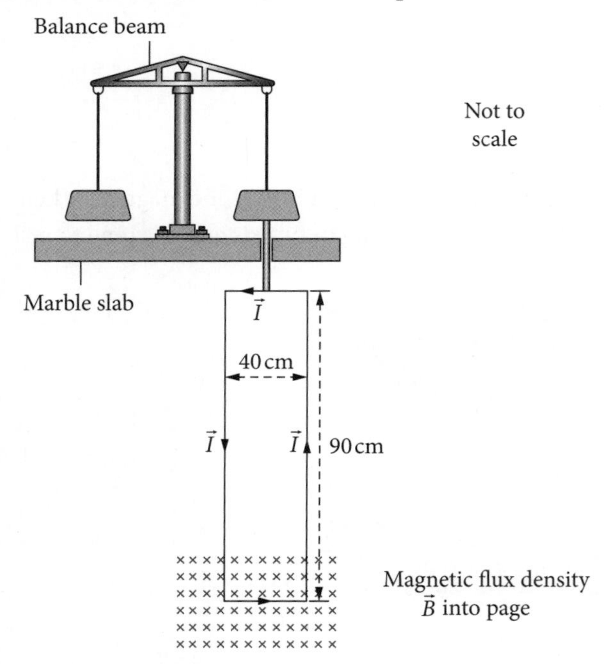

If the direction of the current is reversed but its magnitude stays the same, then 20 grams needs to be added to the left balance pan to restore equilibrium. Calculate the magnetic flux density.

Question 38 (5 marks)

Two parallel current-carrying wires are arranged so that the magnetic field strength and direction at a point P can be measured. P is 20 cm from wire M and a perpendicular distance of d from wire N.

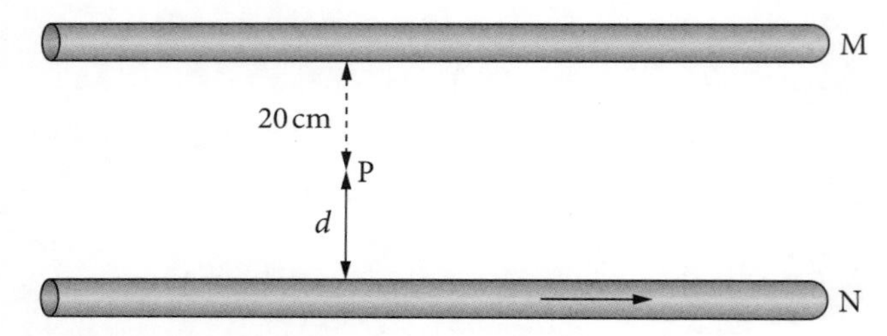

The current in wire M is kept constant and the current in wire N is varied. Data is recorded in the table below.

Current in wire N (A)	Magnetic field direction at P	Magnetic field strength at P (T)
0.3	Out of page	1.3×10^{-6}
0.7	Out of page	1.8×10^{-6}
1.2	Out of page	2.2×10^{-6}
1.8	Out of page	2.7×10^{-6}
2.5	Out of page	3.5×10^{-6}
3.4	Out of page	4.5×10^{-6}

Plot a graph to determine the direction and magnitude of the current in wire M and the distance d.

Electromagnetic induction

Question 39 (14 marks)

A transformer is constructed of two coils of copper wire of diameter 2 mm wound around a solid iron core.

An experiment is performed using the transformer in which current and voltage data is collected for both coils and tabulated below.

$V_{primary}$ (V)	$I_{primary}$ (A)	$V_{secondary}$ (V)	$I_{secondary}$ (A)	Efficiency (%)
10.5	1.6	4.2	3.2	
20.4	3.1	8.1	6.2	79.4
29.9	5.0	12.0	9.3	74.6
40.1	6.9	16.1	12.2	71.0
49.4	8.9	20.2	13.9	

a Is this a step-up or step-down transformer? 1 mark

b Plot a graph of $V_{primary}$ against $V_{secondary}$ and draw a line of best fit through the data. 3 marks

c There are 100 turns in the primary coil. Use the graph to estimate the number of turns in the secondary coil. 3 marks

d Calculate the values of efficiency for the first and last data set in the table. 2 marks

e Why does the efficiency change in the way seen in the table? 2 marks

f Justify two changes that could be made to this transformer to improve its efficiency. 3 marks

Question 40 (6 marks)

A rectangular loop of wire is able to rotate about its axis and is arranged so that the plane of the loop is initially parallel to the uniform magnetic field between two magnetic poles.

The wire is rotated with a frequency of 1 revolution per second for 2 seconds and then at twice that frequency for 2 seconds.

a Sketch a graph showing the change in flux through the coil over time using an appropriate numerical scale on the horizontal axis. 3 marks

b Sketch an emf versus time graph with an appropriate numerical scale on the horizontal axis (no specific numerical scale is needed on the vertical axis). 3 marks

Question 41 (4 marks)

A powerful magnetic field is used to separate metal rubbish from other waste as the waste slides down a ramp at a disposal depot.

Suppose that the ramp is frictionless and that two of the objects that are released simultaneously are a small square loop of copper wire and a small square of copper sheet. The two squares have the same dimensions and mass.

With the aid of diagrams, compare the electromagnetic processes that occur as each object slides down the ramp, entering, moving into a position of being completely within and leaving the magnetic field. Your response should include details of processes within the copper and of the motion of the objects.

9780170465304

Question 42 (8 marks)

A conducting rod is moved at a constant velocity of 5.0 m s^{-1} along two long, frictionless metal rails within a magnetic field of strength 0.04 T. The rails are 25 cm apart and connected at the end with a 0.25 Ω resistor.

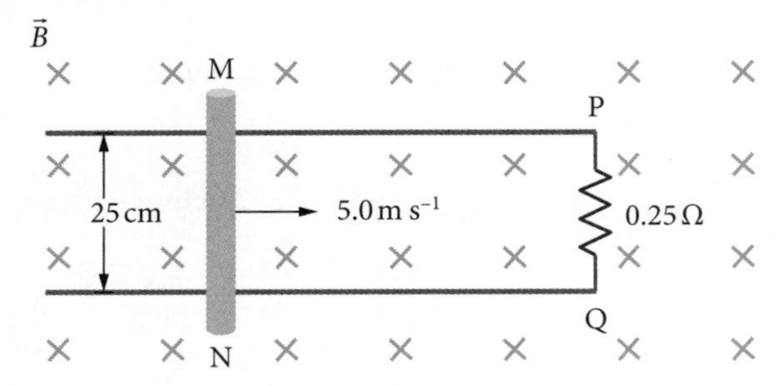

a Determine the direction and magnitude of the current flowing through the conducting rod. 4 marks

b Explain any changes that could be made to this set-up that could increase the magnitude of the current and reverse its direction. 4 marks

Applications of the motor effect

Question 43 (4 marks) 

On some exercise bicycles, the design features an electromagnet set-up so that when it is switched on it can establish a magnetic field over part of a spinning disc as seen in the diagram.

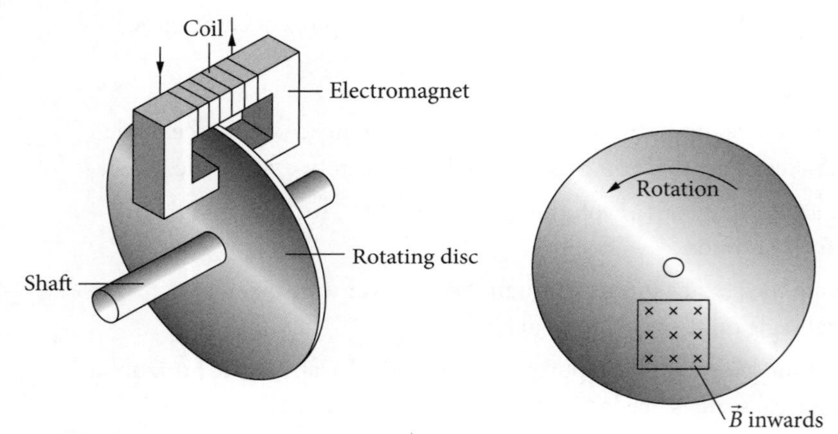

By annotating a sketched diagram similar to the one shown, explain how the electromagnet is able to make it harder to make the wheel spin when cycling.

Question 44 (3 marks)

A simple motor consists of 4 loops of wire made into a rectangle of length 8 cm and width 4 cm. This is situated in a uniform magnetic field of strength 22 mT. A current of 0.14 A runs through the motor and the plane of the coil makes a 25° angle with the field lines.

Calculate the ratio of the torque on the motor in this position to the maximum torque.

Question 45 (4 marks)

Many motors will feature magnets with a curved surface that create a magnetic field with a particular shape as shown.

With reference to the force and torque generated by this motor, explain why it has an advantage over a motor with identical number of coils, current, area and magnetic field strength featuring a uniform field created by flat-ended magnets.

Sketches and/or graphs may assist in the clarity of your response.

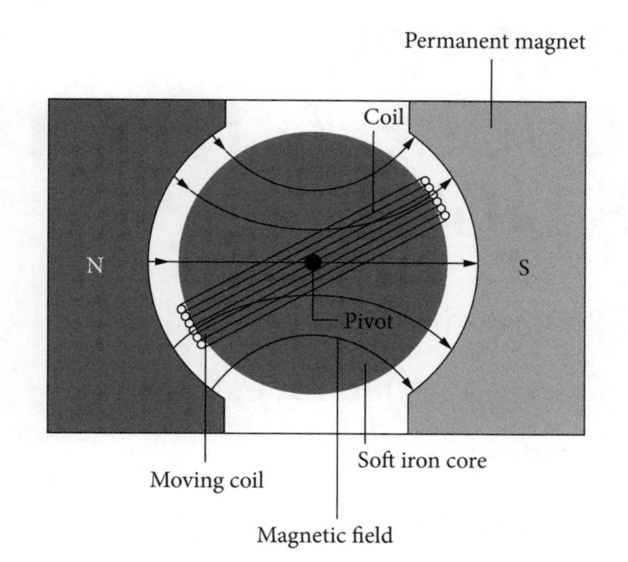

Question 46 (4 marks)

The 'motor' shown on the right was designed by Michael Faraday. Explain the nature and direction of the movement of the wire when a direct current flows through the circuit.

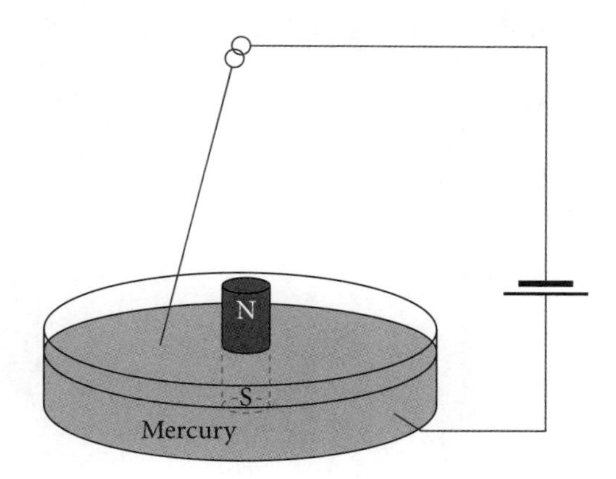

Question 47 (5 marks)

A schematic diagram of an AC induction motor is shown. With reference to the two main components shown on the diagram, explain how an AC induction motor generates torque.

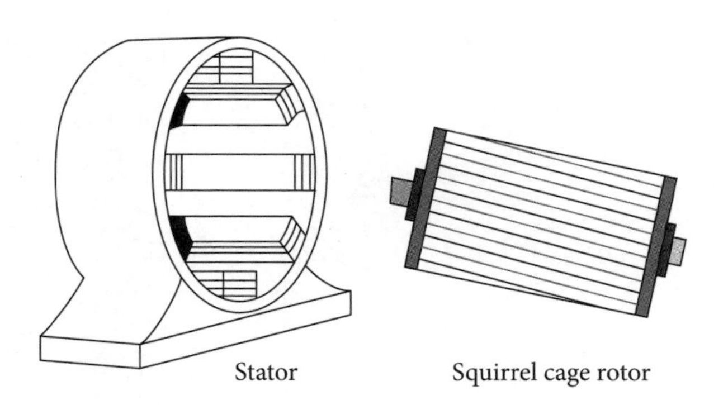

Question 48 (12 marks)

A feature of the operation of all DC motors is back emf.

a With reference to Lenz's law, explain the statement that 'back emf sets a maximum operating speed for motors'. 4 marks

b Sketch a graph of how the current in an ideal DC motor with no load would change over time from the switch being closed until it has stayed at its maximum speed for a short period. 3 marks

c Why can a motor 'burn out' if its rate of rotation is too low? 3 marks

d How is energy conserved in an ideal DC motor with no load? 2 marks

CHAPTER 3 MODULE 7: THE NATURE OF LIGHT

Module summary	67
3.1 Electromagnetic spectrum	69
3.2 Light: wave model	73
3.3 Light: quantum model	77
3.4 Light and special relativity	80
Glossary	85
Exam practice	87

Chapter 3
Module 7: The nature of light

Module summary

Outcomes

On completing this module, you should be able to:

- develop and evaluate questions and hypotheses for scientific investigation
- design and evaluate investigations in order to obtain primary and secondary data and information
- conduct investigations to collect valid and reliable primary and secondary data and information
- select and process appropriate qualitative and quantitative data and information using a range of appropriate media
- communicate scientific understanding using suitable language and terminology for a particular audience or purpose
- describe and analyse evidence for the properties of light and evaluate the implications of this evidence for modern theories of physics in the contemporary world.

Working Scientifically skills

In this module, you are required to demonstrate the following Working Scientifically skills:

- develop and evaluate questions and hypotheses for scientific investigation
- design and evaluate investigations in order to obtain primary and secondary data and information
- conduct investigations to collect valid and reliable primary and secondary data and information
- select and process appropriate qualitative and quantitative data and information using a range of appropriate media
- analyse and evaluate primary and secondary data and information
- solve scientific problems using primary and secondary data, critical thinking skills and scientific processes
- communicate scientific understanding using suitable language and terminology for a specific audience or purpose.

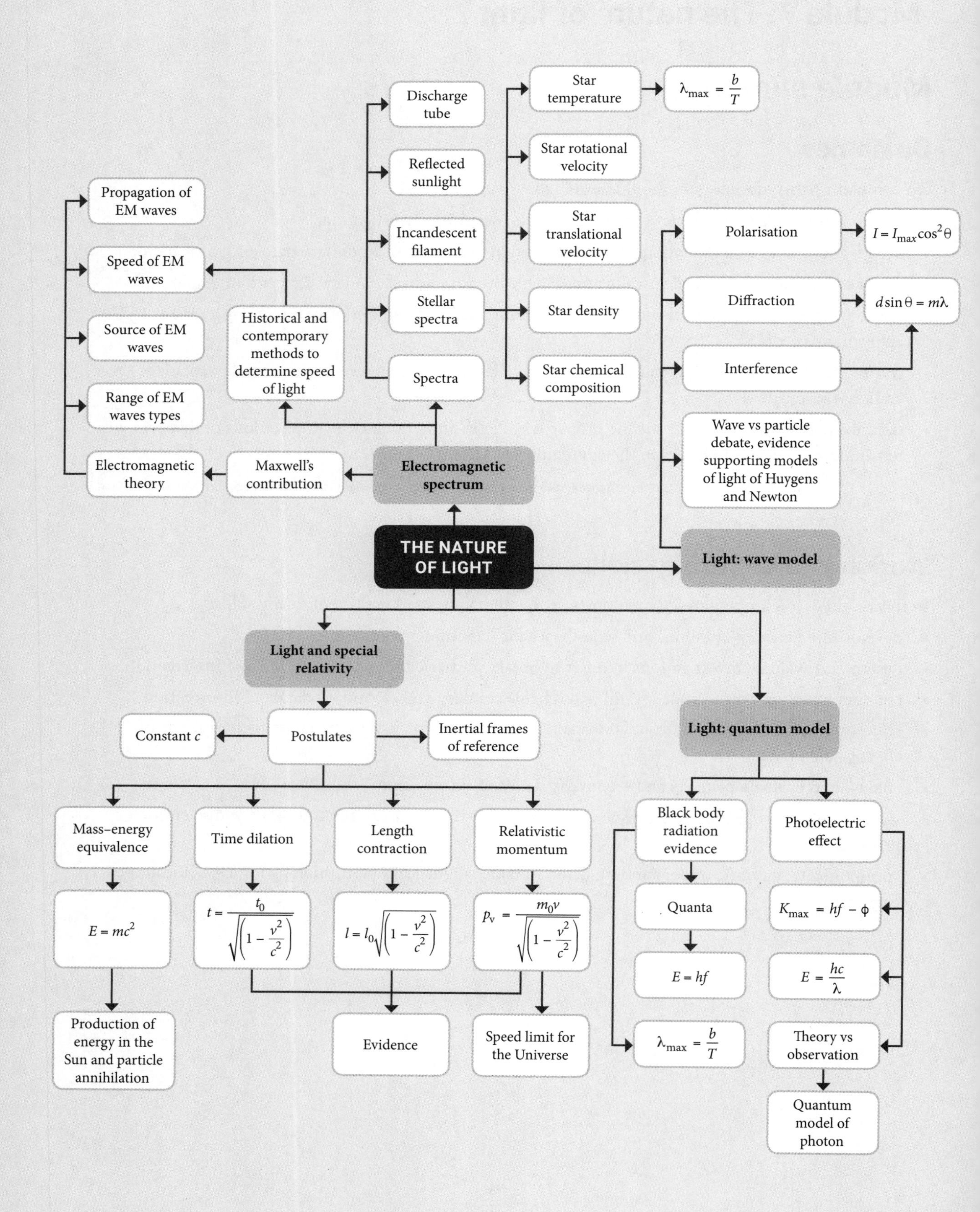

THE NATURE OF LIGHT
Electromagnetic spectrum
Maxwell's contribution
Electromagnetic theory
Propagation of EM waves
Speed of EM waves
Source of EM waves
Range of EM waves types
Historical and contemporary methods to determine speed of light
Spectra
Discharge tube
Reflected sunlight
Incandescent filament
Stellar spectra
Star temperature
$\lambda_{max} = \frac{b}{T}$
Star rotational velocity
Star translational velocity
Star density
Star chemical composition
Light: wave model
Polarisation
$I = I_{max}\cos^2\theta$
Diffraction
$d\sin\theta = m\lambda$
Interference
Wave vs particle debate, evidence supporting models of light of Huygens and Newton
Light and special relativity
Postulates
Constant c
Inertial frames of reference
Mass–energy equivalence
$E = mc^2$
Production of energy in the Sun and particle annihilation
Time dilation
$t = \frac{t_0}{\sqrt{\left(1 - \frac{v^2}{c^2}\right)}}$
Length contraction
$l = l_0\sqrt{\left(1 - \frac{v^2}{c^2}\right)}$
Relativistic momentum
$p_v = \frac{m_0 v}{\sqrt{\left(1 - \frac{v^2}{c^2}\right)}}$
Evidence
Speed limit for the Universe
Light: quantum model
Black body radiation evidence
Quanta
$E = hf$
$\lambda_{max} = \frac{b}{T}$
Photoelectric effect
$K_{max} = hf - \phi$
$E = \frac{hc}{\lambda}$
Theory vs observation
Quantum model of photon

3.1 Electromagnetic spectrum

The word 'light' is most commonly used to describe a narrow range of frequencies of **electromagnetic waves**. This rainbow of colours is often called the visible spectrum. Humans have inbuilt detectors of this range of frequencies – eyes. There are electromagnetic waves with both greater and lower frequencies than the frequencies of the visible spectrum. The full range is called the **electromagnetic spectrum**.

Theoretical and experimental physicists have been working to establish the nature and properties of light and the electromagnetic spectrum for over 300 years.

3.1.1 Maxwell's theory of electromagnetism

James Clerk Maxwell believed that light was related to both electricity and magnetism, rather than all three being independent phenomena. He was able to combine the existing observations, theories and equations into a single coherent theory – the theory of electromagnetism.

Maxwell's theory of electromagnetism is built upon four equations, the details of which fall beyond this course. The equations could be used to make predictions about light.

- Oscillating charges produce changing electric fields that, in turn, create changing magnetic fields, which induce changing electric fields and so forth in a mutually inducing manner. As a consequence, these mutually perpendicular, oscillating fields will radiate from the source as an electromagnetic wave.
- Therefore, electromagnetic waves are self-sustaining and do not require a medium.

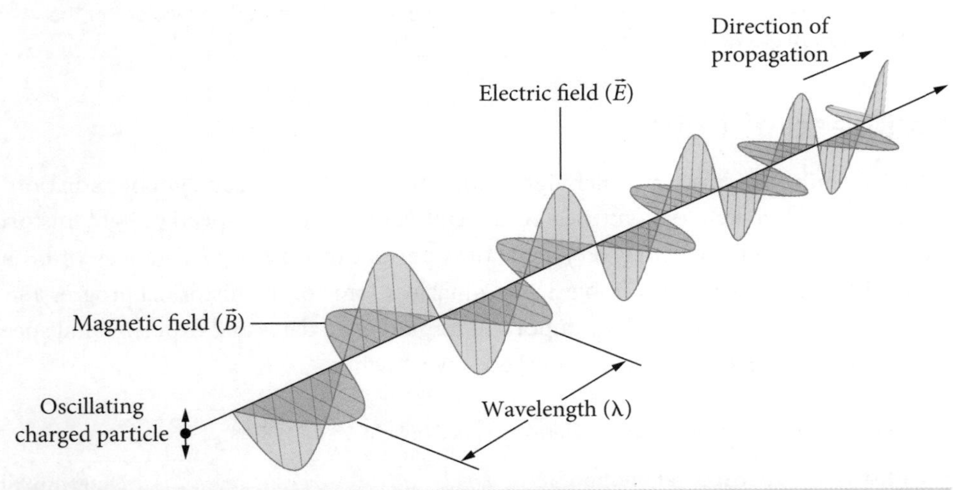

FIGURE 3.1 An electromagnetic wave is produced by an oscillating charged particle.

- All materials restrict the formation of electric and magnetic fields to some degree. This is called **permittivity** (ε) for electric fields and **permeability** (μ) for magnetic fields. The least restriction to electric and magnetic fields occurs in a vacuum (ε_0 and μ_0 respectively), so the speed of an electromagnetic wave would be greatest through a vacuum. Maxwell predicted this speed mathematically to be $3 \times 10^8\ \mathrm{m\,s^{-1}}$, the same as the experimentally determined speed of light.
- There would exist electromagnetic waves of frequencies both higher and lower than those of the visible spectrum and hence the visible spectrum would comprise a small proportion of the electromagnetic spectrum only.

Maxwell's predictions about the speed of light, the production and propagation of electromagnetic waves and the existence of a range of frequencies comprising the electromagnetic spectrum are universally accepted by the scientific community today. Maxwell died at the age of 48, before he was able to see aspects of his theory experimentally confirmed.

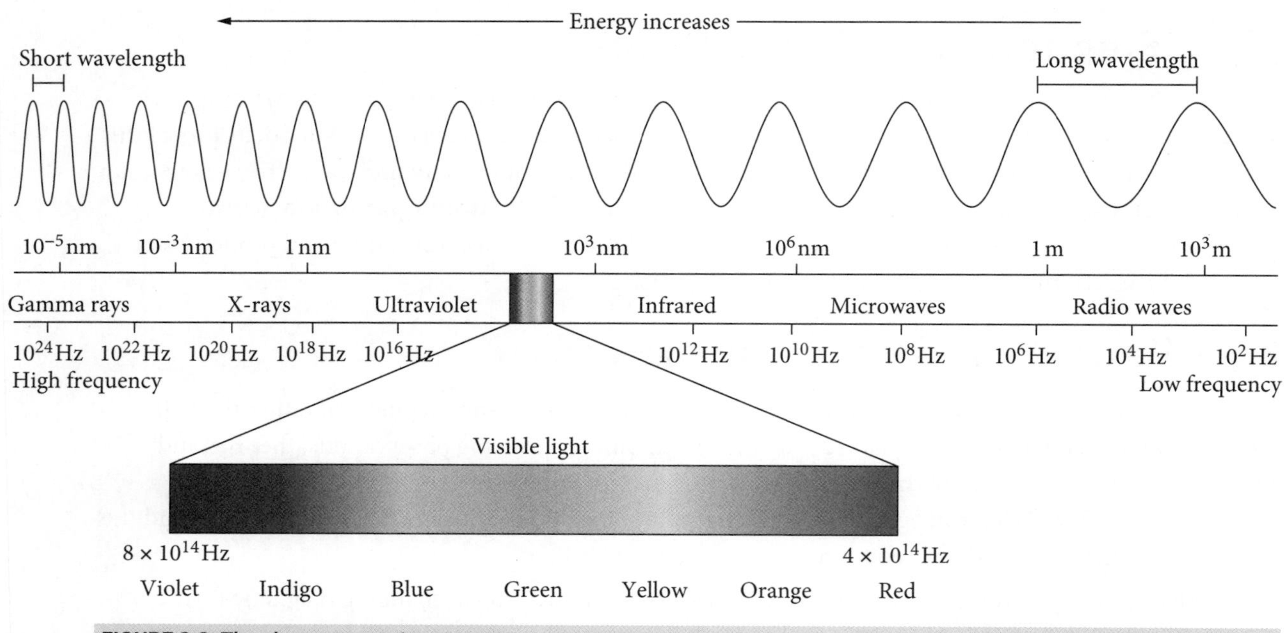

FIGURE 3.2 The electromagnetic spectrum

> **Note**
> There is no need to remember frequency or wavelength values but to know the trends is important. However, the boundary frequency and wavelength values (red and violet) for the visible spectrum are worth knowing.

3.1.2 The speed of light

Experiments to determine the speed at which light and, by extension, electromagnetic radiation propagates have been undertaken for centuries. As a result of the very high speed of light, historically, only sophisticated experiments involving very short time periods or very large distances, or both, were able to establish reasonable values. Table 3.1 summarises some of the historical progression of measurements of the speed of light. It is not important to memorise dates and experimental speeds; but understanding broadly how each experiment worked is worthwhile.

TABLE 3.1 Development of historical methods to determine speed of light

Year	Scientist	Experimental method	Determined speed of light value ($m\,s^{-1}$)
1638	Galileo Galilei	Galileo and his assistant stood on hilltops of known distance apart. Galileo uncovered his lantern, and his assistant opened the shutter on his own lantern when he saw the light. Galileo noted the time he saw the return light.	At least ten times the speed of sound; possibly infinite
1676	Ole Roemer	Recorded times between successive eclipses of one of Jupiter's moons measured at different stages of Earth's orbit around the Sun were compared to determine the time taken for light to traverse the extra distance resulting from Earth's changing position in its orbit.	2.2×10^8
1728	James Bradley	Measured the stellar aberration (change in the perceived position of the stars observed from Earth) of stars caused by Earth's orbital velocity.	3.0×10^8
1849	Armand Fizeau	Light passed through a spinning toothed wheel is reflected from a mirror 17.25 km away and travels back to the source after passing through the adjacent gap.	3.13×10^8
1862	Leon Foucault	Light strikes a spinning multi-faceted mirror and is reflected from a mirror 8 km away and passed back to strike another of the spinning device's mirror faces so that it reaches a detector.	3.09×10^8
1926	Albert Michelson	Similar technique to Foucault with longer distances.	2.998×10^8

Contemporary methods over the last 50 years have utilised wave properties such as interference to obtain increasingly more accurate values. In 1983 the value of the speed of light was finalised using the principle of interference and the universal wave equation as having the value of $2.997\,924\,58 \times 10^{8}\,\mathrm{m\,s^{-1}}$. This value is used to provide the definition of the metre.

3.1.3 Spectra

When light of a variety of frequencies is passed through a device (such as a **diffraction grating** or prism) that can disperse it on the basis of frequency, a spectrum can be produced. There are three types of spectra (Figure 3.3).

- A **continuous spectrum** is produced by a sufficiently hot object such as a filament (**incandescent**) light bulb. It comprises all the visible wavelengths of light. In the Year 11 course, this was referred to as '**black body radiation**'. A rainbow produced when sunlight passes through water droplets can be considered a simple continuous spectrum.
- An **emission spectrum** is produced when gaseous elements are excited (given energy) by an electrical discharge in a discharge tube or a flame. An emission spectrum features lines of light of specific wavelengths called **emission lines** or spectral lines, which are characteristic of the element, against a black background. The wavelengths of spectral lines in emission spectra can be used to identify elements in a sample of mixed gases.
- An **absorption spectrum** is produced when a continuous spectrum is passed through a gas. An absorption spectrum appears as a continuous spectrum with black lines (**absorption lines**, also called spectral lines) at specific wavelengths corresponding to light that has been absorbed by the gas. The wavelengths of spectral lines in absorption spectra can also be used to identify elements in a sample of mixed gases.

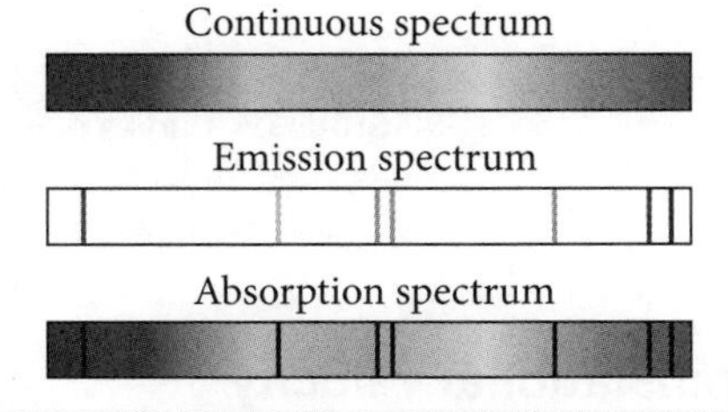

FIGURE 3.3 Continuous spectrum and emission and absorption spectra for the same gas sample

Note
These three spectral types will be referenced again and used in Module 8.

3.1.4 Using spectra

The study of spectra, known as spectroscopy, can be used to provide a range of information. Stellar spectra are particularly useful for providing information about the nature, motion and composition of stars, which would be impossible to gain in any other way.

Chemical composition of stellar atmosphere

The core of a star emits light with a continuous spectrum. That light passes through gases of the star's atmosphere, where some light of specific wavelengths is absorbed. Analysis of the proportion and the wavelengths of absorbed light can be used to identify the relative abundance of elements in the star's atmosphere.

9780170465304

Surface temperature

Recall from Year 11 that the wavelength of maximum intensity in a continuous spectrum can be used to predict the temperature of a **black body** using Wien's law, $\lambda_{max} = \frac{b}{T}$. A star can be considered to emit light as a black body and, subsequently, its surface temperature can be predicted.

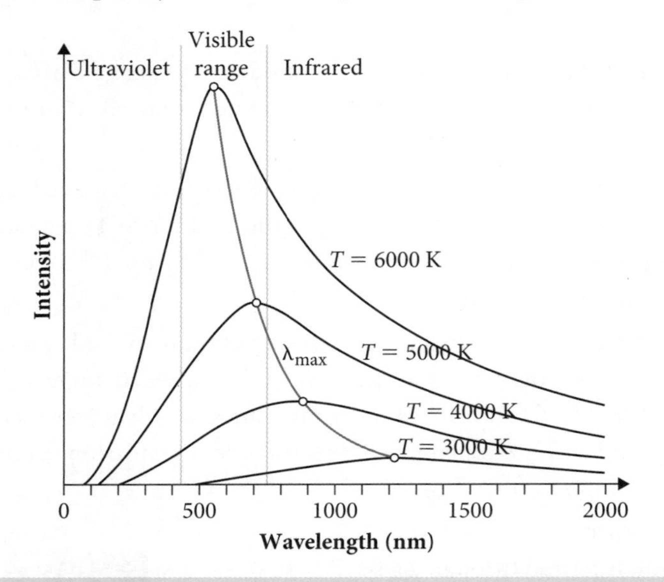

FIGURE 3.4 Curves of intensity of emitted radiation vs wavelength of that radiation measured for a black body at a range of temperatures. Note the line labelled λ_{max}, which shows how the wavelength of maximum intensity varies with temperature.

Translational velocity

Recall from Year 11 that the wavelength, and frequency, of sound changes when there is relative movement between the observer and the source. This same Doppler effect occurs with light waves. When the source of light is moving away from the observer, the wavelength of the observed light will be longer. Since this means the observed wavelengths are closer to the red end of the spectrum, this is called **red shift**. Using similar logic, when the source of light is moving towards the observer, the observed light will be blue shifted. The extent to which red shift or **blue shift** occurs can be used to determine translational velocity.

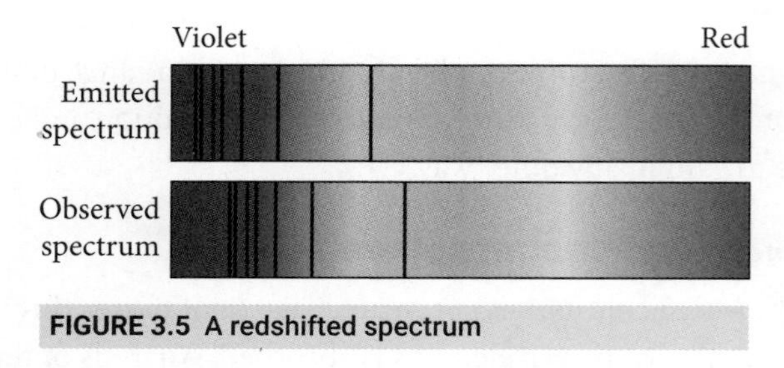

FIGURE 3.5 A redshifted spectrum

The concept of red shift is important again in Module 8. Make sure you have a clear understanding of it.

Rotational velocity

Depending on the axis of rotation of a star, when the star is rotating, one side may be approaching Earth while the opposite side is moving away. Consequently, a simultaneous red shift and blue shift will occur, which results in broader spectral lines. The extent of the broadening can be used to determine rotational velocity.

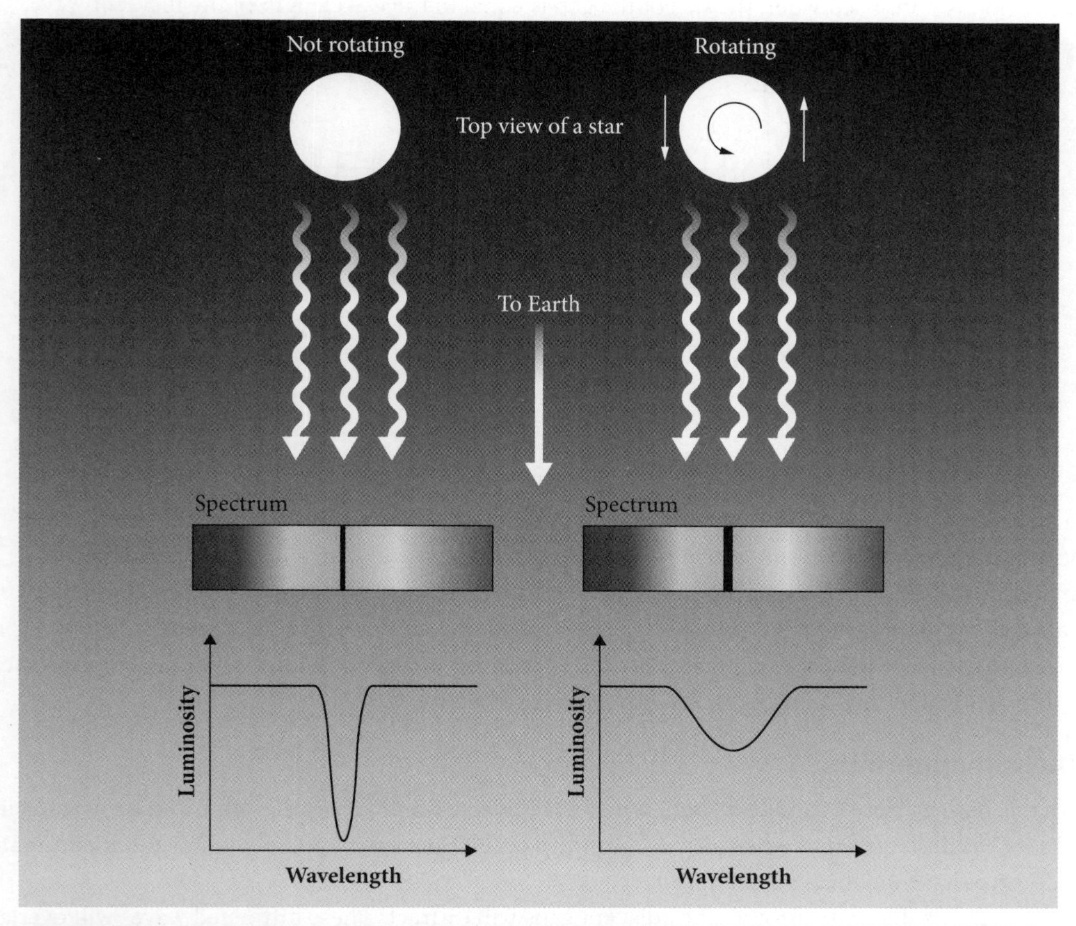

FIGURE 3.6 Broadened spectral lines are a consequence of star rotation.

Density

The density of gases in the outer (atmospheric) layers of the star will play a role in determining the appearance of the spectral lines. Stars with a low-density atmosphere will have sharper, finer spectral lines than stars with a high-density atmosphere.

3.2 Light: wave model

Before Maxwell's theory of electromagnetism was developed in the 1860s, there had been debate throughout the late 17th century about the nature of light. There were two competing models of light (wave and particle) at the time, and experimental evidence supporting the models was gathered.

3.2.1 Diffraction of light

In the Year 11 course, you studied the behaviour of waves when they encounter an obstacle or a gap in a barrier (Figure 3.7). This process in which the waves spread out beyond the obstacle or gap is called **diffraction**. Light can be shown to diffract, and this can be seen as evidence for a wave model of light.

Diffraction is optimised when the size of the gap is equal to the wavelength of the incident wave.

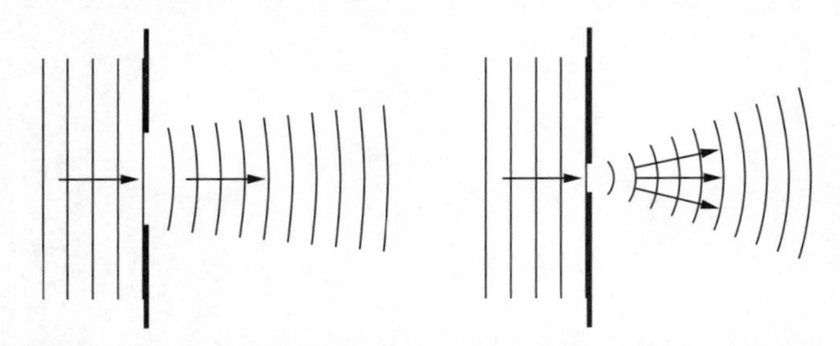

FIGURE 3.7 Significant diffraction occurs when the wavelength approximates the slit width.

3.2.2 Interference of light

When two waves of the same type encounter each other in a medium, a process called **interference** will occur. Interference will result in temporary changes to the wave according to the **principle of superposition** studied in Year 11. At any point, interference can be constructive when the two waves are in **phase** at that point, or destructive when the two waves are out of phase at that point.

Waves, particularly light waves, create observable patterns as a result of diffraction and interference. These diffraction and interference patterns are mainly of two types.

Double-slit apparatus

Light that is both **coherent** and **monochromatic** is projected onto a barrier with two narrow and closely spaced slits. (Light is coherent when all the light wave oscillations are in phase. Light is monochromatic when all waves have identical wavelength.)

The wavefronts that pass through the adjacent gaps will diffract. These diffracted waves will overlap and, therefore, interfere. At some places the waves from the two openings will be in phase and interfere constructively to create a bright point of light. In other places the waves from the two openings will be out of phase and interfere destructively to create a dark region (no light).

A series of light and dark bands will be observed (sometimes called maxima and minima respectively or bright and dark fringes). Moving away from the central bright band ($m = 0$), these alternating bands are numbered as $m = 1, 2, 3 \ldots$

Constructive interference will occur when the difference in the length of the path travelled by the light from the source to the point is an integer multiple of the wavelength of the light.

Destructive interference will occur when the difference in the length of the path travelled by the light from the source to the point is an integer plus a half multiple of the wavelength of the light.

Trigonometry can be used to relate features of the double-slit interference pattern to the apparatus, as seen in the equation $d \sin \theta = m\lambda$. The distance between adjacent slits is measured as d. The angle θ is the angle between the central maximum and the point of interest and can be determined from the experimental results as seen in Figure 3.8.

A diffraction and interference pattern occurs with a single slit too, which might be a little counterintuitive. The double-slit pattern in Figure 3.8 is in part a product of the single-slit patterns created by each slit. It is worth being familiar with it but it is not in the syllabus.

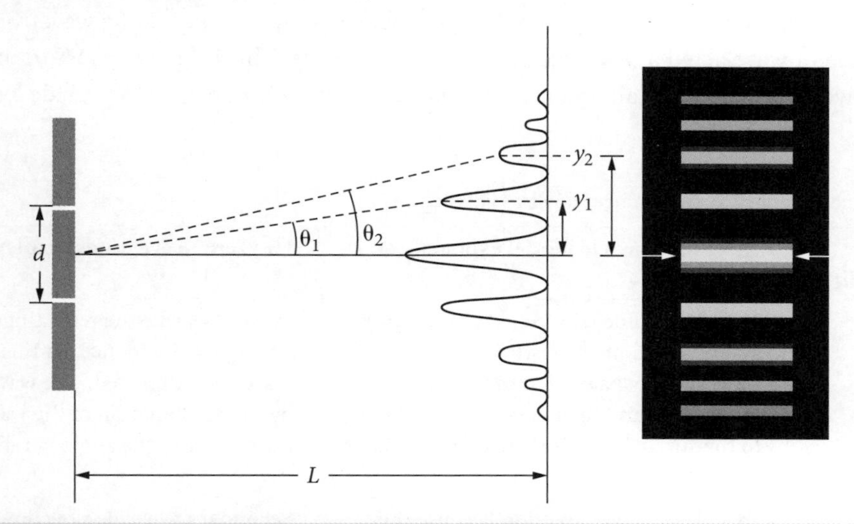

FIGURE 3.8 The interference pattern created by light incident on a double slit is represented on the right side with a bright central maximum and alternating bands of subsequent minima and maxima distributed symmetrically on either side. A graphical representation of intensity aligns with the pattern itself. Values of the angle to maxima (or minima) can be determined using trigonometry and the distance from slit to screen (here shown as L) and from central maximum to the point of interest (here shown as y_1 and y_2).

Diffraction gratings

Multiple slits can be used to produce brighter and sharper maxima that result from the superposition of light from many slits rather than just two. Typically, diffraction gratings feature thousands or tens of thousands of slits over a small space created by etching a large number of narrow parallel lines on a glass plate.

The equation used to relate features of the interference pattern of a diffraction grating is the same as the one for double slits, but typically the distance d between adjacent slits must be determined from data about how many slits there are per unit length and the value of d is much less for a diffraction grating.

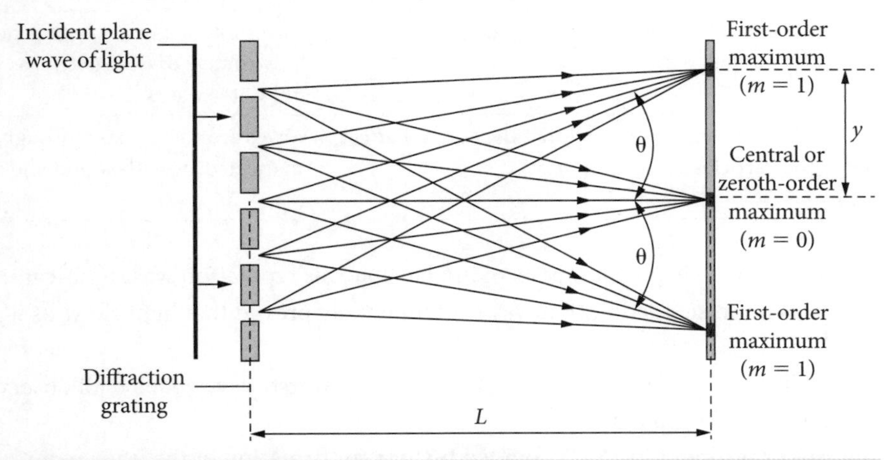

FIGURE 3.9 The rays representing the diffracted light waves are shown constructively interfering at the central maximum and two adjacent maxima.

9780170465304

3.2.3 Competing models of light

The two dominant models of light existing in the late 17th century were those of Christiaan Huygens, who proposed a longitudinal wave model of light, and Isaac Newton, who proposed a particle (**corpuscule**) model of light.

According to Isaac Newton, light comprised extremely small, very fast-moving particles that had mass and obeyed all the laws of physics governing the motion of objects.

Christiaan Huygens, conversely, proposed light is propagated by each point of a wavefront and that each point on a light wave acted as a source of secondary wavelets. This is known as **Huygen's principle**. Neither model was able to explain all observed phenomena, yet both were able to provide explanations for most.

TABLE 3.2 Evidence supporting competing models of light

Phenomenon observed of light	Newton's particle model explanation	Huygens' wave model explanation
Reflection	Particles collide with reflecting surfaces and rebound in an elastic collision in which their velocity parallel to the surface is unchanged and their speed perpendicular to the surface is the same before and after the collision.	Wavelets strike the reflecting surface at an angle of incidence. At this point, new wavelets are created. The wavefront created by the combination of the wavelets leaves the surface at the angle of reflection.
Refraction	Particles change direction because they speed up when they move into the denser medium as a consequence of an attractive force. Different colours have particles of different mass and hence are refracted at different angles.	Secondary wavelets slow down when they move into the denser medium. A new wavefront formed by joining the fronts created by secondary wavelets will be at an angle to the incident wavefront, since secondary wavelets were formed in the new medium at different times.
Rectilinear motion (moving in straight lines)	Particles travel in a straight line unless acted on by a force.	A new wavefront is formed by the joining of the fronts created by the secondary wavelets.
Diffraction	Particles bump into each other at edges and corners and as a result of the impact spread around the corner.	Secondary wavelets formed at the edge of the opening will spread in a circular wavefront that spreads behind the edge of the barrier.
Polarisation	Particles of light have flat sides and can only pass through a filter in certain orientations.	Could not be explained by Huygens' theory in which he considered light to be a longitudinal wave.
Interference	Could not be explained by the particle model.	Waves interact in a way that can be constructive or destructive, as with sound and water waves.
Light rays can pass through one another unaffected	Particles are so small that they do not affect particles of the other ray.	Waves are able to pass through each other unaffected, as with sound and water waves.

Although some of Newton's explanations seem implausible, his reputation was significant, and his corpuscular theory was not discarded until it was experimentally proved that light slows as it moves into a denser medium.

Modifications to Huygens' theory considering light as a transverse wave enabled all observed phenomena to be satisfactorily explained.

Young's double-slit experiment in 1803 provided further confirmation of the wave nature of light and Huygens' model became the accepted model of light.

3.2.4 Polarisation of light

Unpolarised light emitted by the Sun or a light globe, for example, has oscillations of its electric field in all possible directions perpendicular to the direction of propagation.

When this light is passed through a polarising filter, only light with its electric field oscillating in the plane parallel to the plane of polarisation of the filter is transmitted. This is called **polarised** light. The fact that light can be polarised is significant evidence for light as a wave. The intensity of the polarised light will be half of the intensity of the unpolarised light before it passed through the filter.

If this polarised light is then passed through another polarising filter (typically called an analyser), then only the component of the electric field parallel to the plane of polarisation of this filter is transmitted (Figure 3.10).

The transmitted intensity is described by **Malus' law**, which has the equation $I = I_{max} \cos^2 \theta$, where I_{max} describes the intensity of the polarised light that is incident on the analyser.

In the process explained above, the first polarising filter is typically called the polariser and the second is called the analyser.

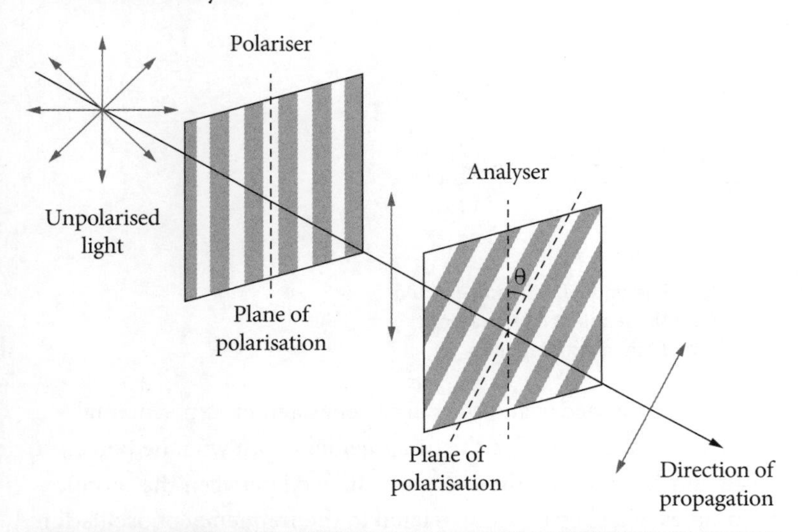

FIGURE 3.10 Light passes through the polariser and the analyser.

> **Note**
> A common source of confusion and error in polarisation questions is misunderstanding what I_{max} refers to. Ensure you are clear on this point.

3.3 Light: quantum model

By the late 19th century, the scientific community was very satisfied with an electromagnetic wave model of light that explained almost all observed phenomena. Soon, however, a radical rethinking of energy, and light, was needed to explain some observations and a new branch of physics developed.

3.3.1 Black body radiation and quanta

Recall from work in Year 11 that black body radiation describes energy emitted by an object as a consequence of its temperature and that Wien's law relates the wavelength of maximum emission with temperature in the equation

$$\lambda_{max} = \frac{b}{T}$$

where λ_{max} is the wavelength of the peak of the curve, T is the absolute temperature in K, and b is Wien's constant, $b = 2.898 \times 10^{-3}$ m K.

A perfect black body is so named because it absorbs *all* incident electromagnetic radiation, raising the kinetic energy of its particles (temperature). Energy detected radiating from a black body represents the temperature of the body alone. (There is no component due to energy reflected from an external source.) Heating the black body to a particular temperature causes the black body to emit radiation with a spectrum that is characteristic of that temperature. Examples of black bodies are the Sun and other stars, light bulb filaments, and the element in a toaster.

At the end of the 19th century, the problem regarding black body radiation was that the existing classical theory regarding how hot objects radiate energy predicted that an infinite amount of energy would be emitted at short wavelengths, which would defy the laws of energy conservation. Small increases in heat energy supplied would result in increasingly large amounts of heat energy radiated. This problem was known as the ultraviolet catastrophe.

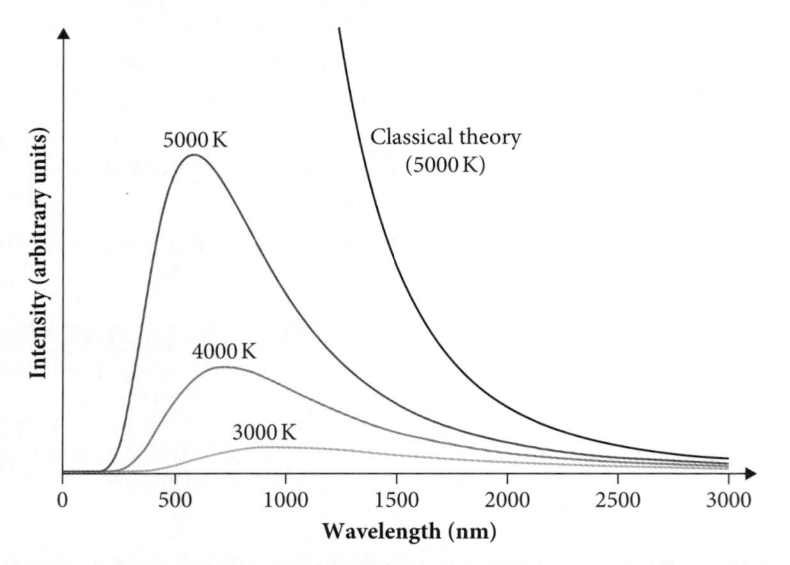

FIGURE 3.11 A graph showing the experimental data of intensity of radiation emitted from a hot object vs the wavelength of that radiation for three object temperatures and the prediction made by **classical physics** at one of these temperatures.

> **Note**
> This concept appears again in Module 8. Keep in mind the possibility of cross-modular questions using these ideas from Modules 7 and 8.

The disagreement between the curve predicted by theory and the curve generated by experimental data posed a significant problem. Max Planck was able to resolve the disagreement with what he perceived as a mathematical trick. This 'trick' involved considering that the energy exchanged between the 'atomic oscillators' of the black body must be of multiples of a specific value related to the frequency of oscillation expressed in the equation $E = hf$.

> **Note**
> Don't overlook the fact that this equation not only indicates the existence of quanta but also that the *size* of each **quantum** varies with frequency. The higher the frequency (lower wavelength), the larger the quantum. The intensity curve shows that there are few very large quanta and few very small quanta – most are in between, tending towards larger values as temperature rises, like a skewed bell curve.

This was the starting point for a new branch of physics – **quantum physics** – for which Planck was awarded the Nobel Prize in Physics in 1918.

This gave rise to the concept of quantisation. If a quantity is **quantised**, it can take on only certain allowed values. Charge is an example of something that is quantised, with only integer multiples of the fundamental electronic charge (1.602×10^{-19} C) possible.

3.3.2 Photoelectric effect and the wave model of light

In the late 19th century, observations by Heinrich Hertz suggested a relationship between light and electricity – a **photoelectric effect**. Essentially, Hertz noticed that electrons were emitted from a metal surface when it was illuminated by incident light. The electromagnetic wave model of light, based on the theories of Maxwell and others, stated the following.

- The energy transferred by light is evenly distributed across the wavefront.
- The intensity of light is related to the square of the **amplitude** of the wave.
- The frequency of light is the number of waves arriving per second.

Consequently, classical physics makes the following predictions about the photoelectric effect.

1 Electrons should be emitted at any frequency of the incident light.
2 The number of electrons ejected from the metal should depend on the frequency of the incident light.
3 The kinetic energy of the ejected electrons should be a function of light intensity, not frequency.
4 When light intensity is low, a delay in emission of electrons should occur while absorbed energy accumulates to a sufficient level.

3.3.3 Photoelectric effect and the particle model of light

Experiments using apparatus such as the one schematically illustrated in Figure 3.12 were designed to test the effect of changing the independent variables of intensity and frequency of incident light on the number of electrons (indicated by the current or **photocurrent**) and maximum kinetic energy of electrons emitted (measured by the stopping voltage).

The experimental observations significantly contrasted with the predictions made by the classical wave model.

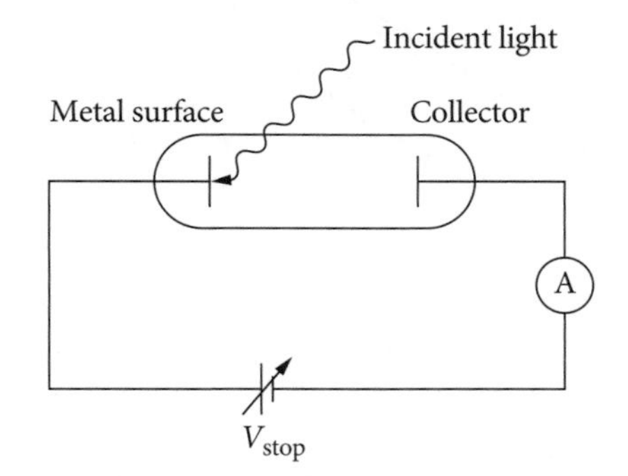

FIGURE 3.12 The type of apparatus that could be used to collect photoelectric effect data

1 Electrons were only emitted at frequencies above a certain **threshold frequency** (f_0) and this value was characteristic of the metal surface.
2 Above this threshold frequency the number of electrons emitted was determined only by the intensity of the light.
3 **Photoemission** was immediate above the threshold frequency regardless of the light intensity.
4 For any metal emitter, the kinetic energy of emitted electrons was dependent only on the frequency of the incident light.

> **Note**
> Good online simulations exist for the photoelectric effect and can be useful in consolidating understanding.

Five years after Planck had solved the black body radiation problem with a radical proposal, Einstein suggested an extraordinary solution to the photoelectric effect problem, which incorporated Planck's work.

There were four key features of Einstein's solution.

1 Light consisted of a stream of packets of energy (later named **photons**) that carried energy given by Planck's equation $E = hf$.
2 One photon interacts with one electron in the metal.
3 A specific minimum amount of energy, called the **work function** (ϕ), is required to remove an electron from the metal. Different metals are characterised by different work functions.
4 Intensity of light is a function of the number of photons.

The value of frequency used in Planck's equation ($E = hf$) in the first point is related to the wavelength of the incident light by the universal wave equation $c = f\lambda$ and, therefore, can be written as $E = \frac{hc}{\lambda}$. The energy of the photon can be seen to be related to the wavelength of incident light.

The solution proposed by Einstein enabled the principle of conservation of energy to be applied to the interaction between the photon, the metal and the emitted electron to explain the observed phenomena.

1 Electrons will only be emitted from the metal if the energy of the photon is equal to or exceeds the metal's work function; that is, $hf \geq \phi$. This threshold frequency is, therefore, the frequency such that $hf_0 \geq \phi$.
2 The number of electrons emitted is proportional to light intensity. If the incident light is of a frequency above the threshold frequency, then each photon can liberate one electron so the photocurrent will increase as light intensity increases.

3 There is immediate photoemission. No accumulation of energy is required because each photon will instantly deliver the energy required to liberate an electron, providing the photon energy is greater than the work function.

4 Electron energy is a function of light frequency. The conservation of energy applied to the interaction between the photon and the electron in the metal surface can be expressed as
energy of photon = energy provided to metal to liberate electron + energy of liberated electron.

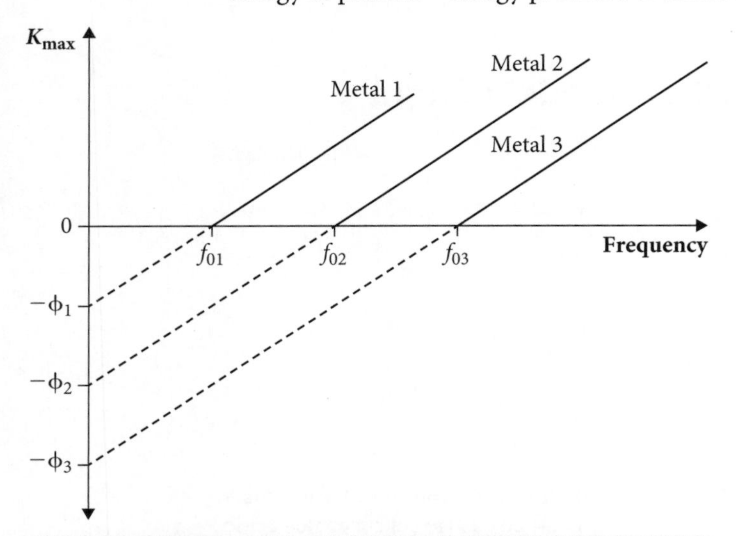

FIGURE 3.13 A graph of maximum kinetic energy of photoelectron vs incident light frequency for three different metals

Consequently, the equation for the maximum kinetic energy of an ejected electron can be written as $K_{max} = hf - \phi$.

Data collected can be used to present a range of styles of graph. One of the more significant types of graph can be seen in Figure 3.13.

This graph can be used to understand the elegant solution to the disparity between theory and observation for the photoelectric effect offered by Einstein. Analysing the straight lines on the K_{max} vs frequency graph from the perspective of $y = mx + b$, it follows that the intercept on the y-axis is the metal's work function, ϕ, and the gradient is **Planck's constant**, h.

It is important to note that Einstein was not suggesting the wave model of light was wrong, but rather that light had wave properties that were observable in some situations and particle properties that were observable in other circumstances. This is often called the **wave–particle duality** of light.

There will be electrons that require more energy to be liberated than the amount described by the work function and so emitted electrons have a range of kinetic energies. Those with the maximum energy are the electrons of interest. The relationship between the maximum electron kinetic energy and the **stopping voltage** is seen through the work done to 'stop' the liberated electron in the equation $K_{max} = q_e V_{stop}$.

3.4 Light and special relativity

In the latter part of the 19th century, the scientific community had great confidence in their knowledge in many areas of science. Motion, as described by Newton's work, and the interactions between matter and energy were thought to be well understood. Time, distance and mass, under these models, were absolute quantities – they were the same when viewed by observers in different frames of reference. In 1905, Einstein published his theory of special relativity, which challenged these models and forced changes to occur on the basis of two postulates (propositions). These changes had broad-reaching implications.

3.4.1 Postulates of special relativity

Einstein's theory of special relativity is based on two **postulates**:

1 All **inertial frames of reference** are equivalent.

2 The speed of light in a vacuum is an absolute constant.

The specific wording of the postulates varies. An alternative wording that is equally acceptable is:

- The laws of physics are the same in all inertial frames of reference.
- The speed of light has the same value, c, in all inertial frames. It does not depend on the speed of either the source or the observer.

First postulate

A **frame of reference** described as inertial will be one in which a free body (one with no net external forces acting on it) will not accelerate – one in which Newton's first law will be obeyed. Any frame of reference moving at a constant velocity with reference to another inertial frame will itself be an inertial frame of reference.

This first postulate extends a principle of relativity proposed by Galileo and supported by Newton, which indicated that all inertial frames were equivalent for laws of mechanics to apply to all laws of physics.

The consequence of the first postulate is that no inertial frame of reference is preferred over another and, so, there is no **absolute frame of reference**. Furthermore, no experiment can be performed entirely within an inertial frame that can determine whether the frame is moving or stationary. Experimentally, for a passenger in a plane flying in a straight line at 700 km h^{-1}, the situation is indistinguishable from the same plane parked in a hangar.

Second postulate

The second postulate means that, regardless of the relative motion of observer and source, the speed of light will be measured to be the same by all inertial observers.

The second postulate seems to defy logic and does defy the straightforward appreciation of relative motion understood in Module 1 in Year 11. The recorded speed of anything would be expected to be determined by the motion of the source and/or observer. Einstein's second postulate, however, indicates that the notion of relative motion does not apply to light and the speed of light is constant no matter the motion of the observer relative to the source.

The unvarying nature of the speed of light is a logical consequence of the first postulate. Einstein suggested that, according to Galileo's principle of relativity, an observer travelling at the speed of light in the direction of a light wave would see a stationary electromagnetic wave but, if Maxwell's theory of electromagnetism is correct, the electromagnetic wave should consist of changing electric and magnetic fields. Einstein held Galileo's and Maxwell's work in high regard, but concluded that the only rational conclusion to draw was that the beam of light must be the same no matter the inertial frame of reference of the observer.

The prevailing thinking at the time, despite Maxwell's work, was still that light must travel in a medium and that the medium, called aether, permeated all space and matter. Earth would travel through the aether as it orbited the Sun. Both of Einstein's postulates were inconsistent with the concept of the aether, as it would be a preferred frame of reference and the speed of light would be different depending on its motion relative to the aether. The aether was deemed by Einstein to be irrelevant to his theory.

First postulate evidence

Albert Michelson and Edward Morley designed a very sensitive experiment to determine the rate at which Earth moved through the aether. The experiment featured a beam of light that was split into two perpendicular beams – one of which would travel with and against the theoretical aether wind and another that would travel across the aether wind on each part of its return journey (Figure 3.14). They calculated that the two beams would travel at different speeds and the time difference between the two journeys would be detectable by a change in the interference pattern caused by the beams being out of phase when the two beams were recombined. Extensive experimentation revealed no change in the interference pattern and the experiment failed to detect the aether. The existence of the aether would have suggested a preferred frame of reference – the lack of evidence of the aether supports Einstein's first postulate.

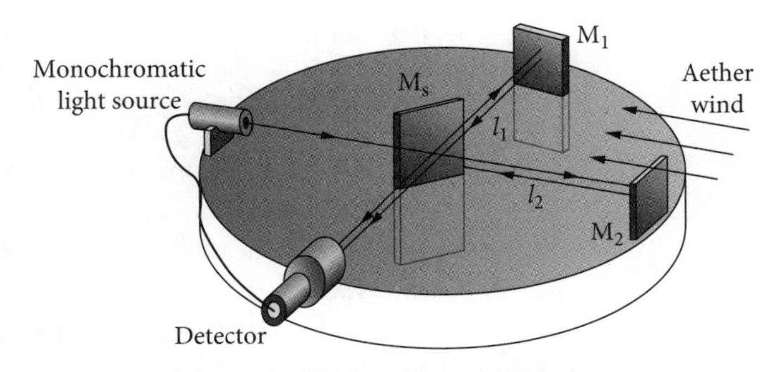

FIGURE 3.14 In this diagram of the Michelson–Morley experimental apparatus, a single beam of monochromatic light is split into two perpendicular beams by a half-silvered mirror (M_s). These two beams each travel a return journey over the distance of l_1 and l_2 respectively, reflecting off mirrors M_1 and M_2, and are then recombined at M_s to arrive at the detector.

Second postulate evidence

- The Michelson–Morley experiment supports the notion of the speed of light being unchanging.
- Subatomic particles moving at close to the speed of light have been observed undergoing a spontaneous decay with a gamma ray as one product. Experimental evidence confirms that regardless of the direction of the emitted gamma ray relative to the subatomic particle, the speed of the gamma ray relative to a stationary observer is constant.
- Cosmological studies of light emitted by matter blasted from a supernova explosion have shown that the light reaches Earth simultaneously regardless of the direction of motion of the matter.
- Light from binary star system pairs reaches Earth at the same speed irrespective of the motion of the star relative to Earth.

3.4.2 Time dilation

Einstein was able to use a simple **thought experiment** (Gedanken) to show quantitatively that two observers in different inertial frames of reference would measure the time taken for an event to occur to be different.

> **Note**
> This is a peculiar idea and difficult to comprehend. Some outcomes of this can seem quite bizarre but it is important to become 'comfortable' with them.

Einstein considered a moving train carriage in which a beam of light was projected from the floor (A) to a mirror on the ceiling (B) and back (Figure 3.15). This device can be considered to be a light clock. One observer is in the carriage with the light clock while another observer is standing outside. The observer in the carriage will see the beam move straight up and down, travelling a distance of $2L$. If the observer on board measures a time t_0 for the whole light journey, then it can be said that light travels a distance of $2L = ct_0$. The observer outside the carriage will see the light travel a greater distance – equivalent to the sum of the hypotenuses of the two right-angled triangles formed ($2D$).

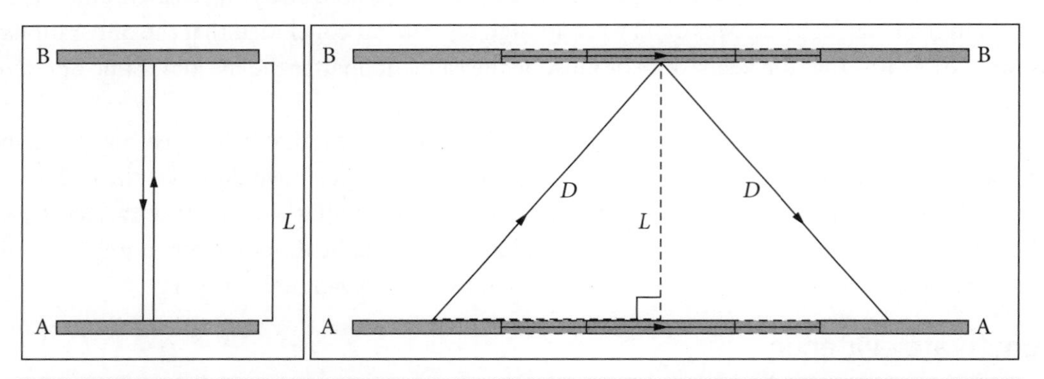

FIGURE 3.15 On the left, the observer on board will see the beam of light travel straight up and down, a distance of $2L$. On the right, it can be seen that the light travels a distance of $2D$ on its two diagonal journeys and $2D$ is clearly greater than $2L$.

Since the speed of light is constant, if the light travels a greater distance ($2D > 2L$), then it must take more time to make the journey. The observer outside the carriage will measure a time t for the journey. Consequently, the two observers will measure different time periods for the same event from their different inertial frames of reference. This is known as **time dilation**. The equation

$$t = \frac{t_0}{\sqrt{\left(1 - \frac{v^2}{c^2}\right)}}$$

can be derived and describes the relationship between the time measured in the frame of reference of the event, t_0 (known as the **proper time**), and the time measured by an observer in a different inertial frame of reference t (known as the **relativistic** or dilated time).

> **Note**
> Since the denominator of this fraction must always be less than 1, t is always greater than proper time. Hence, the phrase 'dilation'. Symmetrically, however, if the event had occurred in the frame of reference of the observer outside the train, then the passenger would record a longer time for the event – by the same factor.

9780170465304

Note

Einstein's time dilation thought experiment relied on two observers, in different frames of reference that are in motion relative to one another, observing a single event *and* knowing that the speed of light is the same for both observers. You should be able to reproduce this explanation with a diagram.

3.4.3 Length contraction

Einstein was also able to use a simple thought experiment to show quantitatively that the measurement of length by observers in different inertial frames of reference would be different. **Length contraction** is a logical consequence of time dilation. All lengths measured in the direction of travel in other inertial frames of reference are contracted by a factor described in the equation $l = l_0\sqrt{\left(1 - \frac{v^2}{c^2}\right)}$ where l_0 is known as **proper length** and is the length measured by an observer in the same frame of reference as the 'object' measured.

3.4.4 Evidence for special relativity

A range of experiments and observations support relativity.

Cosmic origin muons

Muons are subatomic particles created by interactions of cosmic rays with gas molecules in the upper atmosphere. These muons travel towards Earth at a speed close to the speed of light (0.995*c*). This high speed makes them ideal for observing relativistic effects. In the laboratory, stationary muons are observed to have an average lifetime of 2.2 μs.

Newtonian physics predicts that, on average, a muon would travel only 660 m. Muons are created kilometres above sea level and it would therefore be expected that no muons would reach the surface of Earth. Muons have been detected at Earth's surface and so must exist for much longer, when measured in the moving frame of reference, than classical physics predicts.

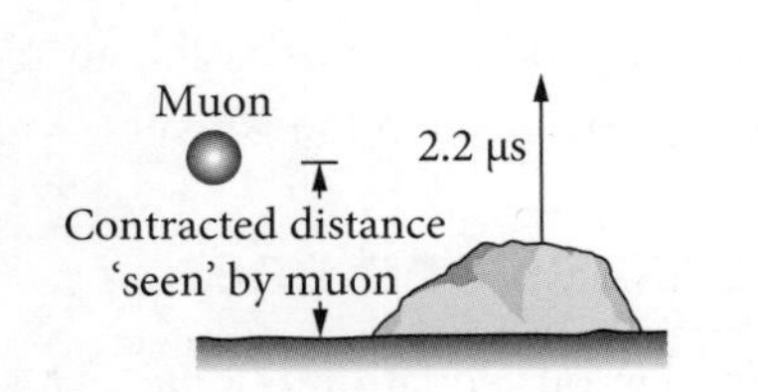

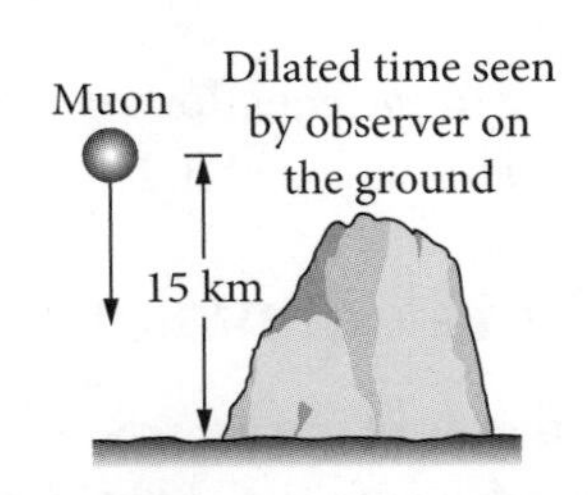

FIGURE 3.16 In the muon's frame of reference (diagram on left), its lifetime is the proper time of 2.2 μs, but the distance to Earth is significantly contracted compared with the altitude as measured by an observer on the ground. From Earth's frame of reference (diagram on right), the proper altitude is measured, but the muon's lifetime is significantly dilated. Each scenario gives some muons sufficient time to reach Earth.

Note

Notice that the frame of reference of the Earth observer and of the muon can both be used to reach the same conclusion here and, thus, this evidence supports both time dilation and length contraction and shows the consistency of the theory.

This observation can be explained if the lifetime of the muon was increased significantly. By using special relativity, the observation can be explained. Special relativity closely predicts the number of muons making it to the surface of Earth. An observer on Earth would see the muons' time run much slower – time dilation. This means that the muons will exist long enough to be observed on the surface of Earth in the same frame of reference as the observer.

From a muon's frame of reference, the distance to the surface of Earth is greatly reduced by the effects of length contraction so, although its average life span is unchanged, it has a much shorter distance to travel at high speed and reach Earth's surface.

Atomic clocks

Atomic clocks are extremely precise, to around 1 second per 100 000 000 years. They use the frequency of radiation emitted during electron transitions within atoms to measure time. This level of precision is ideal for measuring very small time differences.

In 1971, an experiment was conducted – the Hafele–Keating experiment – using three atomic clocks. A clock was placed in each of two aeroplanes, one of which flew east and the other west around Earth twice. The times recorded by the clocks for the journey were compared with each other and with the third clock left at the airport. The clocks that had flown around Earth were found to have different times from the stationary clock – the differences were predicted by relativity.

Particle accelerators

In 2014, a team of physicists measured the rates of transitions between energy states of atoms moving at high speed in a particle accelerator. These values were compared with the rates of transitions of stationary atoms. The difference in rates corresponded with values predicted by special relativity.

3.4.5 Relativistic momentum

It can be shown that, as a consequence of other features of special relativity and of conservation of momentum, an observer in an inertial frame of reference in relative motion to another frame will conclude that masses in the other frame are greater – this is mass dilation. The dilation of mass means that momentum is a relative quantity like time and distance. This relativistic momentum is related to the classical Newtonian momentum by the equation

$$p_v = \frac{m_0 v}{\sqrt{\left(1 - \frac{v^2}{c^2}\right)}}$$

Note
Both sides of this equation can be divided by velocity to show that 'mass dilation' (not mentioned in the syllabus) occurs.

where m_0 represents the **rest mass** (mass measured in the frame of reference of the observer) and, thus, $m_0 v$ is the Newtonian momentum.

It can be seen from the equation above that as an object approaches the speed of light, the momentum of the object approaches infinity. Further increases to the momentum of the object would require a force approaching infinity according to the equation $\Delta p = Ft$. Consequently, it is impossible to accelerate an object to the speed of light and the speed of light is the maximum speed of the Universe.

3.4.6 $E = mc^2$ and the equivalence of mass and energy

A consequence of Einstein's theory of special relativity is the famous equation $E = mc^2$. This equation means that energy and mass are two forms of the same entity, the **mass–energy equivalence**, much like ice and water are two forms of H_2O, and that conversion of one to the other is possible.

Extensive experimental evidence has been accumulated supporting this aspect of Einstein's theory and it is accepted that all nuclear and chemical reactions involving a release or absorption of energy are associated with a change in mass.

Examples of such situations include:

- stars converting mass to energy in nuclear fusion reactions
- combustion reactions yielding energy and having products with slightly less mass than the reactants (although this difference is likely to be immeasurably small)
- nuclear fission reactions in power plants converting small proportions of mass to energy
- **annihilation** of particles that encounter their **antiparticle**, such as an electron and a positron, producing energy corresponding to the mass that was annihilated
- production of a pair of particles, one matter and the other its **antimatter** equivalent, from electromagnetic radiation.

Note
Again, this is important throughout Module 8, and cross-modular questions are common.

Glossary

absolute frame of reference A frame of reference against which all others could be measured

absorption lines Black lines appearing on a continuous spectrum, indicating an absence or deficit of light of that wavelength

absorption spectrum A spectrum featuring black lines at certain wavelengths, indicating an absence or deficit of light of that wavelength

amplitude The maximum extent of a vibration or oscillation, measured from the position of equilibrium

annihilation The event that occurs when a matter particle and its antimatter equivalent meet, resulting in their entire masses being converted to energy

antimatter Matter consisting of elementary particles that are the antiparticles of those making up normal matter

antiparticle A particle with the opposite charge to its matter equivalent

black body An object that absorbs all wavelengths of electromagnetic radiation and emits radiation at a peak intensity related to the object's temperature

black body radiation Electromagnetic radiation emitted from a black body

blue shift The shift in the wavelengths of light towards the blue end of the spectrum as an object moves towards us

classical physics Physics that does not use quantum mechanics or special relativity

coherent light A light ray consisting of electromagnetic waves that are in phase with one another

constructive interference An interaction of two waves that results in a wave of greater amplitude

continuous spectrum A spectrum made up of all wavelengths

corpuscle Small particles that comprise light according to Newton's theory

destructive interference An interaction of two waves that results in a wave of lesser amplitude

diffraction The spreading of a wave after it passes an obstacle or goes through a gap

diffraction grating A device that diffracts light to result in an interference pattern using a large number of parallel lines etched into a transparent object

electromagnetic spectrum The collection of frequencies of electromagnetic radiation

electromagnetic wave A wave featuring perpendicular oscillating electric and magnetic fields

https://get.ga/aplus-hsc-physics-u34

A+ DIGITAL FLASHCARDS
Revise this topic's key terms and concepts by scanning the QR code or typing the URL into your browser.

emission lines Bright lines of certain wavelengths on a black background showing emission of those wavelengths

emission spectrum A black background with bright lines of particular wavelengths

frame of reference A place (point of view) from which an observation is made

Huygens' principle The wave theory that suggests each point on a wave acts as a source of secondary wavelets

incandescent Light of a range of wavelengths generated by the motion of particles as a result of heat

inertial frame of reference A frame of reference that is not accelerating; i.e. one in which Newton's first law is obeyed

interference The process whereby two waves occupying the same point in space interact with each other

length contraction The process whereby a length appears shorter when the observer is in motion relative to the object

Malus' law A law that describes the intensity of light emerging from an ideal polarising filter in terms of the intensity of incident polarised light

mass–energy equivalence The statement that mass and energy are two forms of the same quantity, illustrated by $E = mc^2$

monochromatic light Light of a single wavelength

muon A type of subatomic particle formed when cosmic rays interact with the upper atmosphere

permeability A measure of the magnetisation that a material or substance obtains in response to an applied magnetic field

permittivity A measure of the ability of a material to store electrical potential energy under the influence of an electric field

phase A term describing the comparative stage of an oscillating cycle

photocurrent The current created by the emission of electrons from a metal surface as a consequence of incident light

photoelectric effect The emission of electrons when a material, such as a metal, is exposed to light

photoelectron An electron liberated from a metal surface as a consequence of incident light

photoemission The process whereby electrons are emitted from a metal surface as a consequence of incident light

photon A particle or quantum of electromagnetic radiation with specific energy given by $E = hf$

Planck's constant A constant, h, that relates photon energy to light frequency according to $E = hf$

polarised Light waves with aligned electric fields

postulate A proposition assumed to be true in order to develop subsequent ideas

principle of superposition The quantitative principle by which the amplitude of waves in an interference situation can be determined

proper length The length of an object measured by an observer in the same frame of reference as the object

proper time The time measured for an event by an observer in the same frame of reference as the event

quantised Existing in discrete amounts (quanta) that cannot be divided into smaller quantities

quantum A discrete unit of a quantity

quantum physics The study of the smallest units of the physical world

red shift The shift in the wavelengths of light towards the red end of the spectrum as an object moves away from us

relativistic Moving at a velocity such that there is a significant change in properties (such as mass) in accordance with the theory of relativity

rest mass The mass measured in an inertial frame of reference in which the object is at rest

stopping voltage The reverse bias voltage required to completely stop the flow of photoelectrons in a photoelectric effect apparatus

thought experiment A theoretical experiment used to investigate a theoretical position that may be difficult to test practically

threshold frequency The minimum frequency of incident light required to liberate electrons from a metal surface

time dilation The slowing of time in a moving frame of reference as measured by an observer in a stationary frame

wave–particle duality The concept that radiation or matter can be both a particle and a wave simultaneously

work function The minimum amount of energy required to eject an electron from a metal surface

9780170465304

Exam practice

Multiple-choice questions

Solutions start on page 177.

Electromagnetic spectrum

Question 1

Maxwell's theory of electromagnetism is based on

A experiments using light and mirrors to establish a value for the speed of light.

B experimental observations of oscillating charged particles.

C one equation relating quantised energy to frequency.

D four equations relating charged particles, magnetic fields and electric fields.

Question 2

The spectrum from a newly found star and laboratory spectra for four elements expected to be present are shown.

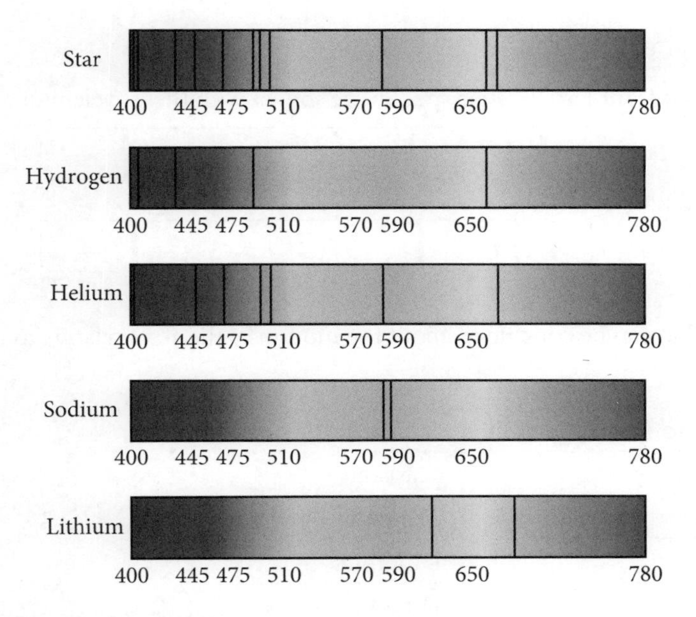

A trainee astronomer successfully identifies that two elements definitely not found in the new star are

A hydrogen and lithium.

B helium and hydrogen.

C sodium and helium.

D lithium and sodium.

Question 3

The image below illustrates one of the historical methods, using a rotating mirror wheel and a number of stationary mirrors, to determine an experimental value for the speed of light.

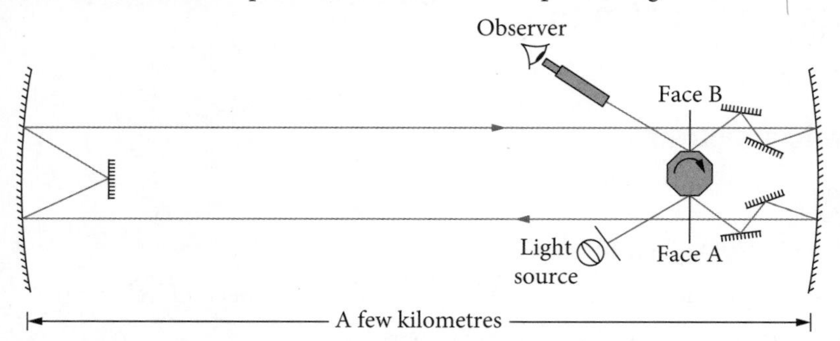

In order to gain data useful for determining the speed of light, as the wheel is gradually rotated at an increasing rate, the observer would be required to report only when

A no light reaches them.

B an interference pattern can be seen.

C a continuous spectrum can be seen.

D light reaches them.

Question 4 ©NESA 2019 SI Q2

Two stars were observed from Earth. Their spectra are shown with the wavelength in nanometres.

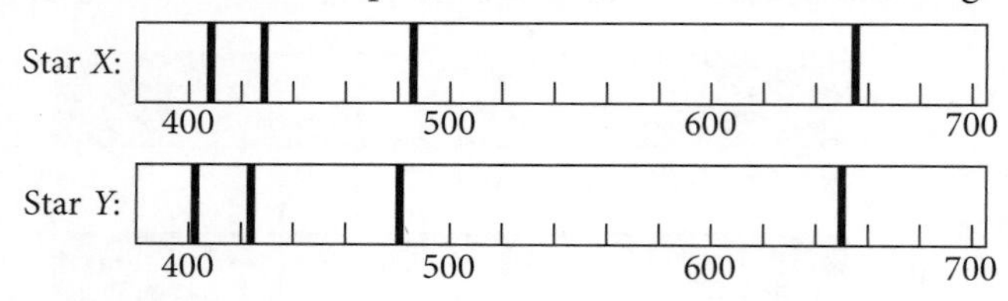

Using their spectra, what can be concluded about the motion of the stars relative to Earth and their chemical compositions?

	Motion relative to Earth	Chemical composition
A	The same	The same
B	Different	The same
C	The same	Different
D	Different	Different

Question 5 ©NESA 2020 SI Q18

An observer sees Io complete one orbit of Jupiter as Earth moves from P_1 to P_2, and records the observed orbital period as t_P. Similarly, the time for one orbit of Io around Jupiter was measured as Earth moved between the pairs of points at Q, R and S, with the corresponding measured periods of Io being t_Q, t_R and t_S.

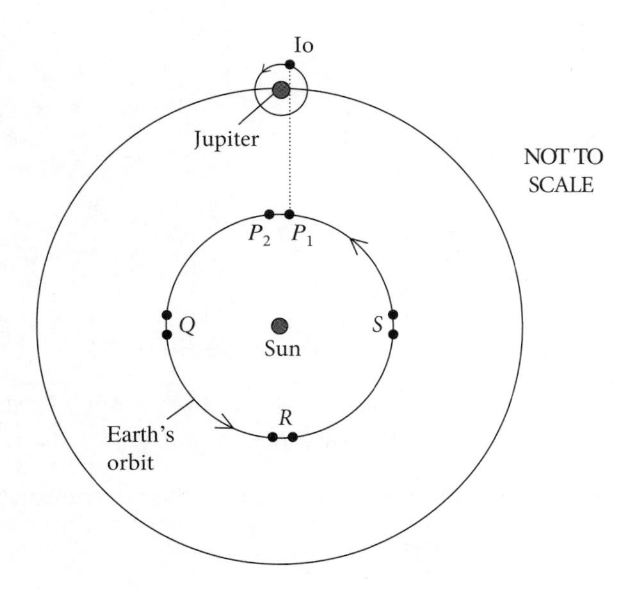

Which measurement of the orbital period would be the longest?

A t_P

B t_Q

C t_R

D t_S

Question 6

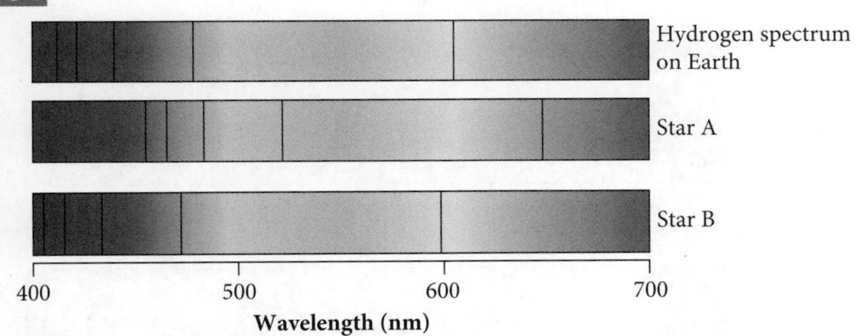

From these spectra it can be assumed that

A star A is moving away from Earth at a greater rate than star B is moving towards Earth.

B star A is moving towards Earth at a greater rate than star B is moving away from Earth.

C star B is moving away from Earth at a greater rate than star A is moving towards Earth.

D star B is moving towards Earth at a greater rate than star A is moving away from Earth.

Question 7 ©NESA 2021 SI Q5

The spectrum of an object is shown.

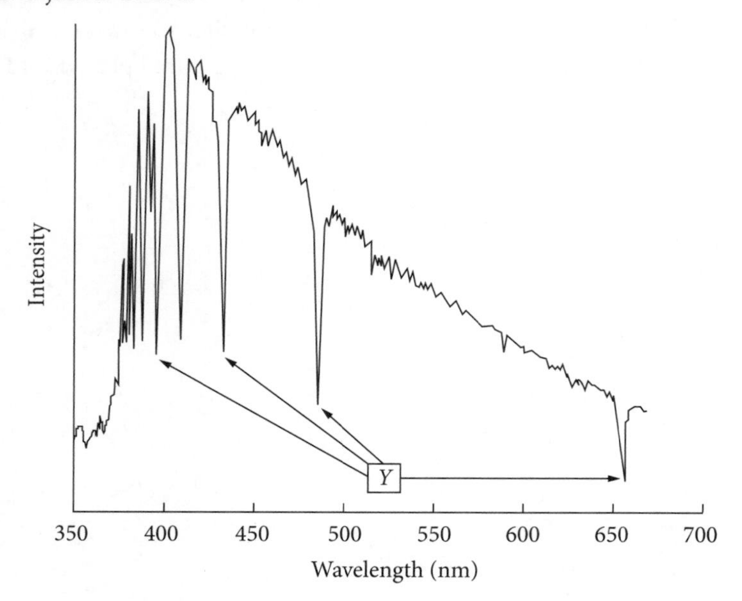

Which row of the table correctly identifies the most likely source of the spectrum and the features labelled *Y*?

	Source of spectrum	Features labelled *Y*
A	Star	Absorption lines
B	Discharge tube	Absorption lines
C	Star	Emission lines
D	Discharge tube	Emission lines

Question 8

The image below shows spectral data collected from two stars in the same galaxy.

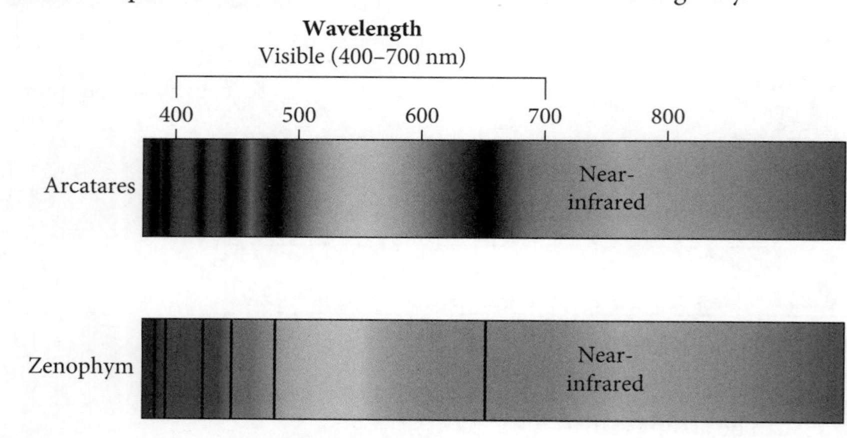

This information could be used to conclude that Arcatares

A was formed much earlier than Zenophym.

B has a denser atmosphere than Zenophym.

C is moving away from Earth faster than Zenophym.

D is rotating at a lower rate than Zenophym.

9780170465304

Light: wave model

Question 9 ©NESA 2020 SI Q1

The diagram shows a model used to explain the refraction of light passing from medium *X* into medium *Y*.

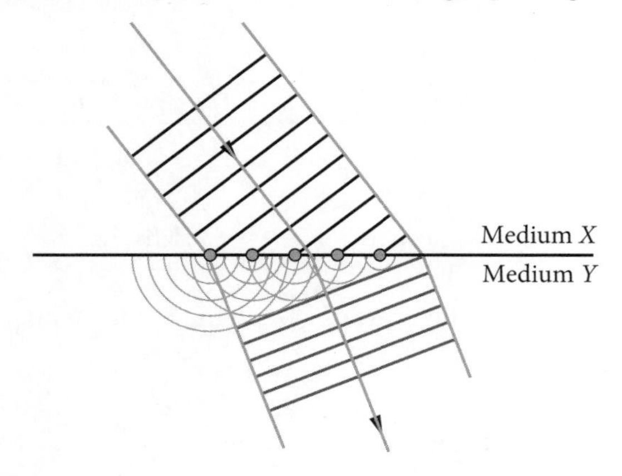

Who proposed this model?

A Malus

B Planck

C Newton

D Huygens

Question 10

Infrared cameras are used by rescue services and police to 'see' people in situations where there is no natural or artificial light. The external surface temperature of a clothed person is 25°C. What will be the wavelength of infrared radiation detected at greatest intensity by the camera?

A 0.116 m

B 9.72×10^{-3} m

C 1.16×10^{-4} m

D 9.72×10^{-6} m

Question 11 ©NESA 2021 SI Q8

Light from a point source is incident upon a circular metal disc, forming a shadow on a screen as shown. A bright spot is observed in the centre of the shadow.

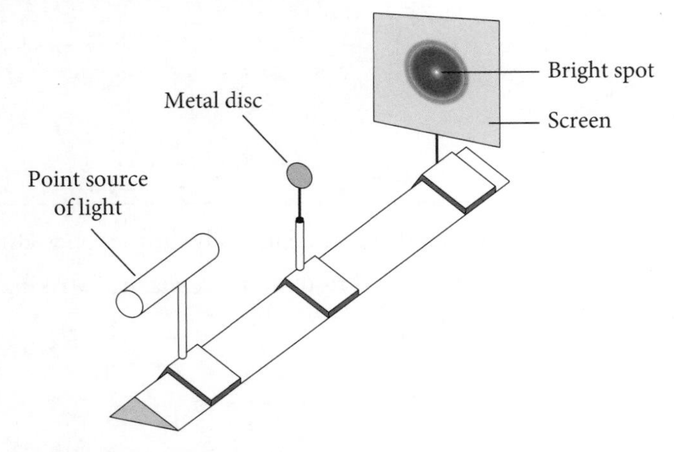

The bright spot is caused by a combination of

A interference and refraction.

B refraction and polarisation.

C polarisation and diffraction.

D diffraction and interference.

Question 12 ©NESA 2019 SI Q10

A beam of light passes through two polarisers. The second polariser has a transmission axis at an angle of 30° to that of the first polariser. The intensity of the beam before and after the second polariser is I_0 and I_B respectively.

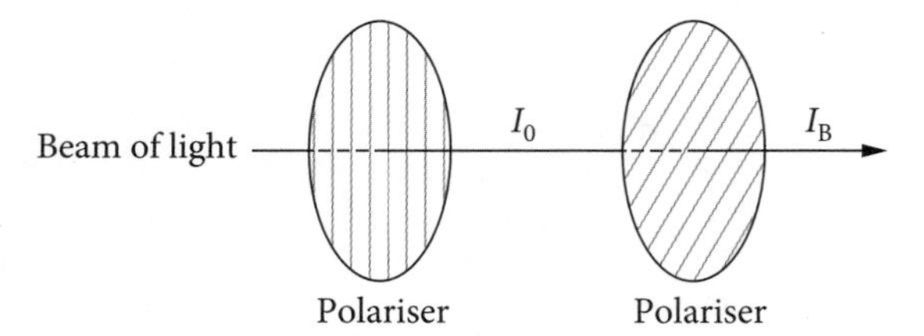

Which row of the table correctly identifies the value of $\frac{I_B}{I_0}$ and the model of light demonstrated by this investigation?

	Value of $\frac{I_B}{I_0}$	Model of light demonstrated
A	0.750	Wave model
B	0.750	Particle model
C	0.866	Wave model
D	0.866	Particle model

Question 13

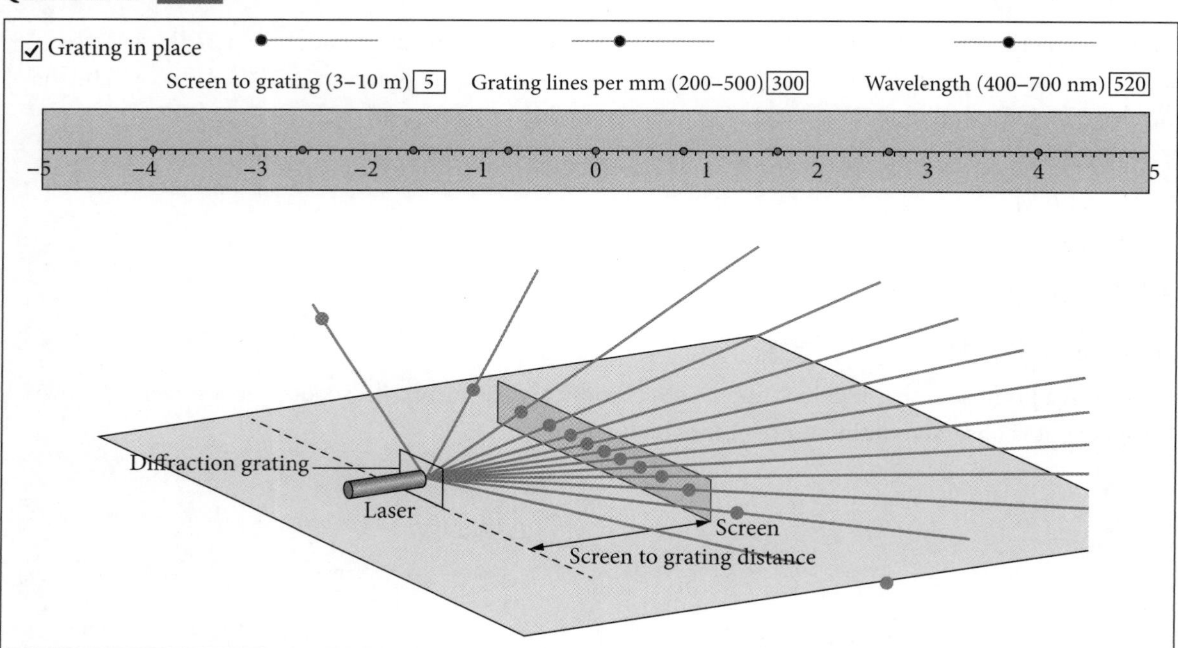

The image above is a screenshot of a simulation that is used to study diffraction gratings.

Using the information provided, which change would result in more maxima striking the screen if all other factors remain constant?

A Using a blue laser

B Increasing the number of grating lines per millimetre

C Increasing the screen to grating distance

D Replacing the diffraction grating with a double slit with a slit separation the same as the one featured in the diffraction grating

Question 14 ©NESA 2019 SI Q15

Monochromatic light passes through two slits 1 μm apart. The resulting diffraction pattern is measured at a distance of 0.3 m.

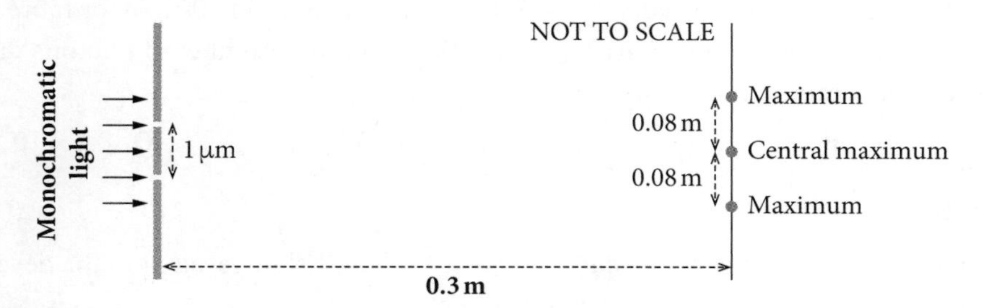

This diffraction pattern can be analysed using the equation $d \sin \theta = \lambda$.

What values of d and θ should be used in the equation?

	d	θ
A	0.3 m	$\tan^{-1}\left(\frac{0.08}{0.3}\right)$
B	0.3 m	$\sin^{-1}\left(\frac{0.08}{0.3}\right)$
C	1 μm	$\tan^{-1}\left(\frac{0.08}{0.3}\right)$
D	1 μm	$\sin^{-1}\left(\frac{0.08}{0.3}\right)$

Question 15

When white light is incident on a diffraction grating, a series of continuous spectra can be seen on either side of the central maximum. Which description is most accurate for the observed pattern?

A The central maximum will be white, and the continuous spectra will each have violet furthest from the central maximum.

B The central maximum will be white, and the continuous spectra will each have red furthest to the right.

C The central maximum will be white, and the continuous spectra will each have red furthest to the left.

D The central maximum will be white, and the continuous spectra will each have red furthest from the central maximum.

Question 16 ©NESA 2021 SI Q15

Unpolarised light is incident upon two consecutive polarisers as shown. The second polariser has a fixed transmission axis which cannot be rotated. I_1 is the intensity of light after the first polariser, and I_2 is the intensity of light after the second polariser.

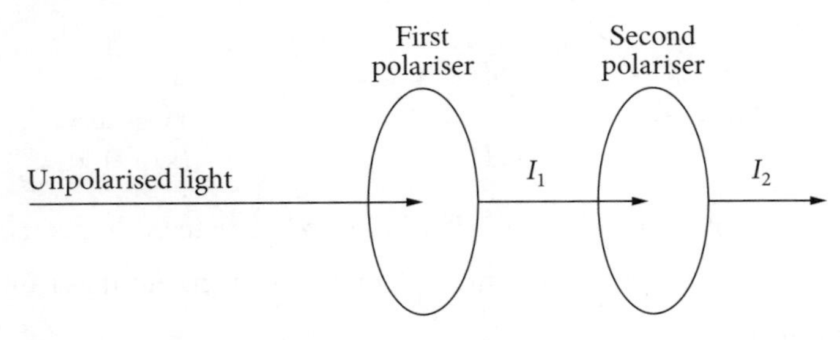

How would I_1 and I_2 be affected if the transmission axis of the first polariser was rotated?

A Both would change.

B Only I_1 would change.

C Only I_2 would change.

D Neither would change.

Light: quantum model

Question 17

According to the classification system used, a class IV laser has an output of 500 mW or more. If the laser is emitting green light of a wavelength of 550 nm, what is the minimum number of photons that will be emitted from the laser in a minute?

A 1.4×10^{18} **B** 8.3×10^{19} **C** 8.3×10^{22} **D** 8.2×10^{40}

Question 18

Experimental data that was used to generate curves similar to those below resulted in the development of a revolutionary concept because of the work of one scientist.

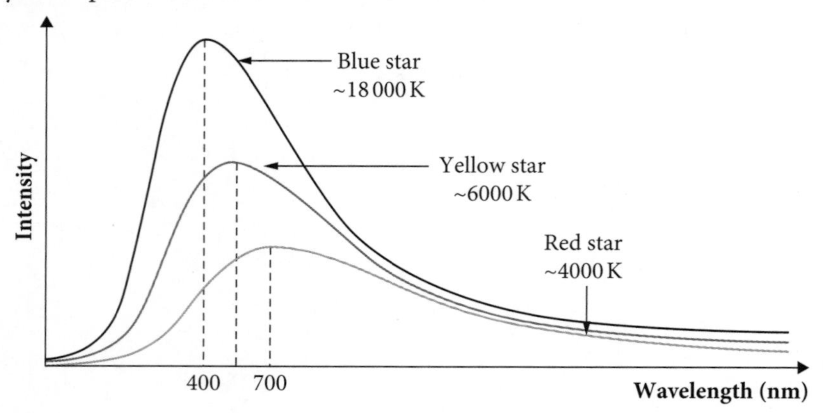

Identify the scientist and the revolutionary concept.

	Scientist	Concept
A	Galileo	The speed of light is infinite.
B	Einstein	Time and distance are relative quantities.
C	Maxwell	Light is an electromagnetic wave.
D	Planck	Energy is exchanged in quantised amounts.

Question 19 ©NESA 2020 SI Q13

The graph shows the relationship between the frequency of light used to irradiate two different metals, and the maximum kinetic energy of photoelectrons emitted.

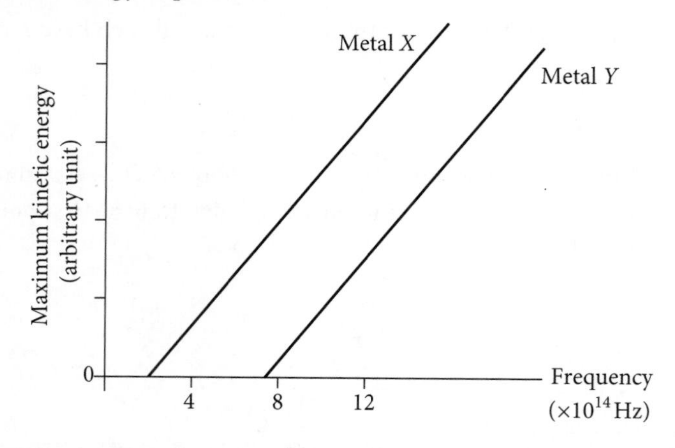

Suppose that light having a frequency of 8×10^{14} Hz is used to irradiate both metals.

Compared to the photoelectrons emitted from metal *X*, photoelectrons emitted from metal *Y* will

A have a lower maximum velocity.

B have a higher maximum velocity.

C take a longer time to gain sufficient energy to be ejected.

D take a shorter time to gain sufficient energy to be ejected.

Questions 20–22 are based on the following diagram, which shows apparatus to test the photoelectric effect.

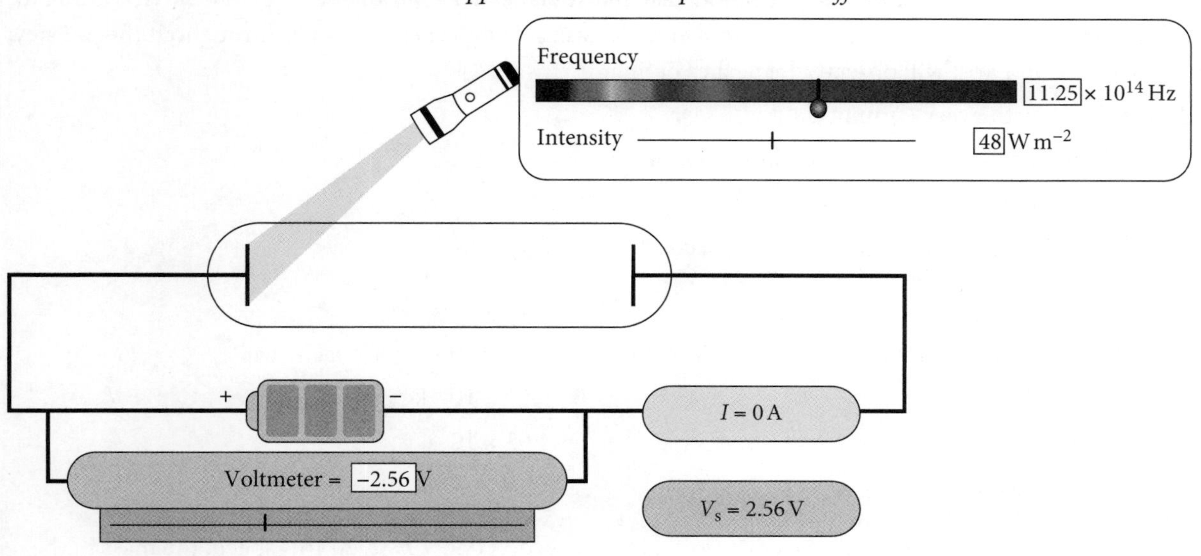

Question 20

Based on the information provided, incident photons striking the emitter plate each have energy of

A 7.45×10^{-19} J
B 48 W
C 2.56 eV
D 0 A

Question 21

Based on the information provided, in order to reach the collector from the emitter plate, photoelectrons would need to leave the emitter plate with a velocity of at least

A 3.37×10^{-19} m s^{-1}
B 4.11×10^{-19} m s^{-1}
C 9.49×10^{5} m s^{-1}
D 1.28×10^{6} m s^{-1}

Question 22

Based on the information provided, the greatest current will flow if the following change is made.

A Caesium (work function 2.1 eV) is replaced as the emitter plate by calcium (work function 2.9 eV).
B The intensity is increased to 60 W m^{-2}.
C Frequency is increased to 11.26×10^{14} Hz.
D V_{stop} is decreased to 2.5 V.

Light and special relativity

Question 23

The two postulates used by Einstein to develop the special theory of relativity can be stated as

A I: The laws of physics are the same in all frames of reference.
II: The speed of light is a constant irrespective of the relative motion of the source.

B I: The laws of physics are the same in all inertial frames of reference.
II: The speed of light is a constant irrespective of the medium in which it travels.

C I: The laws of physics are the same in all frames of reference.
II: The speed of light is a constant irrespective of the medium in which it travels.

D I: The laws of physics are the same in all inertial frames of reference.
II: The speed of light is a constant irrespective of the relative motion of the source.

Question 24

A spaceship passes Earth and takes a trip to a distant exoplanet at a constant velocity of $0.9c$. According to the astronaut on the spaceship, his spaceship emits a flash of light every 6 minutes throughout the journey. An observer on Earth will observe the flashes of light

A reaching Earth every 6 minutes.

B reaching Earth every 13 minutes and 45 seconds.

C travelling at $0.1c$.

D reaching Earth every 2 minutes and 37 seconds.

Question 25

The total power of the Sun is 3.83×10^{26} W. What is the decrease in its mass each year?

A 3.73×10^{13} kg

B 5.59×10^{15} kg

C 1.34×10^{17} kg

D 4.03×10^{25} kg

Question 26

The Large Hadron Collider can accelerate protons to a speed of $0.999\,998c$. At this speed, the mass of the proton, as observed by a scientist at the LHC, would have increased by a factor of

A 500

B $2.999\,999\,4 \times 10^8$ m s^{-1}

C 0.02

D 250 000

Question 27 ©NESA 2020 SI Q17

In a thought experiment, observer X is on a train travelling at a constant velocity of $0.95c$ relative to the ground. Observer Y is standing on the ground outside the train. As observer X passes observer Y, observer X sends a short light pulse towards the sensor.

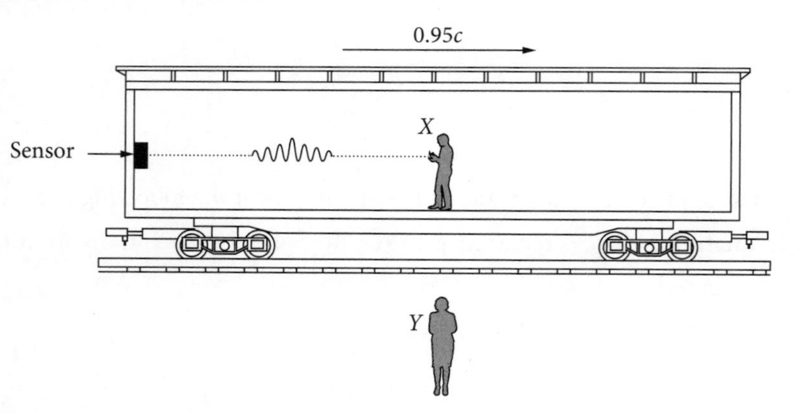

Which statement about the light pulse is correct as observed by X or Y in their respective frames of reference?

A Its velocity observed by Y is $0.05c$.

B X sees it travel a shorter distance to the sensor than Y.

C X sees it take a longer time to reach the sensor than Y.

D Both X and Y see it travel the same distance in the same amount of time.

Question 28

A carriage in which observer A sits passes observer B at a speed of $0.75c$.

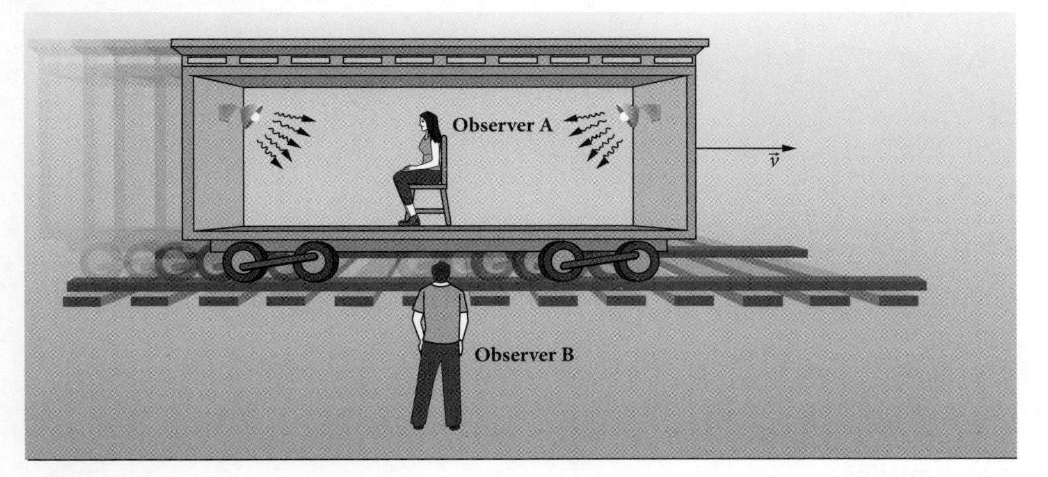

Observer A has measured the distance from the wall in front of her to the wall behind, which she calls x, to be 8 m and the distance from roof to floor, which she calls y, to be 2.4 m.

What values will observer B measure for x and y?

	x (m)	y (m)
A	5.3	1.6
B	5.3	2.4
C	8	1.6
D	8	2.4

Short-answer questions

Solutions start on page 181.

Electromagnetic spectrum

Question 29 (4 marks)

Account for the production of each of the three spectra illustrated.

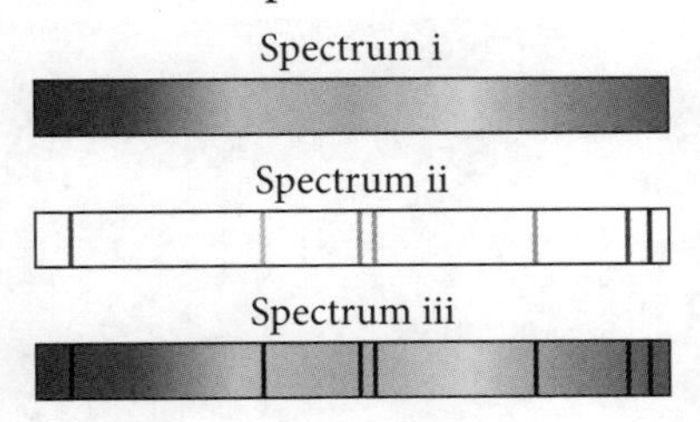

Question 30 (3 marks)

Explain how red shift occurs and how it can be used to determine the speed of stars relative to us.

Question 31 (6 marks)

Our understanding of a phenomenon requires the development of theories that are subsequently proved correct, and which enable the development of additional related theories.

Discuss this statement in the context of Maxwell's contribution to our understanding of light.

Question 32 (5 marks)

Describe how spectra can be used to determine various properties of stars.

Question 33 (5 marks)

The diagram is a schematic representation of one of the methods used to experimentally determine the speed of light.

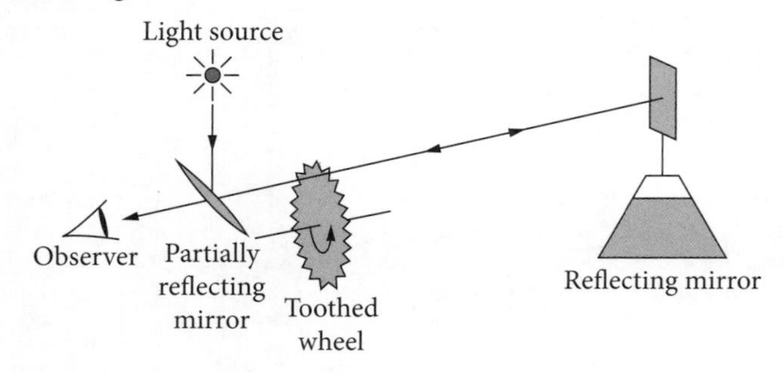

Explain what measurements would need to be made and how this apparatus could be used to determine a value for the speed of light.

Light: wave model

Question 34 (4 marks)

Describe how the wavelength of an unknown monochromatic light source can be determined using Young's double-slit apparatus.

Question 35 (4 marks)

A light source is placed a short distance from a rotating polarising filter. A detector is used to measure the intensity of light that emerges from the polarising filter. In the first part of the experiment, the light from the source is polarised; in the second part of the experiment, the light from the source is unpolarised but of equal intensity. With the aid of sketched graphs on a single set of axes, compare intensity of light measured by the detector in each part of the experiment. Justify your response.

Question 36 (7 marks)

The image below was created when a beam of coherent, monochromatic light of wavelength 585 nm was incident on a diffraction grating. The distance from the diffraction grating to the projection screen was 6.10 m. The distance from the central maximum to the third-order maximum was measured to be 3.62 m.

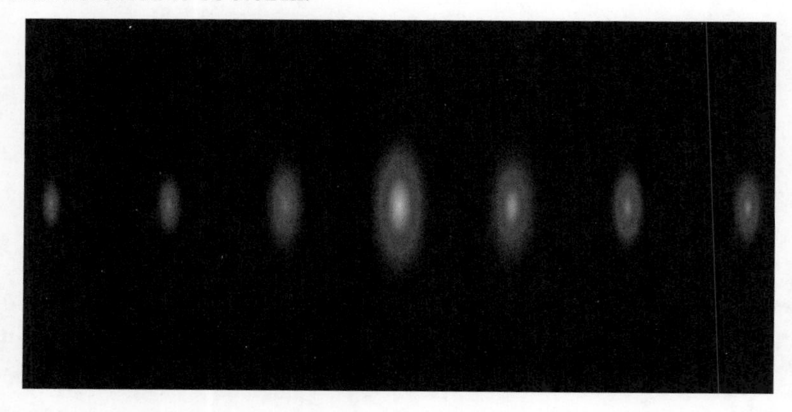

a Explain how the third-order maximum is created in terms of diffraction and interference. 2 marks

b Determine the number of slits per centimetre of the diffraction grating to the nearest thousand. 3 marks

c If the wavelength of light is increased gradually to 700 nm, how will the observed pattern be affected? 2 marks

Question 37 (4 marks)

Compare the way in which the models of light proposed by Newton and Huygens explain refraction.

Light: quantum model

Question 38 (4 marks)

Analyse Planck's contribution to the quantum model of light.

Question 39 (6 marks)

Analyse the evidence from photoelectric effect experiments that demonstrated inadequacies of the wave model of light.

Question 40 (15 marks)

A group of students performed an investigation of the photoelectric effect using apparatus similar to that in the diagram.

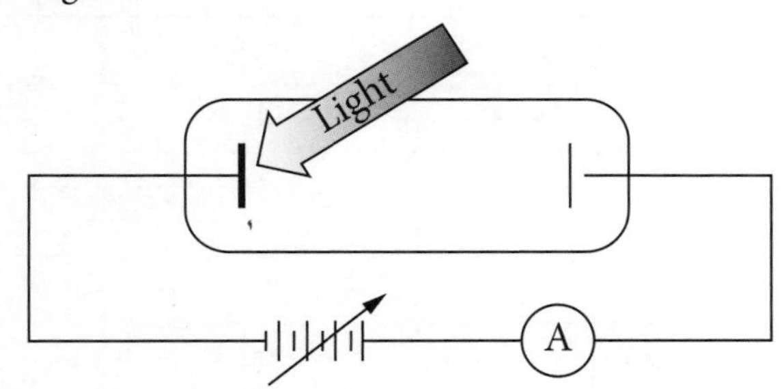

Using a caesium metal surface, the voltage was varied until no photocurrent flowed for each of the incident light frequencies. The metal was then replaced by another unknown metal and the process repeated. The data collected is given in the following table.

Frequency of light (Hz) × 10^{14}	Stopping voltage caesium (V)	K_{max} caesium (J)	Stopping voltage unknown metal (V)	K_{max} unknown metal (J)
4.5	0.60		0.15	
5.5	0.95		0.50	
6.5	1.30		0.85	
7.0	1.55		1.10	
7.5	1.90		1.45	

a Complete the table by calculating the maximum kinetic energy values for photoelectrons for each metal at the given frequencies. 2 marks

9780170465304

b Plot the data appropriately on the axes provided and draw lines of best fit for each metal. 3 marks

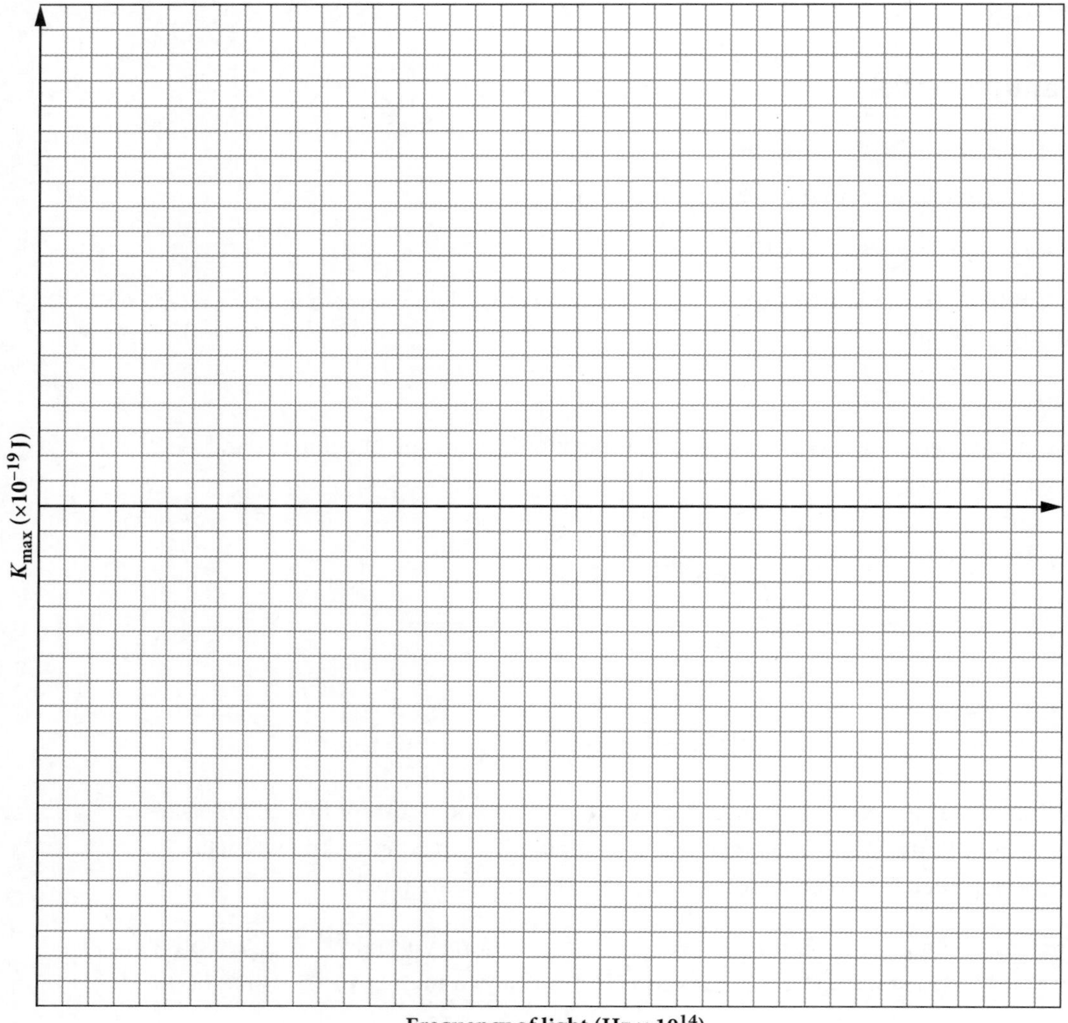

c Use the graph to determine the threshold frequency of each metal. 2 marks

d Which metal has the greater work function? Justify your response. 2 marks

e Use the graph to determine an experimental value for Planck's constant. Show all working. 3 marks

f How does the straight line of best fit plotted for either of the metals suggest that the law of conservation of energy applies to the photoelectric effect investigated? 3 marks

Light and special relativity

Question 41 (7 marks)

Describe Einstein's thought experiment for time dilation and a piece of subsequent experimental evidence that validated his theory.

Question 42 (4 marks)

A proton is accelerated in a particle accelerator to a speed of 2.66×10^8 m s^{-1}.

a Calculate the relativistic momentum of this proton. 2 marks

b With reference to your work in part **a**, explain how this provides evidence for the notion that the speed of light is the maximum speed for the Universe. 2 marks

Question 43 (3 marks)

Calculate the frequency of the two identical gamma ray photons that are produced when an electron and a positron annihilate each other.

Question 44 (5 marks)

An observer on the ground is able to view and measure a train when it is stationary within a glass-sided tunnel at a point where the centre of the train is exactly at the centre of the tunnel. The train is 60 m in length, and it is measured that 2.0 m of the train extends past the end of the tunnel at each end.

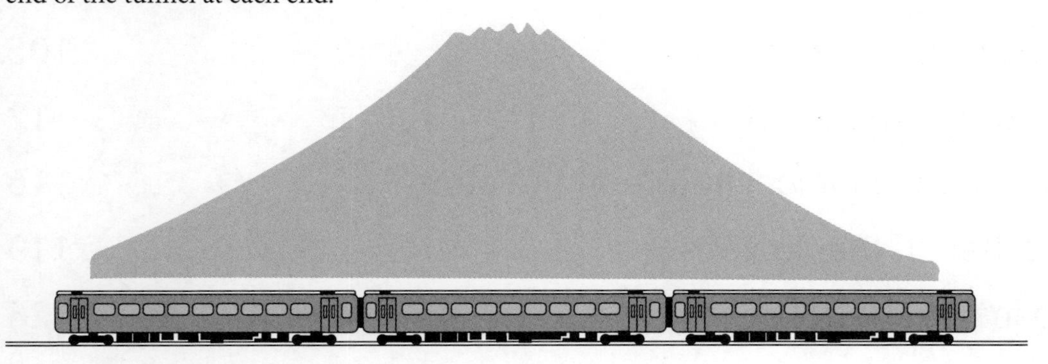

a How fast would the train need to be travelling so that the same observer sees the train fit exactly within the tunnel? 3 marks

b At that instant, what would an observer on the train, at the centre of the train, notice about the train's fit within the tunnel? 2 marks

Question 45 (3 marks)

Discuss a piece of evidence supporting or denying one of Einstein's two postulates.

CHAPTER 4
MODULE 8: FROM THE UNIVERSE TO THE ATOM

Module summary 103

4.1 Origins of the elements 105

4.2 Structure of the atom 112

4.3 Quantum mechanical nature of the atom 116

4.4 Properties of the nucleus 119

4.5 Deep inside the atom 126

Glossary 129

Exam practice 131

4

Chapter 4
Module 8: From the Universe to the atom

Module summary

Outcomes

On completing this module, you should be able to:

- analyse and evaluate primary and secondary data and information
- solve scientific problems using primary and secondary data, critical thinking skills and scientific processes
- communicate scientific understanding using suitable language and terminology for a specific audience or purpose
- explain and analyse the evidence supporting the relationship between astronomical events and the nucleosynthesis of atoms and relate these to the development of the current model of the atom.

Working Scientifically skills

In this module, you are required to demonstrate the following Working Scientifically skills:

- develop and evaluate questions and hypotheses for scientific investigation
- design and evaluate investigations in order to obtain primary and secondary data and information
- conduct investigations to collect valid and reliable primary and secondary data and information
- select and process appropriate qualitative and quantitative data and information using a range of appropriate media
- analyse and evaluate primary and secondary data and information
- solve scientific problems using primary and secondary data, critical thinking skills and scientific processes
- communicate scientific understanding using suitable language and terminology for a specific audience or purpose.

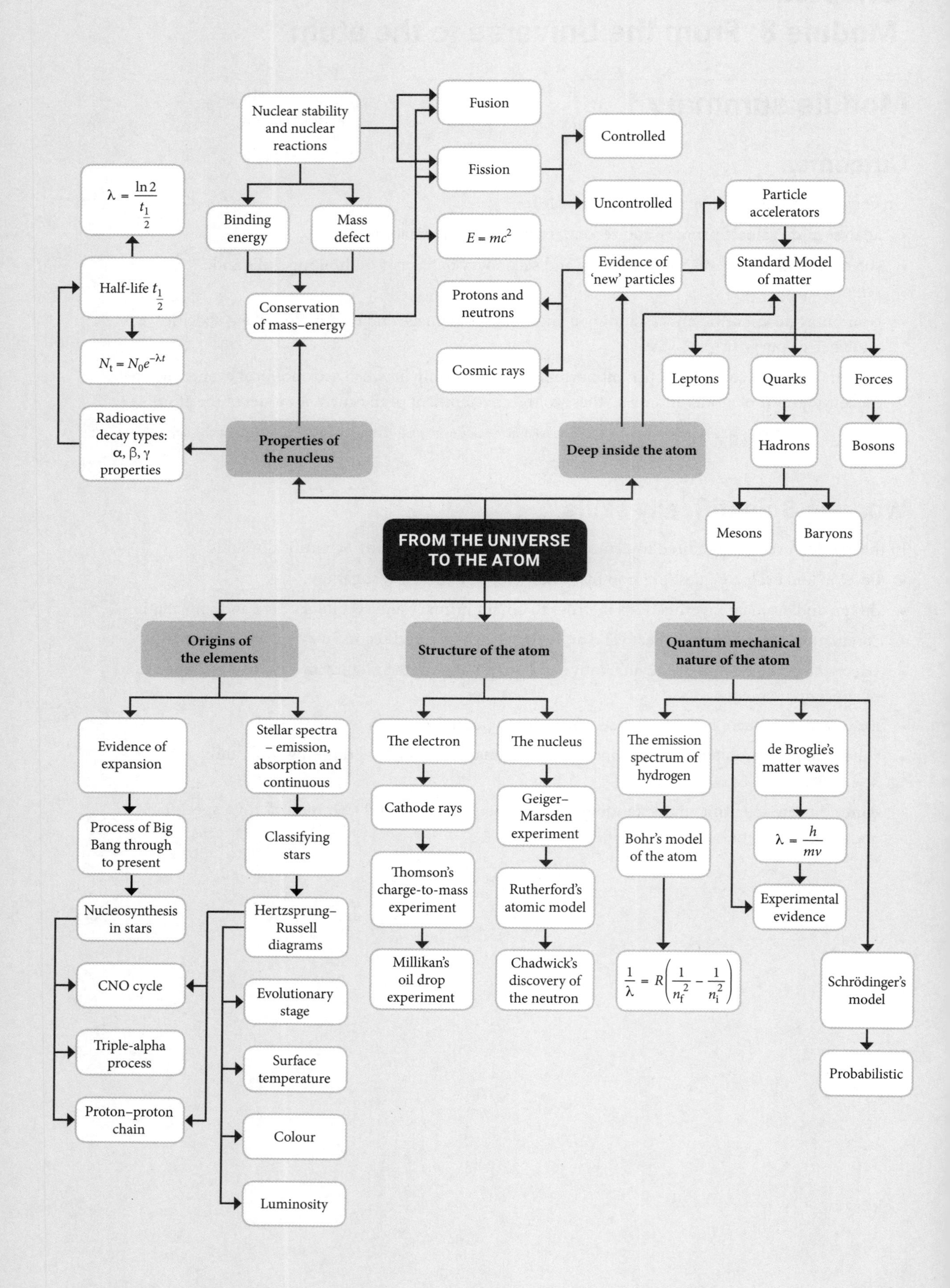
FROM THE UNIVERSE TO THE ATOM
Properties of the nucleus
Deep inside the atom
Origins of the elements
Structure of the atom
Quantum mechanical nature of the atom
Nuclear stability and nuclear reactions
Fusion
Fission
Controlled
Uncontrolled
$E = mc^2$
Binding energy
Mass defect
Conservation of mass–energy
$\lambda = \frac{\ln 2}{t_{\frac{1}{2}}}$
Half-life $t_{\frac{1}{2}}$
$N_t = N_0 e^{-\lambda t}$
Radioactive decay types: α, β, γ properties
Protons and neutrons
Cosmic rays
Evidence of 'new' particles
Particle accelerators
Standard Model of matter
Leptons
Quarks
Forces
Hadrons
Bosons
Mesons
Baryons
Evidence of expansion
Process of Big Bang through to present
Nucleosynthesis in stars
CNO cycle
Triple-alpha process
Proton–proton chain
Stellar spectra – emission, absorption and continuous
Classifying stars
Hertzsprung–Russell diagrams
Evolutionary stage
Surface temperature
Colour
Luminosity
The electron
Cathode rays
Thomson's charge-to-mass experiment
Millikan's oil drop experiment
The nucleus
Geiger–Marsden experiment
Rutherford's atomic model
Chadwick's discovery of the neutron
The emission spectrum of hydrogen
Bohr's model of the atom
$\frac{1}{\lambda} = R\left(\frac{1}{n_f^2} - \frac{1}{n_i^2}\right)$
de Broglie's matter waves
$\lambda = \frac{h}{mv}$
Experimental evidence
Schrödinger's model
Probabilistic

4.1 Origins of the elements

For many centuries, we have had a fascination with the origin of all that can be observed. As the enormity of galaxies and the distances that separate them became understood, that fascination became questions about concepts such as the origin of the Universe, whether it was finite or infinite, and the origin of all the matter that makes up the Universe. As technology has evolved to collect more and better scientific evidence and knowledge has accumulated, theories have been developed and discarded. The Big Bang theory, the idea of an expanding Universe, is now widely accepted on the basis of the scientific evidence available today from investigations that use modern technologies. Evidence gathered on the nucleosynthesis reactions in stars allows scientists to understand how elements are made in the nuclear furnace of stars.

4.1.1 The Big Bang theory and an expanding Universe

Up until the start of the 20th century, the scientific community believed that the Universe was stable, or in a steady state, rather than evolving. In 1922, Russian mathematician Alexander Friedmann, using Einstein's theory of general relativity, solved equations that indicated an expanding Universe – a bizarre concept.

In 1923, Edwin Hubble began collecting spectral data from galaxies at known distances. The spectra were all redshifted, which revealed that the galaxies were moving away.

Determining the distance to another galaxy relied on the discovery by Henrietta Leavitt of a relationship between **luminosity** and period for a type of star with variable luminosity called a Cepheid variable.

By studying spectra from distant galaxies, Hubble was able to show an approximately linear relationship between the distance to the galaxy and the amount by which the spectral lines are redshifted. This relationship illustrated that the more distant a galaxy, the faster it is moving away from Earth. It became known as Hubble's law.

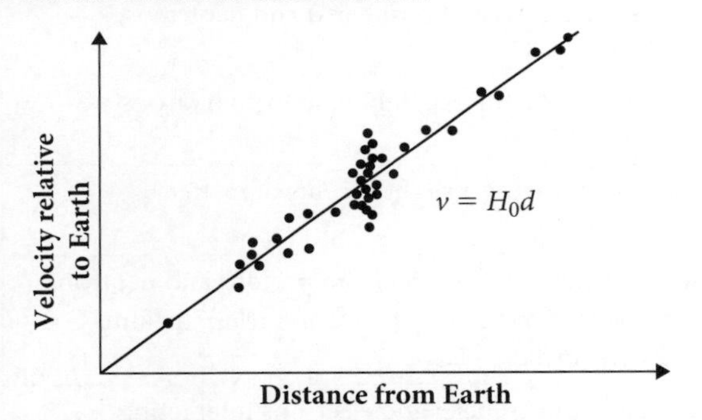

FIGURE 4.1 Data collected by Hubble shows a linear relationship between distance to galaxy and speed at which galaxy is moving away.

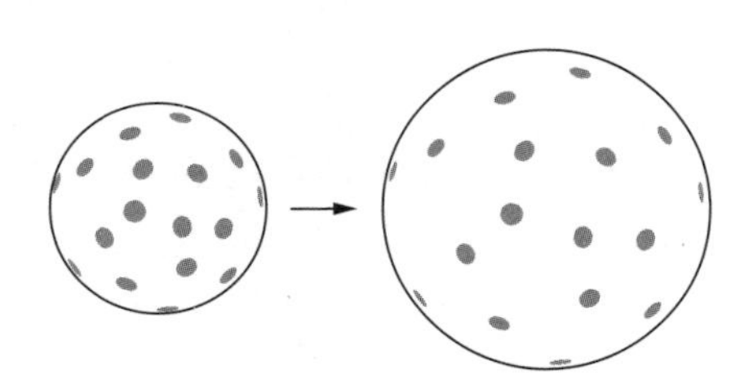

FIGURE 4.2 As the Universe expands, the distance between any point and all other points in the Universe increases.

It can be inferred from this that all galaxies are moving away from all other galaxies. By incorporating Einstein's general theory of relativity, it can be shown that the galaxies are not moving through space further from each other, but rather the space between them is stretching. The conclusion from that concept is that the Universe is expanding, as proposed by Friedmann.

The gradient of the graph shown in Figure 4.1 is known as the Hubble constant (H_0) and gives an estimate of the rate of expansion of the Universe. From the graph, $H_0 = \frac{v}{d}$ and $v = \frac{d}{t}$, so $H_0 = \frac{1}{t}$. Therefore, the inverse of Hubble's constant can be used to estimate the age of the Universe. Based on current data, the estimate stands at approximately 13.7 billion years.

9780170465304

The concept of the expanding Universe, with a beginning, gave birth to the **Big Bang** theory. The Big Bang theory asserts that the Universe began from an extremely hot, infinitely small point of extraordinary energy called a **singularity**. Time began at the Big Bang and space started being created by the expansion. The theory made predictions about the proportions of the light elements found in the Universe, as well as the existence and wavelength of radiation (the cosmic microwave background radiation) left over from the beginning of the Universe, predictions that were subsequently proven correct.

The Big Bang theory proposes a timeline for the transformation of energy into matter as the singularity expanded and time began, along with the creation of the forces that govern the processes occurring in the Universe in its current form.

Note

Of course, the inability to replicate the energy levels and the lack of knowledge about the laws of physics that governed the early Universe result in conjecture and speculation, but an approximate timeline of highlights is given in Table 4.1. There is considerable variation in information from different sources. The order of events and very approximate values are needed. Specific times and temperatures are not important.

TABLE 4.1 Timeline of the Big Bang

Time period following Big Bang	Key events
10^{-43} to 10^{-12} seconds	Universe expands dramatically (period of cosmic inflation). Four forces of nature separate out from a single unified force – first gravity, then strong nuclear force followed by the weak nuclear force and electromagnetic force. First basic matter–antimatter pairs of particles created from energy in form of quarks and antiquarks.
10^{-12} to 10^{-6} seconds	A surplus of matter particles results in only matter quarks remaining after annihilation, along with leptons and antileptons (such as electrons and positrons) and neutrinos, making up most of the matter of the Universe.
10^{-6} to 1 second	Temperature decreases sufficiently for quarks to begin to combine to form hadrons (such as protons and neutrons).
1 second to 3 minutes	Leptons and antileptons annihilate each other, leaving a slight surplus of matter particles (electrons).
3 to 20 minutes	Temperature becomes sufficiently low for formation of nuclei from protons and neutrons to occur through nuclear fusion. Isotopes of hydrogen (about 75% of nuclei), helium (about 25% of nuclei) and lithium (trace) are formed.
20 minutes to 240 000 years	The Universe is a hot, seething collection of atomic nuclei and electrons interacting repeatedly with vast numbers of photons – the Universe is said to be opaque.
240 000 to 300 000 years	Temperature decreases to 3000 K and electrons are able to combine with nuclei to form the first atoms, making the Universe transparent. This recombination results in the release of huge numbers of photons (the source of the cosmic microwave background radiation).
300 million to 500 million years	Temperature is sufficiently low for gravitational attraction to begin to create pockets of dense matter that continue to collapse under their own gravitational forces. Stars begin to form as matter is converted into energy in nuclear fusion reactions according to $E = mc^2$ and the Universe creates light. Gravity clusters stars together into the first galaxies.

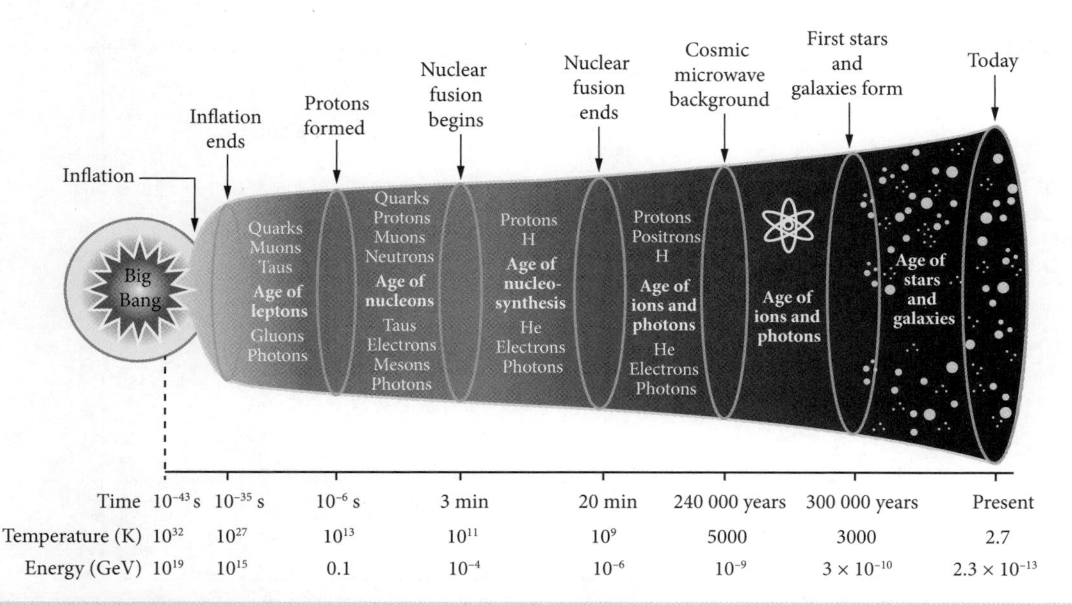

FIGURE 4.3 As the Universe expands, the forces are separated and matter is created from energy. The temperature changes as the expansion occurs and the forces transform the building blocks of basic matter into the Universe that is observed now.

4.1.2 Classification of stars from stellar spectra

Recall from section 3.1.3 that when light of a variety of frequencies is passed through a device that can disperse it on the basis of frequency (such as a diffraction grating or prism), a spectrum can be produced. There are three types of spectra: continuous, emission and absorption.

Stellar spectra are particularly useful for providing information about the nature, motion and composition of stars.

The most significant aspects of stellar spectra are the wavelengths of peak intensity, which can be used to calculate surface temperature, and will determine colour; and the location and radiant intensity of absorption lines, which can be used to determine relative proportions of elements and ions that comprise the star's atmosphere.

Using this knowledge, a stellar classification system was developed (and subsequently refined) that classifies stars into seven **spectral classes**: O, B, A, F, G, K and M (Table 4.2). Astronomers use a system of subdivisions of each of these classes into ten subcategories for greater detail.

TABLE 4.2 Spectral classification of stars

Spectral class	Temperature range (K)	Colour
O	>30 000	Blue
B	11 000–30 000	Blue-white
A	7500–11 000	White
F	6000–7500	Yellow-white
G	5000–6000	Yellow
K	3500–5000	Orange
M	<3500	Red

4.1.3 The Hertzsprung–Russell diagram and stellar evolution

In 1911 two astronomers, Hertzsprung and Russell, independently plotted spectral class against luminosity for the known stars. The graph, known as the **Hertzsprung–Russell (H–R) diagram**, revealed some significant patterns and enabled stars to be classified into three main groups: main sequence stars, red giant and supergiant stars, and white dwarfs. This has become a vital tool in astrophysics.

Note
The H–R diagram is a powerful tool, and since much about a star can be concluded from its position on it, there is a lot of opportunity for questions based on it from most of section 4.1.

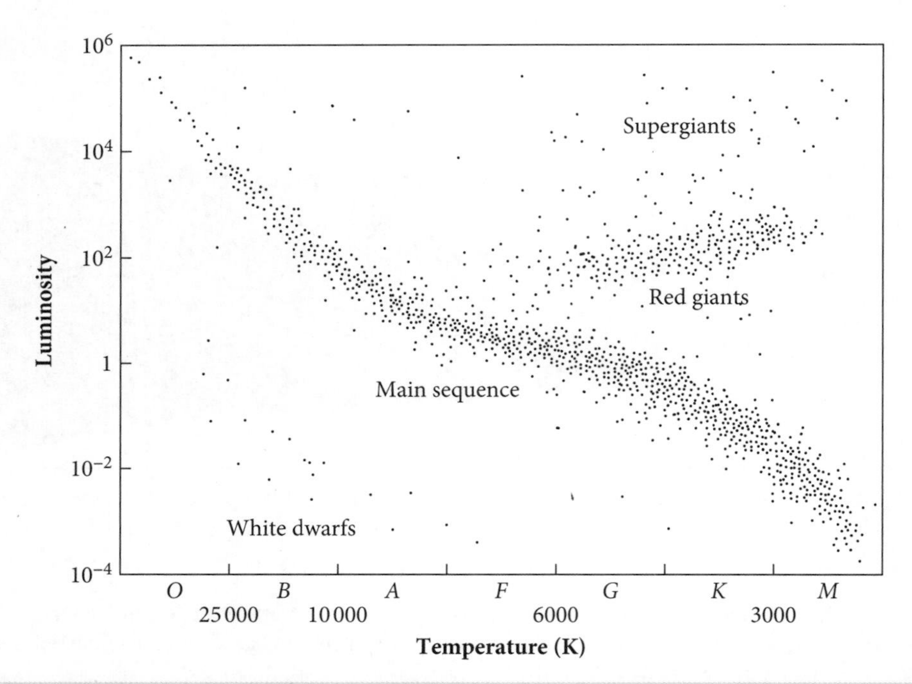

FIGURE 4.4 The Hertzsprung–Russell diagram shows the three main groups of star types. Note that the four extremes of the diagram represent: top right = cool and bright; bottom right = cool and dim; bottom left = hot and dim; and top left = hot and bright. Understanding the relationships that link temperature with luminosity and luminosity with surface area can enable the conclusions that stars at top right are very large, bottom right are small, bottom left are very small and top left are large.

The three star types identified represent the three long-term visible stages of a star's evolutionary life cycle.

Main sequence stars

- Approximately 90% of stars belong in this category.
- Stars spend most of their active 'life' as **main sequence** stars from the time at which they begin emitting light.
- Nuclear **fusion** in the core converts hydrogen to helium.
- Mass ranges from about 0.1 solar masses to 200 solar masses.
- They fuse hydrogen at a rate related to mass, so stars at top left will be main sequence stars for about 10 million years, stars in the middle will be main sequence stars for about 10 billion years, and stars at the bottom right will be main sequence stars for about 10 trillion years.
- Our Sun is a main sequence star.

Red giant and supergiant stars

- They comprise about 1% of stars.
- The most common are 'red giants'.
- Red giants have masses between one-quarter and eight times that of the Sun, while supergiants have masses between eight and 200 times that of the Sun but are hundreds to thousands of times larger.
- They have fused most of the hydrogen in the core and so are at a later stage of their evolution. They will fuse heavier elements in the core and hydrogen to helium in their outermost shell.

White dwarf stars

- They comprise nearly 10% of all stars.
- They are no longer undertaking any fusion and are at the last stage of their star life.
- They radiate energy as a consequence of residual heat.
- They are very dense and small (approximately Earth-sized on average).
- They gradually become dull.

All stars begin as a gas cloud, called a nebula, that will predominantly consist of hydrogen and helium, the most common elements of the Universe. Other elements may be present depending on the source of the material from which the star is formed. Gravity causes the cloud to contract and aggregate other matter. If the mass is sufficiently large (>0.1 solar mass), then the temperature (about 10^7 K) and pressure will facilitate nuclear fusion of hydrogen in the core. Hydrogen will fuse first because less energy is required to overcome electrostatic repulsive forces than for larger nuclei.

In most stars, at some point an equilibrium will be reached between the inwards force of gravity and the outwards pressure resulting from the fusion reactions. The star is stable at this point and begins its time on the main sequence.

At some point, helium will accumulate in the core to a level that causes hydrogen fusion to stop. Gravitational collapse ensues. Astrophysicists have determined that if the star has less than 0.23 solar masses it will become a white dwarf, although there is no evidence of this yet because the main sequence life of such small stars is extremely long. If the mass is greater than 0.23 solar masses and less than 8 solar masses, hydrogen fusion around the helium core will commence. The outermost layers expand significantly and cool and the star becomes a red giant. If the star has sufficient mass, fusion of helium will continue in the core (stars above 0.5 solar masses), producing heavier elements sequentially (stars above 5 solar masses). Depending on the mass of the star, fusion will cease at some stage. The largest stars will fuse elements up to the production of iron. Fusion beyond iron does not occur in stars because these fusion reactions require energy rather than release energy.

Three possible futures lie ahead for the star at this stage, depending on its mass.

If the mass is less than 8 solar masses, stars will lose their outer layers (and create a planetary nebula) as the core shrinks to become a white dwarf.

Larger stars will collapse dramatically, and a huge **supernova** explosion will result. Most of the star's mass will be blasted into space in a supernova, and elements heavier than iron will be created in the process. (The abundance of energy and copious free neutrons enable massive fusion reactions beyond iron.)

If the remaining mass is between 1.4 and 3 solar masses, the electrons in the core will be forced into the protons in nuclei that comprise the core, and a lump of very tightly packed neutrons will result. This is a **neutron star** – a very small (radius around 10 km) and dense object (a handful of a neutron star will have mass exceeding 10^7 tonnes).

If the core mass exceeds 3 solar masses, the remnants will collapse to form a **black hole** – an object so dense that the escape velocity exceeds the speed of light.

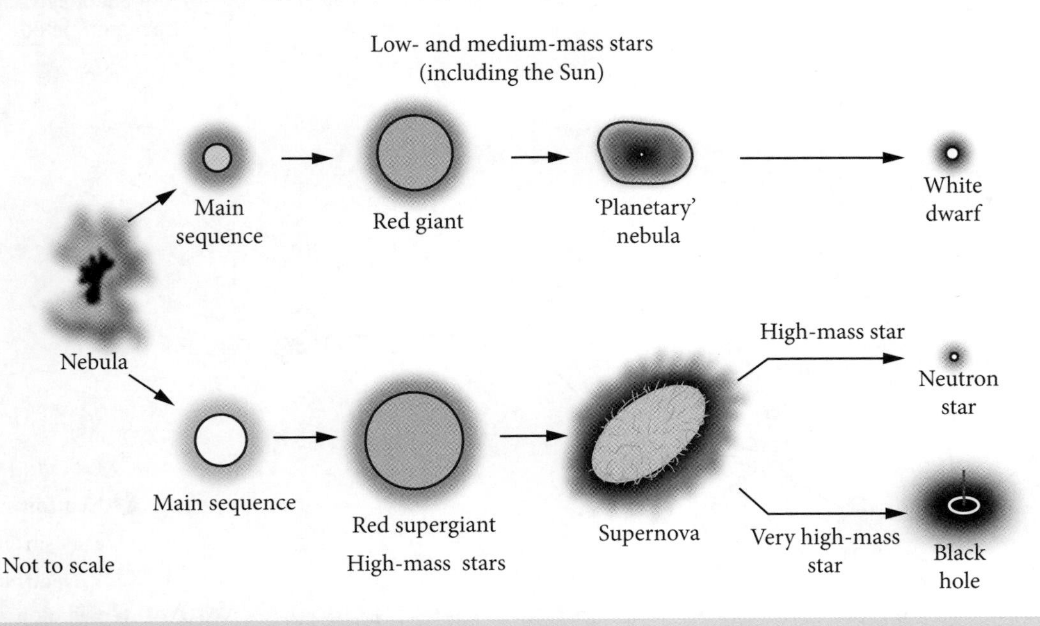

FIGURE 4.5 Mass determines the evolutionary path of stars.

Some of the evolutionary stages of stars after their time on the main sequence can be represented on the Hertzsprung–Russell diagram (Figure 4.6).

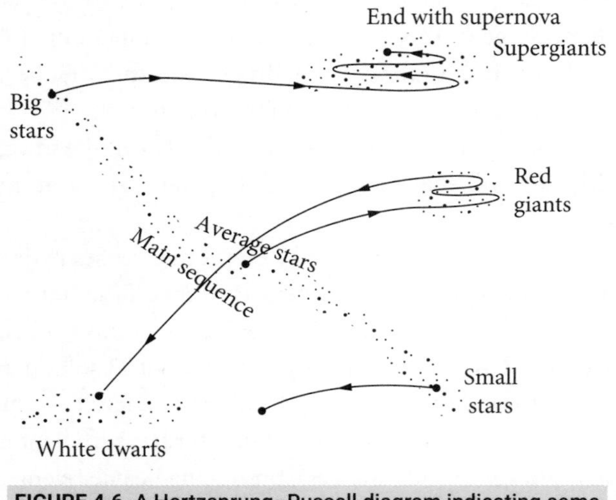

FIGURE 4.6 A Hertzsprung–Russell diagram indicating some evolutionary path options

4.1.4 Nucleosynthesis in stars and mass–energy equivalence

Main sequence stars produce energy by converting mass into energy, according to $E = mc^2$, in the process of fusing hydrogen to helium by one of two nuclear fusion reactions. The fusion reaction converts four hydrogen nuclei into a helium nucleus. The creation of a new nucleus in this manner is called **nucleosynthesis**.

The main fusion reaction in relatively low mass stars requires lower temperatures (10^7 K) and is called the **proton–proton (p–p) chain** reaction. Our Sun produces most of its energy in this way. The p–p chain releases energy in the form of two **gamma ray** photons.

The sequence of equations is as follows and can be represented in Figure 4.7.

$2\,{}^{1}_{1}\text{H} \xrightarrow{\text{yields}} {}^{2}_{1}\text{H} + {}^{0}_{+1}\text{e} + \nu$ (two of these reactions required)

${}^{2}_{1}\text{H} + {}^{1}_{1}\text{H} \xrightarrow{\text{yields}} {}^{3}_{2}\text{He} + {}^{0}_{+1}\text{e} + \nu$ (two of these reactions required)

$2\,{}^{3}_{2}\text{He} \xrightarrow{\text{yields}} {}^{4}_{2}\text{He} + 2\,{}^{1}_{1}\text{H}$

Note
The syllabus does not indicate that this chain of reactions needs to be remembered.

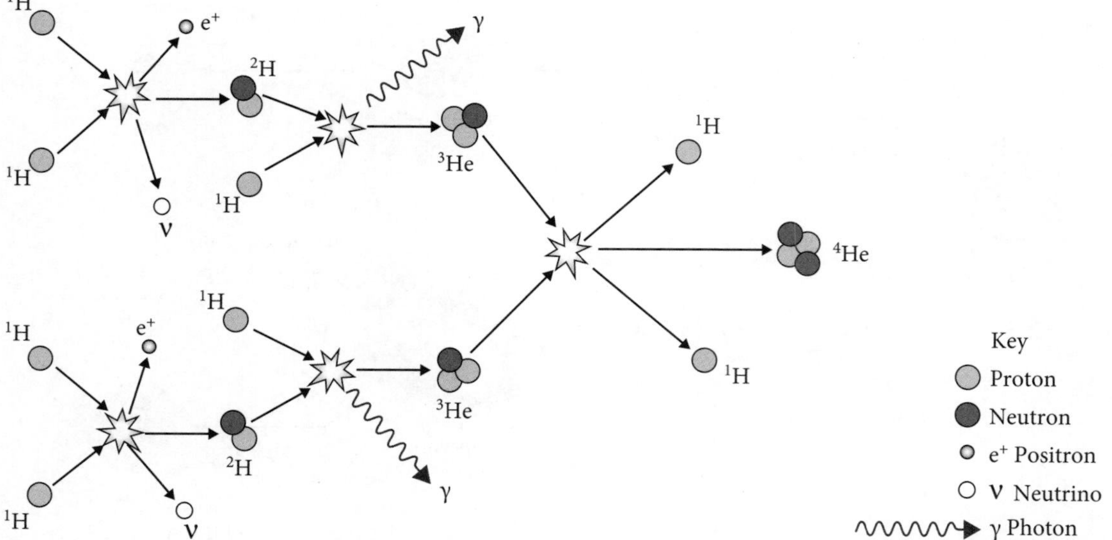

FIGURE 4.7 Steps in the proton–proton chain

The **carbon–nitrogen–oxygen (CNO) cycle** is a series of reactions with the same reactants and products as the p–p chain, but requiring temperatures of greater than 1.5×10^7 K. In this cycle, carbon-12 acts as a catalyst in the reaction in that it is required to enable the reaction but is produced in the final step. In a similar fashion, as can be seen in the diagram, nitrogen and oxygen are produced and consumed during the process.

The CNO cycle produces three gamma ray photons but the same amount of energy as the proton–proton chain.

$^{12}_{6}\text{C} + ^{1}_{1}\text{H} \xrightarrow{\text{yields}} ^{13}_{7}\text{N} + \gamma$

$^{13}_{7}\text{N} \xrightarrow{\text{yields}} ^{13}_{6}\text{C} + ^{0}_{+1}\text{e} + \nu$

$^{13}_{6}\text{C} + ^{1}_{1}\text{H} \xrightarrow{\text{yields}} ^{14}_{7}\text{N} + \gamma$

$^{14}_{7}\text{N} + ^{1}_{1}\text{H} \xrightarrow{\text{yields}} ^{15}_{8}\text{O} + \gamma$

$^{15}_{8}\text{O} \xrightarrow{\text{yields}} ^{15}_{7}\text{N} + ^{0}_{+1}\text{e} + \nu$

$^{15}_{7}\text{N} + ^{1}_{1}\text{H} \xrightarrow{\text{yields}} ^{12}_{6}\text{C} + ^{4}_{2}\text{He}$

Note
The syllabus does not indicate that this cycle of reactions needs to be remembered.

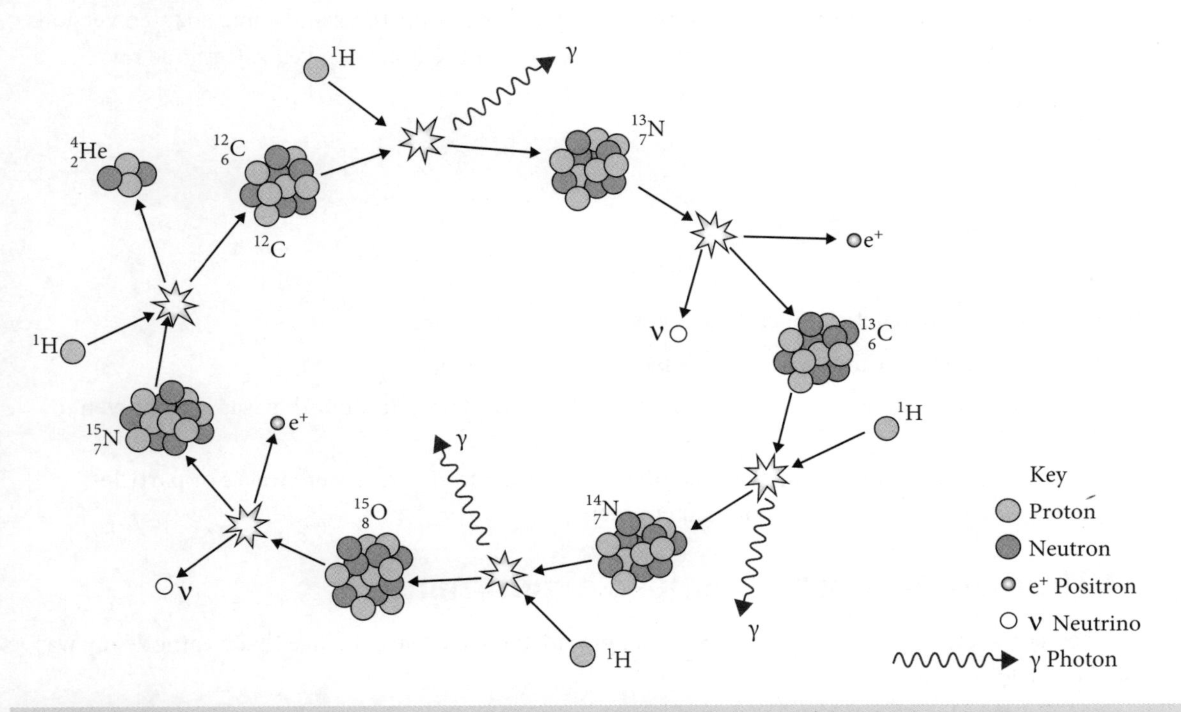

FIGURE 4.8 The CNO cycle shows the steps in the fusion of hydrogen to helium nuclei.

Both the p–p chain and the CNO cycle have a net reaction as follows:

$4^{1}_{1}\text{H} \xrightarrow{\text{yields}} ^{4}_{2}\text{He} + 2\text{e}^{+} + 2\nu + \text{energy}$

Note that e^+ is a positron (the antimatter electron particle) and ν is a neutrino.

A variety of fusion reactions is possible in post-main sequence stars and will occur depending on star mass and evolutionary stage. In the core of sufficiently massive stars, a reaction called the triple-alpha (nucleosynthesis) reaction occurs as follows:

$^{4}_{2}\text{He} + ^{4}_{2}\text{He} \xrightarrow{\text{yields}} ^{8}_{4}\text{Be}$

$^{8}_{4}\text{Be} + ^{4}_{2}\text{He} \xrightarrow{\text{yields}} ^{12}_{6}\text{C} + 2\gamma$

Depending on star mass, this may be followed by reactions such as:

$^{12}_{6}\text{C} + ^{4}_{2}\text{He} \xrightarrow{\text{yields}} ^{16}_{8}\text{O} + \gamma$

9780170465304

4.2 Structure of the atom

Beginning in the late 19th and early 20th centuries, experimental discoveries revolutionised the accepted understanding of the nature of matter on an atomic scale. Observations of the properties of matter and light inspired the development of better models of matter that, in turn, have been modified or abandoned as a result of further experimental investigations.

By studying the development of the atomic models through the work of Thomson and Rutherford, who established the nuclear model of the atom – a positive nucleus surrounded by electrons – you will further your understanding of the limitations of theories and models.

4.2.1 Cathode rays

Around the middle of the 19th century, the development of vacuum pumps, high-voltage sources and glassblowing had reached a stage that enabled the development of a device that became known as a **cathode ray** tube.

Scientists observed that a ray is emitted by the cathode (the electrode connected to the negative terminal of the power source) in this tube. Using the simple cathode ray tube, and some adapted versions of it, scientists made a series of observations of cathode rays over the following years. Cathode rays:

- expose photographic plates
- travel in straight lines
- cast a shadow
- can cause **fluorescence**
- pass through thin metal foil
- can be deflected by magnetic and electric fields
- can carry and transfer momentum (cause a paddle wheel to turn).

Initially, cathode rays were thought not to be deflected by electric fields, but this was later proven to be incorrect.

Debate in the scientific community centred on whether the cathode rays were waves or particles because some evidence seemed to support each model.

4.2.2 Thomson's charge-to-mass experiment

J.J. Thomson conducted an experiment in 1897 that provided conclusive evidence that a cathode ray was a charged particle.

Thomson's apparatus placed parallel charged plates within the cathode ray tube to subject the cathode rays to a uniform electric field (Figure 4.9). Electromagnets outside the tube created a uniform magnetic field perpendicular to the electric field. Both the magnetic and electric fields could then be switched on and off and adjusted to known field strengths.

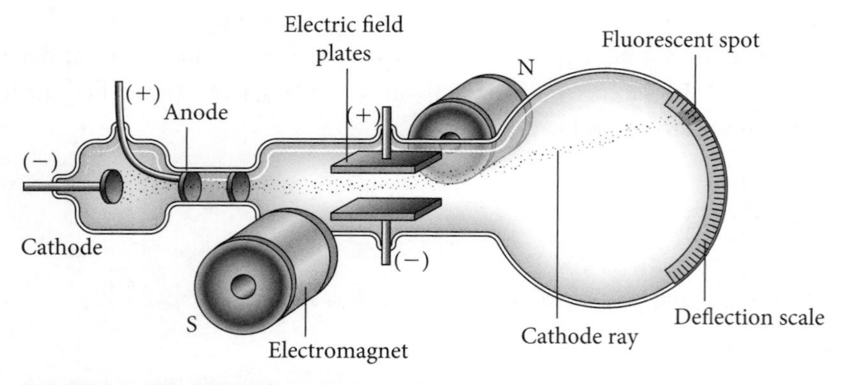

FIGURE 4.9 Thomson's charge-to-mass experimental apparatus

The electric and magnetic fields were established to provide a force on a moving charged particle in opposite directions. By adjusting the electric field strength until the electrons were undeflected, Thomson knew that the electric force and the magnetic force were equal.

$$F_B = F_E$$
$$qvB = qE$$
$$v = \frac{E}{B}$$

This revealed that the velocity of cathode rays was a fraction of the speed of light and suggested that cathode rays were not electromagnetic waves.

By then removing the electric field, Thomson was able to cause the charged particles to move in a circular arc within the magnetic field. In this way the magnetic force was providing a centripetal force. The radius r of the arc described by the cathode ray could be measured:

$$F_B = F_c$$
$$qvB = \frac{mv^2}{r}$$
$$\frac{q}{m} = \frac{v}{Br}$$
$$\frac{q}{m} = \frac{E}{B^2 r}$$

Thomson was able to determine a ratio of charge-to-mass (q/m) for the negatively charged particle that comprised the cathode rays – later known as an electron. This particle was present regardless of the cathode material, and Thomson proposed that it was a subatomic particle present in all atoms.

4.2.3 Millikan's oil drop experiment

Robert Millikan performed an experiment that showed that the electron was the quantum of charge. He balanced the gravitational force acting on an ionised oil droplet against an upwards electrical force.

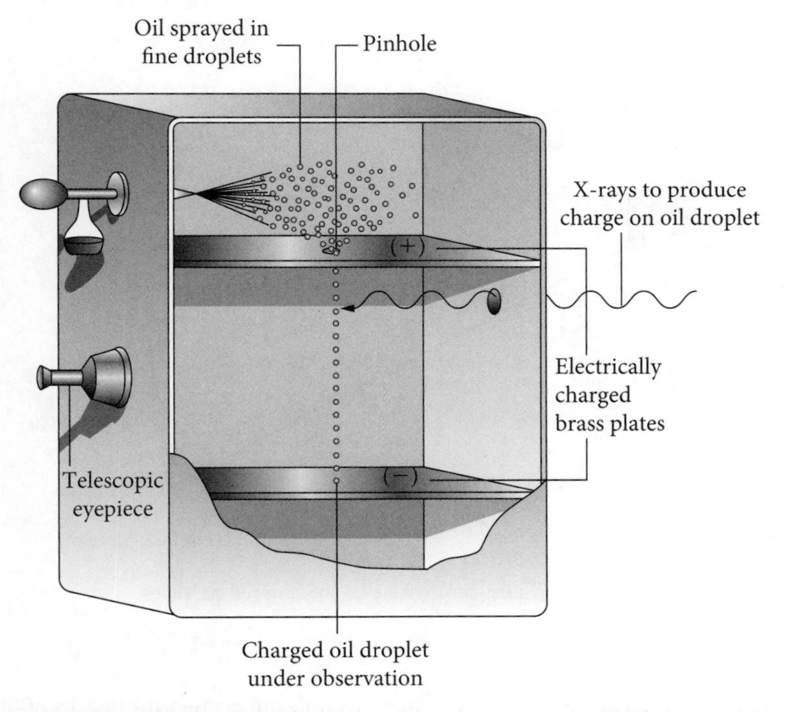

FIGURE 4.10 Millikan's experiment

The electric field strength could be adjusted by changing a variable voltage supply between the plates of known separation until a droplet hung stationary within view.

$$F_g = F_E$$
$$mg = Eq$$
$$q = \frac{mg}{E}$$
$$q = \frac{mgd}{V}$$

The mass of the oil droplet could be determined using its terminal velocity with no electric field.

Millikan experimented with thousands of oil droplets, which provided a large range of charge values. He was able to determine a highest common factor and concluded that this was a unit charge, the charge of an electron. Once the electron charge was known, its mass could be determined from Thomson's $\frac{q}{m}$ ratio.

4.2.4 Geiger–Marsden experiment and Rutherford's atomic model

Thomson proposed a model of the atom that he called the '**plum pudding model**' in which electrons were embedded within a positive atomic mass like the plums of a plum pudding (Figure 4.11).

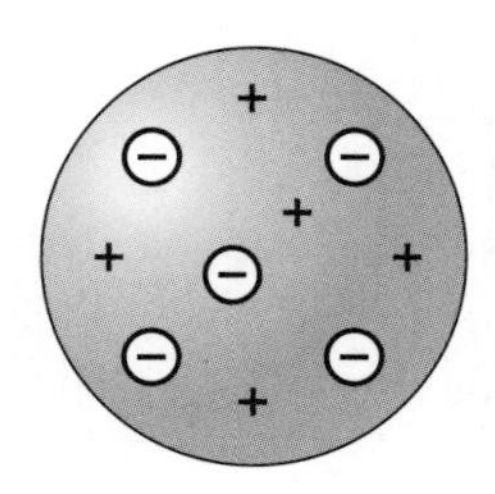

FIGURE 4.11 Thomson's plum pudding model of the atom

Ernest Rutherford designed an experiment, performed by Geiger and Marsden, that aimed to confirm the model proposed by Thomson. Rutherford's experiment involved directing a beam of the recently discovered **alpha particles**, which were emitted by some **radioisotopes**, at a thin gold foil target. The model of the atom suggested by Thomson indicated that the density of atoms was very low and so Rutherford reasoned that the positively charged alpha particles should pass through undeflected or with minimal deflection as a result of interactions with the electrons.

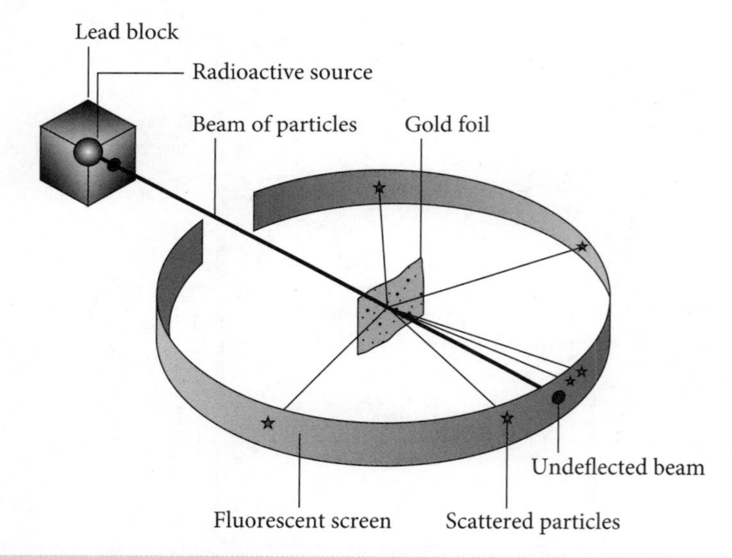

FIGURE 4.12 The Geiger–Marsden experiment

Most alpha particles did pass straight through the gold foil, but some (about 1 in 8000) were found to scatter at large angles. Thomson's model of the atom was incompatible with this observation.

Rutherford suggested a new atomic model that featured a small, dense, positively charged centre, called the nucleus, with electrons orbiting the nucleus (Figure 4.13). The electrons orbited under the influence of an electrostatic attractive force and at a relatively large radius, such that the vast majority of the atom was empty. This model, called the **planetary model** of the atom, was not well regarded as it defied accepted physics that the orbiting electrons are accelerating and so should emit radiation and spiral into the nucleus.

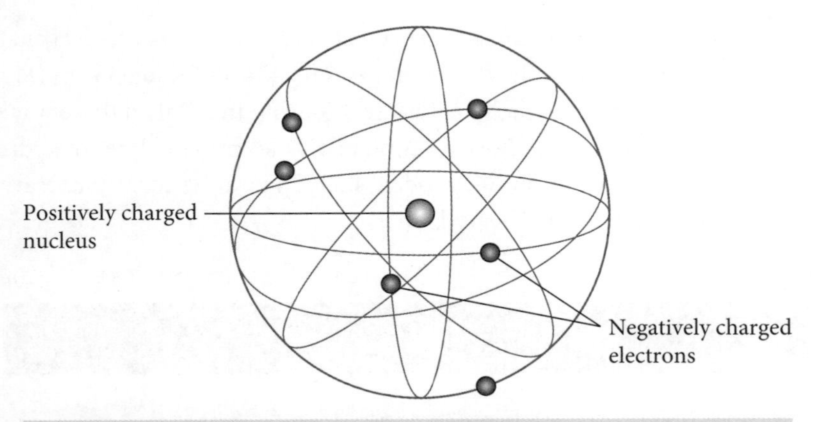

FIGURE 4.13 Rutherford's planetary model of the atom

4.2.5 Chadwick's discovery of the neutron

As a consequence of an experiment that yielded hydrogen ions, Rutherford had proposed the existence of the proton. However, he struggled to reconcile the mass and charge of most nuclei with the mass and charge of the proton. Initially physicists suggested that the nucleus contained nuclear electrons along with protons, but Rutherford hypothesised the existence of a neutral particle in the nucleus that he called the neutron.

Neutrons have no charge and interact with matter infrequently, and so are difficult to detect.

Experiments over several years had struggled to identify a neutral but highly penetrating radiation that was obtained by alpha particle bombardment of beryllium. Eventually James Chadwick, who had been a colleague of Rutherford's, demonstrated that this unknown radiation was in fact made up of neutral particles and not gamma radiation, as had been presumed. These neutral particles were able to 'knock' protons from a paraffin wax target (along with other elemental targets)(Figure 4.14). By using laws of conservation of momentum and energy, Chadwick was able to show that these neutral particles must have a mass marginally greater than that of a proton. Chadwick had confirmed Rutherford's idea of the existence of a neutral nuclear particle – the neutron.

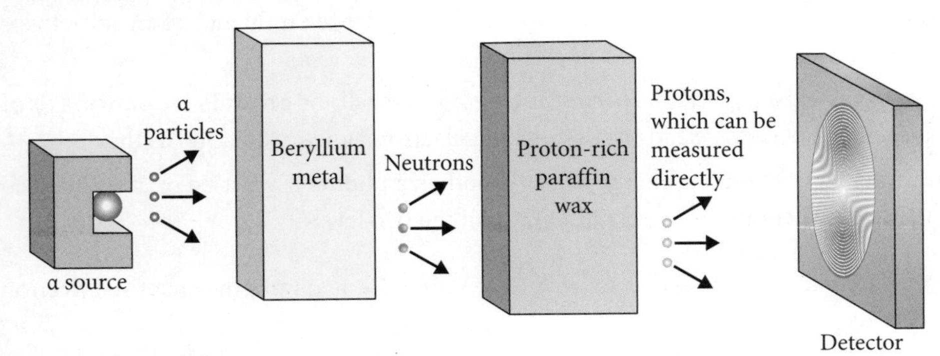

FIGURE 4.14 The experimental apparatus that led to Chadwick's proposal of the existence of the neutron

The nuclear **transmutation** reaction that was occurring when the alpha particles (helium nuclei) were bombarding the beryllium target was

$${}^{9}_{4}\text{Be} + {}^{4}_{2}\text{He} \xrightarrow{\text{yields}} {}^{12}_{6}\text{C} + {}^{1}_{0}\text{n}$$

4.3 Quantum mechanical nature of the atom

The work of Bohr, de Broglie and, later, Schrödinger demonstrated that the quantum mechanical nature of matter was a better way to understand the structure of the atom.

4.3.1 The emission spectrum of hydrogen

When gaseous elements are energised, an emission spectrum is produced. For hydrogen, the emission spectrum features visible spectral lines with wavelengths of 410, 434, 486 and 656 nm (Figure 4.15). Swiss mathematician Johann Balmer derived a mathematical relationship that linked these wavelengths with a constant and integer values 3, 4, 5 and 6. The collection of visible emission lines of hydrogen became known as the Balmer series. The existing atomic models, including Rutherford's planetary model, were unable to explain the emission spectrum observed.

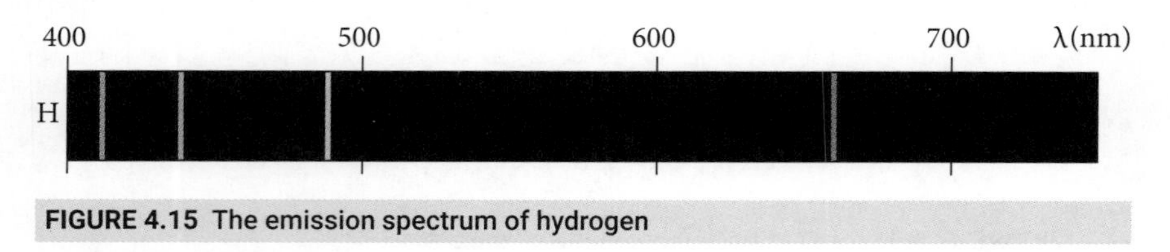

FIGURE 4.15 The emission spectrum of hydrogen

4.3.2 Bohr's atomic model

Rutherford's atomic model of 1911 successfully included a small, dense, positive, central nucleus with electrons orbiting at a relatively large distance and the vast majority of each atom being empty space. However, Rutherford's model had limitations.

- Orbiting electrons are accelerating and should, therefore, emit electromagnetic radiation in accordance with Maxwell's theory and consequently spiral into the nucleus. Yet, they do not.
- It could not explain the internal mechanism of the atom that enabled the production of a characteristic emission spectrum.
- It did not offer any detail of the arrangement of the electrons around the nucleus – simply stating that they were in orbit.
- It did not explain the composition of the nucleus.

In 1913 Niels Bohr, a Danish physicist who had worked with Rutherford, proposed a new model of the atom to explain atomic spectra and overcome the shortcomings of Rutherford's model.

Bohr's model was based on four postulates.

1. An electron moves in a circular orbit about the nucleus under the influence of an attractive electrostatic force.
2. Electrons can only exist in certain, specific energy levels (shells or **orbitals** or **stationary states**). In these specific orbitals, the electrons do not emit electromagnetic radiation as they orbit.
3. When electrons move from one energy level to another, a photon is released or absorbed with an energy that is equivalent to the energy difference of the two levels.
4. This energy difference is expressed as $\Delta E = E_f - E_i = hf = \frac{hc}{\lambda}$ and links the quantised electron energy to Planck's equation.

   The angular momentum is quantised, as indicated by the equation $mvr = \frac{nh}{2\pi}$.

   The following clarifications are noteworthy.

- The first postulate agrees with Rutherford's model.
- The second postulate suggests a quantised nature of electron energy and overcomes a problem with Rutherford's model by saying that the accepted laws of physics don't apply in this case. Bohr didn't

know why the laws wouldn't apply, but he knew that it explained observations. It also addresses some information that Rutherford's model lacked regarding the arrangement of electrons around the nucleus.

- The third postulate explains the emission and absorption spectra observed by using conservation of energy.
- The fourth postulate explains quantised electron energy levels.

Bohr was able to derive an equation, using the work of Balmer and others, to predict the wavelengths of the hydrogen emission spectrum in terms of the shell number in which the electron starts and finishes its transition.

$$\frac{1}{\lambda} = R\left[\frac{1}{n_f^2} - \frac{1}{n_i^2}\right]$$

The constant R is known as Rydberg's constant and the formula as Rydberg's formula. The model provided a feasible mechanism to explain emission and absorption spectra, was compatible with the equation Balmer had derived years earlier, and even predicted the size of the hydrogen atom and its ionisation energy. Furthermore, the model correctly predicted the existence of emission series for hydrogen with wavelengths both longer and shorter than those of the visible spectrum.

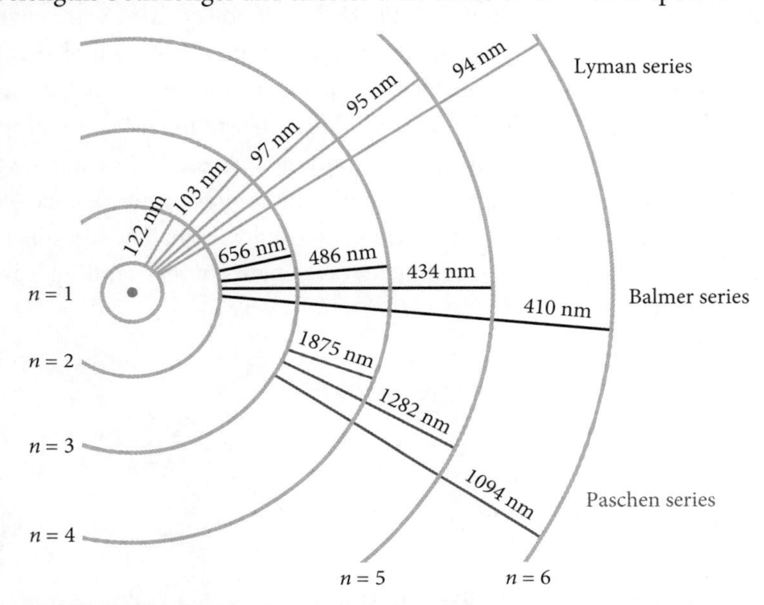

FIGURE 4.16 Some of the spectral series of hydrogen

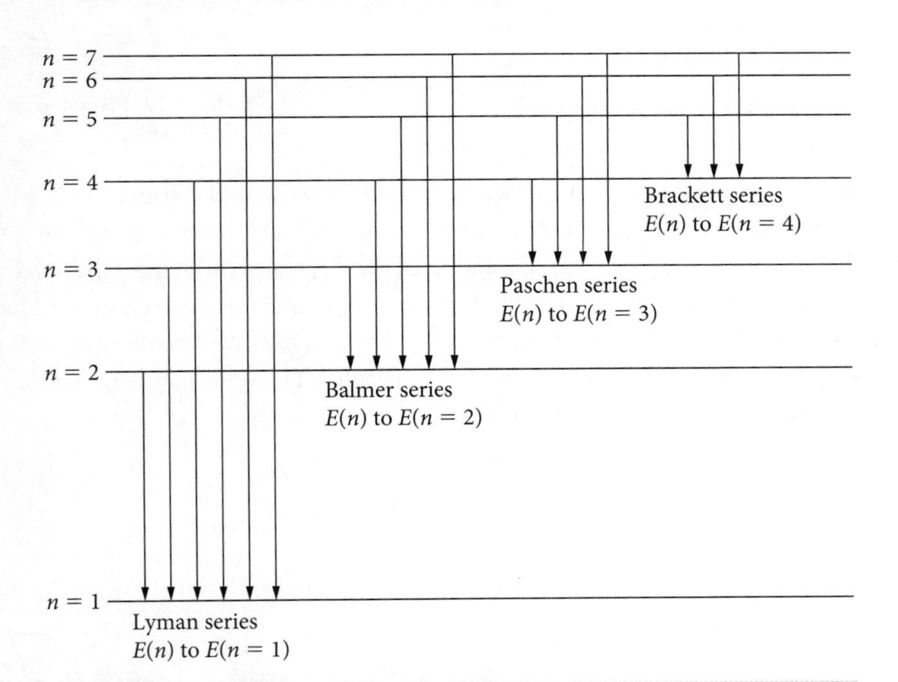

FIGURE 4.17 The energy levels of hydrogen, according to Bohr's model

CHAPTER 4

Bohr's model was incomplete and had some significant limitations. Bohr's model:

- only worked for atoms with one electron
- could not explain the relative intensities of the spectral lines
- could not explain the fine spectral lines that were observed to comprise each spectral line, as technology improved (hyperfine lines)
- could not explain the splitting of lines that occurred when the excited atom was subjected to a magnetic field (**Zeeman effect**)
- seemed to be a convenient mixture of classical physics and the new field of quantum physics.

4.3.3 de Broglie's matter waves

In 1924 Louis de Broglie equated the energy of a photon, as indicated by Planck's equation, with Einstein's mass–energy equivalence equation to derive the equation $\lambda = \frac{h}{mv}$, which suggested a relationship between momentum and wavelength. He proposed that all objects have a wave–particle nature, rather than just light. In other words, de Broglie thought it likely that all objects have wave characteristics, and all waves have particle characteristics. This is referred to as de Broglie's **matter-wave theory**.

This matter-wave theory could be used to explain the quantum nature of the electron energy levels that featured in Bohr's atomic model. de Broglie suggested that the stationary states proposed by Bohr existed as the standing waves of electrons positioned around the nucleus. These standing waves would need to have an integer number of wavelengths to fit within the orbital circumference of the electron, otherwise a complex wave featuring destructive interference would occur. The integers of Bohr's orbitals $n = 1, 2, 3$ etc. corresponded to the number of complete wavelengths of the standing waves within the orbital circumference.

de Broglie was able to:

- explain the quantised electron energy levels proposed but not explained by Bohr's model
- explain the lack of emission of electromagnetic radiation that Rutherford was unable to explain, since waves do not emit radiation
- give a reason for the existence of the stationary states of Bohr's model
- rationalise the uncomfortable blend of quantum and classical physics of Bohr's model
- prove Bohr's equation for quantised angular momentum using standing wave equations.

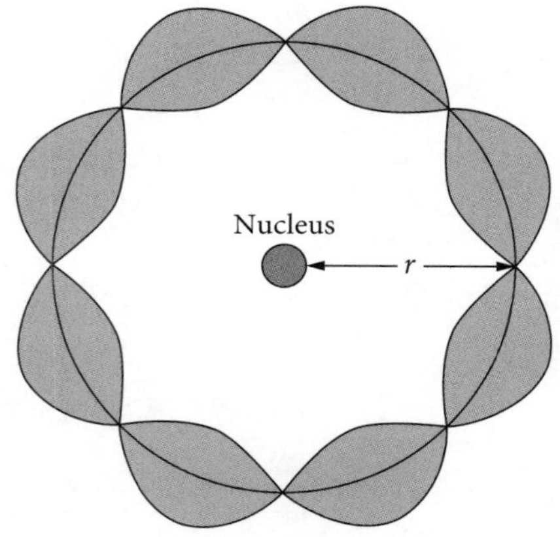

FIGURE 4.18 The standing wave of the $n = 4$ orbital

The concept of wave properties of electrons was tested by Davisson and Germer in 1927 in an experiment in which electrons scattered from the surface of a nickel crystal created a pattern that featured interference maxima. Changes made to the momentum of the electrons, by altering the accelerating voltage, changed the interference pattern in predictable ways. Quantitative analysis of the data found it to be consistent with de Broglie's equation. Subsequent interference experiments have revealed the wave nature of neutrons, protons and small atoms and molecules. The wave nature of matter is utilised scientifically in devices such as electron microscopes.

4.3.4 Schrödinger's contribution to the current atomic model

As the field of quantum mechanics continued to develop the atomic model was updated. In the 1920s, two quantum physicists, Erwin Schrödinger and Werner Heisenberg, independently and using different methods developed a complete quantum model of the atom.

Erwin Schrödinger's method used wave mechanics (based on de Broglie's matter-wave hypothesis) to show that the electron exists as a wave until observed. Hence, Schrödinger's model describes a probability of finding an electron in some region around a nucleus, an electron cloud, rather than stating that the electron is in an orbit of known radius around the nucleus. This is often referred to as a **probabilistic** model rather than the classical, **deterministic** model.

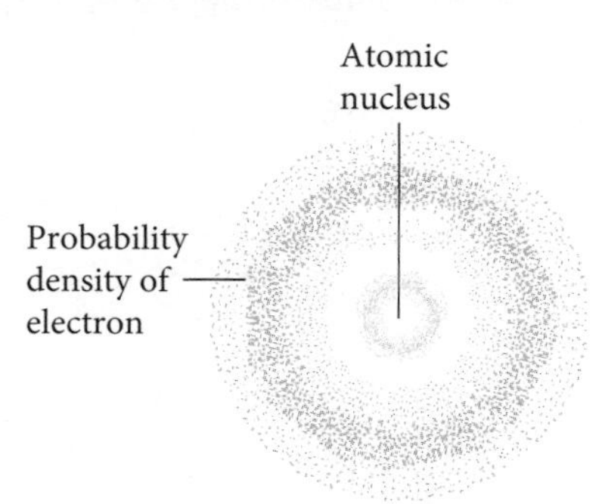

FIGURE 4.19 The quantum physics interpretation of electrons around the nucleus of an atom

Note

Examiners often set questions about the development of the atomic model. You might find it useful to construct a table or flowchart to summarise the contributions of scientists listed in sections 4.2.2–4.2.4 incorporating, where appropriate, the preceding model, process and/or experiment results, model change and limitations.

4.4 Properties of the nucleus

Experimental investigations of the nucleus have led to an understanding of **radioactive** decay, the ability to extract energy from nuclear **fission** and fusion, and a deeper understanding of the atomic model.

4.4.1 Spontaneous nuclear decay

All nuclei except hydrogen feature strong repulsive electrostatic forces acting between protons. The strong nuclear force attracts all **nucleons** to their adjacent nucleons. It is a short-range force that is strongly attractive unless nucleons are too close and then it is repulsive. The combined effect of these forces is relevant for nuclear stability. Some nuclei are inherently unstable and release particles and/or radiation as a consequence of a change that progresses the nucleus towards greater stability. This is called nuclear decay or radioactive decay and is a spontaneous process that results in a nucleus of a different form or different energy state. Unstable **isotopes** of an element are called radioisotopes.

The instability is a consequence of an unsuitable ratio of protons to neutrons in the nucleus, an unsuitable arrangement of protons and neutrons in the nucleus or simply the oversized nature of the nucleus. For stable nuclei, the ratio of number of neutrons to protons forms a clear line. Any **nuclide** not fitting this line will be unstable, as can be seen in Figure 4.20.

Radioactive decay can occur in a number of forms, the nature of which can be predicted by the composition of the nucleus.

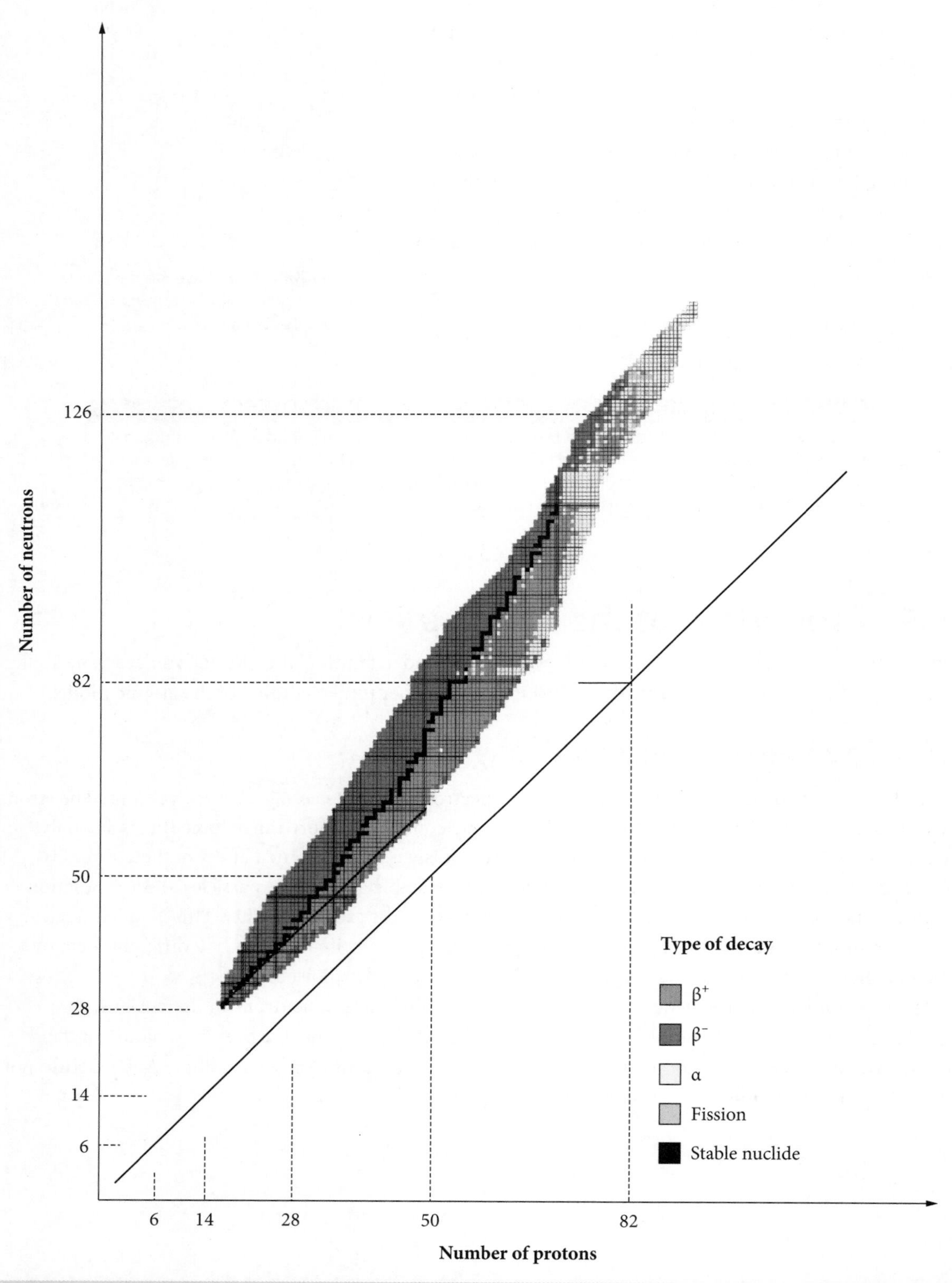

FIGURE 4.20 The relationship between decay type, nuclear size and nuclear composition. Note the black line of stability that approximately follows a 1:1 proton-to-neutron ratio for the first 20 elements.

There are three main types of spontaneous radioactive decay described – alpha, beta and gamma – with beta decay having two distinct forms – beta-plus and beta-minus. They are distinguished by the particles or radiation emitted, the **penetrating ability** and the **ionising ability** of the decay type, charge and mass. They are also distinguished by their motion in electric and magnetic fields as a consequence of their different charge and mass.

TABLE 4.3 Radioactive decay properties

Radiation type	Particle type	Equation symbol	Penetrating ability	Ionising ability	Charge (multiples of electron charge)	Mass (amu) approx.
Alpha	Helium nucleus	$^{4}_{2}He$ or $^{4}_{2}\alpha$	Low (a few cm of air, unable to pass through paper)	High	+2	4
Beta-minus	Electron (co-emitted with an antineutrino)	$^{0}_{-1}e$	Medium (<1 metre of air, able to penetrate several mm of aluminium)	Medium	−1	$\frac{1}{1800}$
Beta-plus	Positron (co-emitted with a neutrino)	$^{0}_{+1}e$	Medium (<1 metre of air, able to penetrate several mm of aluminium)	Medium	+1	$\frac{1}{1800}$
Gamma	Gamma ray photon	γ or $^{0}_{0}\gamma$	High (vast distances in air, able to penetrate several cm of lead)	Low	0	0

Note
In the following examples of reactions, mass number and atomic number are conserved in the nuclear equations.

Alpha decay

In alpha decay, a helium nucleus is ejected from the **parent nucleus** and the mass number (A) is decreased by 4 while the **atomic number** (Z) is decreased by 2.

$$^{A}_{Z}X \xrightarrow{\text{yields}} {}^{(A-4)}_{(Z-2)}Y + {}^{4}_{2}He$$

$$^{238}_{92}U \xrightarrow{\text{yields}} {}^{234}_{90}Th + {}^{4}_{2}He$$

Beta-minus decay

In beta-minus decay, an electron (also called a **beta particle**) is ejected (along with an **antineutrino**) from the parent nucleus and the mass number (A) is unchanged while the atomic number (Z) is increased by 1.

Beta-minus decay occurs when a neutron changes into a proton with the emission of an electron and an antineutrino.

$\overline{\nu}$ is the symbol for an antineutrino (a tiny subatomic antiparticle classified as an antilepton).

$$^{A}_{Z}X \xrightarrow{\text{yields}} {}^{A}_{(Z+1)}Y + {}^{0}_{-1}e + \overline{\nu}$$

$$^{234}_{90}Th \xrightarrow{\text{yields}} {}^{234}_{91}Pa + {}^{0}_{-1}e + \overline{\nu}$$

Beta-plus decay

In beta-plus decay a **positron** (also called a beta particle) is ejected (along with a **neutrino,** ν) from the parent nucleus. The mass number (A) is unchanged while the atomic number (Z) is decreased by 1.

Beta-minus decay occurs when a proton changes into a neutron with the emission of a positron and a neutrino.

ν is the symbol for a neutrino (a tiny subatomic particle classified as a **lepton**).

$${}^{A}_{Z}X \xrightarrow{\text{yields}} {}^{A}_{(Z-1)}Y + {}^{0}_{+1}e + \nu$$

$${}^{195}_{81}Tl \xrightarrow{\text{yields}} {}^{195}_{80}Hg + {}^{0}_{+1}e + \nu$$

Gamma decay

In gamma decay, a gamma ray photon is ejected from the nucleus and the nuclear energy level decreases; no particles are ejected in this decay and there is no change to mass number (A) or atomic number (Z).

$${}^{A}_{Z}X \xrightarrow{\text{yields}} {}^{A}_{Z}X + {}^{0}_{0}\gamma$$

$${}^{99}_{43}Tc \xrightarrow{\text{yields}} {}^{99}_{43}Tc + {}^{0}_{0}\gamma$$

4.4.2 Half-life

A radioactive decay is a random event and for any particular nucleus, irrespective of conditions or data collected, there is no way of predicting if it will decay over a given period of time.

In large populations of radioactive nuclei, however, predictions can be made using probability. Consequently, we are able to predict the time period for half of the nuclei to undergo a radioactive decay. This time period is called the radioisotope's **half-life** ($t_{1/2}$). So, after one half-life, half of the radioactive sample would have decayed and after two half-lives three-quarters of the radioactive sample would have decayed and so forth.

A short half-life indicates a highly radioactive isotope; a long half-life indicates a less radioactive isotope.

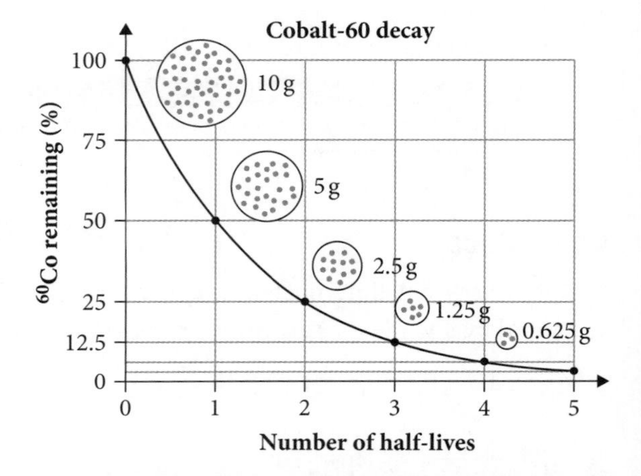

FIGURE 4.21 The population of undecayed nuclei will follow a predictable curve, such as in the example of the decay of radioisotope cobalt-60.

The number of nuclei remaining at time t (N_t) can be calculated using the equation

$$N_t = N_0 e^{-\lambda t}$$

where t is the time elapsed and λ is a constant specific to the radioisotope called the **decay constant**.

The decay constant is related to the half-life by the equation $\lambda = \frac{\ln 2}{t_{\frac{1}{2}}}$.

Alternatively, calculations can be completed using

$$N_t = N_0 \left(\frac{1}{2}\right)^n$$

where n is the number of half-lives elapsed in the time period, and N_0 is the number of nuclei at the start of the time period. It should be noted that this equation is not on the formula sheet.

4.4.3 Nuclear fission

A nuclear fission reaction is one in which a large nucleus splits into two smaller **daughter nuclei**. The products of nuclear fission are random in nature, although splits in which the daughter nuclei are approximately two-fifths and three-fifths of the parent nucleus are more frequently seen.

Fission is usually initiated by the absorption of a neutron, which causes instability in the nucleus. It is frequently accompanied by the emission of neutrons, which can be used to initiate further fission reactions in a process known as a **chain reaction**.

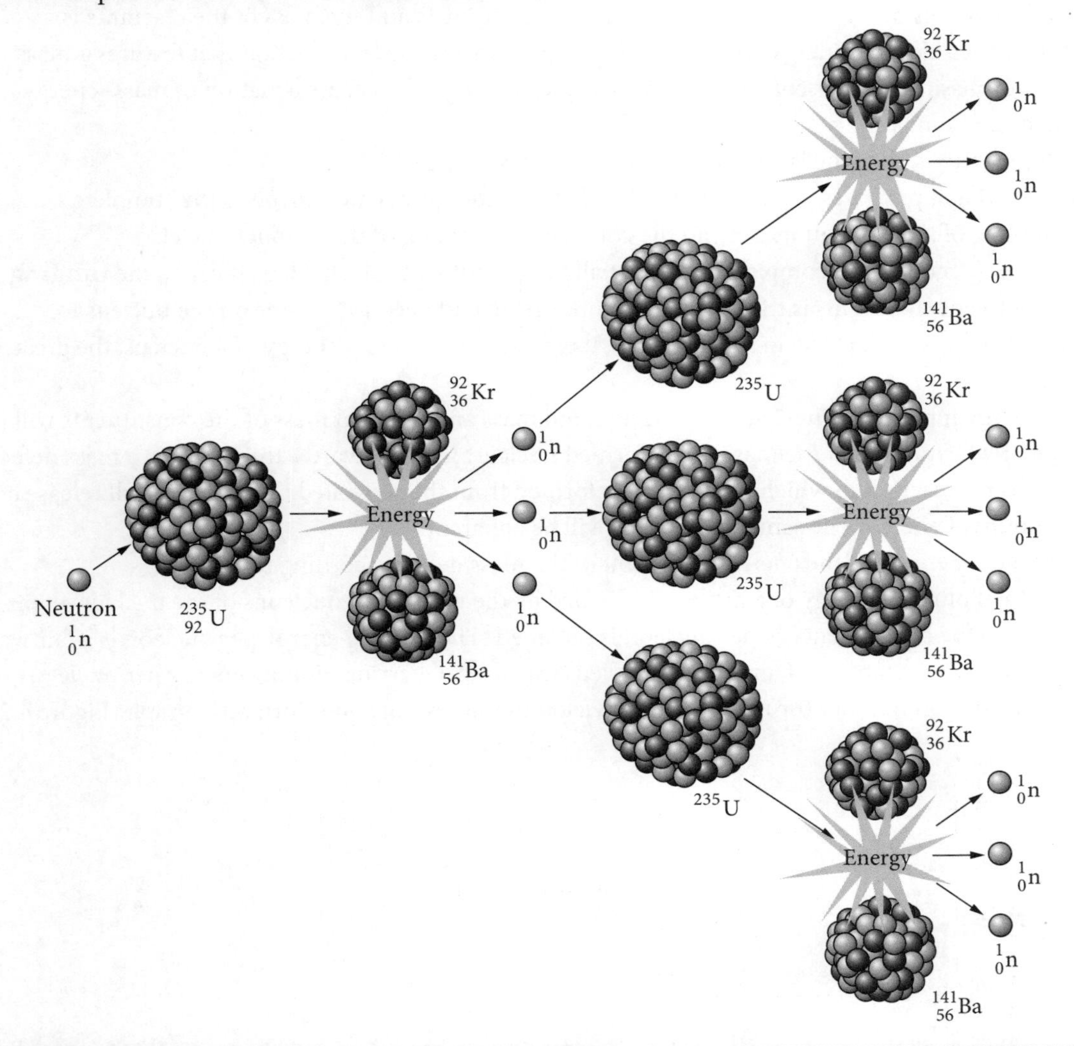

FIGURE 4.22 The nuclear fission chain reaction that can result from a single neutron embedding in the nucleus of a uranium-235 atom. Energy is released along with neutrons that can cause subsequent fission reactions.

This example of a typical nuclear fission reaction describes the process illustrated in Figure 4.22.

$$^{1}_{0}n + ^{235}_{92}U \xrightarrow{\text{yields}} ^{92}_{36}Kr + ^{141}_{56}Ba + 3^{1}_{0}n$$

Uncontrolled fission reactions

If more than one product neutron successfully initiates a subsequent nuclear fission reaction on an ongoing basis, as in the example, then the reaction is said to be an **uncontrolled chain reaction** and the number of reactions will escalate rapidly. Uncontrolled chain reactions are associated with nuclear explosions and nuclear power plant accidents such as meltdowns.

Controlled fission reactions

Controlled chain reactions feature a constant number of fission reactions and are achieved by absorbing sufficient product neutrons so that only one neutron, on average, from each fission successfully initiates a subsequent fission reaction. Controlled reactions are used in nuclear reactors.

4.4.4 Energy from nuclear reactions

Energy is released by a nuclear fission reaction because the total product mass is less than the total reactant mass. This difference between the mass of the products and the mass of the reactants is called the **mass defect**. Energy will, in fact, be released in any nuclear reaction that features a mass defect. The 'missing' mass is converted into energy according to Einstein's equation of mass–energy equivalence, $E = mc^2$.

These reactions obey the law of conservation of mass–energy.

For functional purposes, it can be imagined that any nuclear reaction involves the complete disassembling of the reactant nuclei and the complete assembling of the product nuclei.

The energy required to completely separate all the constituents of a nucleus is called the **binding energy** of the nucleus. This is the amount of work that would need to be done on the nucleus to achieve that outcome. (Hence, in simple terms, the greater the binding energy of a nucleus, the greater its stability.)

The energy input to the nucleus is converted into mass and the total mass of the constituents will consequently be more than the mass of the formed nucleus. In other words, there will be a mass defect.

The reverse reaction in which the nucleus is formed from the separated constituents will release an equal amount of energy. The same mass defect will be apparent.

The binding energy is the energy equivalent of the mass defect according to $E = mc^2$.

If the total binding energy of a nucleus is divided by the number of nucleons of the nucleus, then a value called the binding energy per nucleon is obtained. The binding energy per nucleon is a fair way to judge the nuclear stability of any type of nucleus (nuclide). Charting binding energy per nucleon against the nucleon number (or mass number) yields an interesting and informative graph (Figure 4.23).

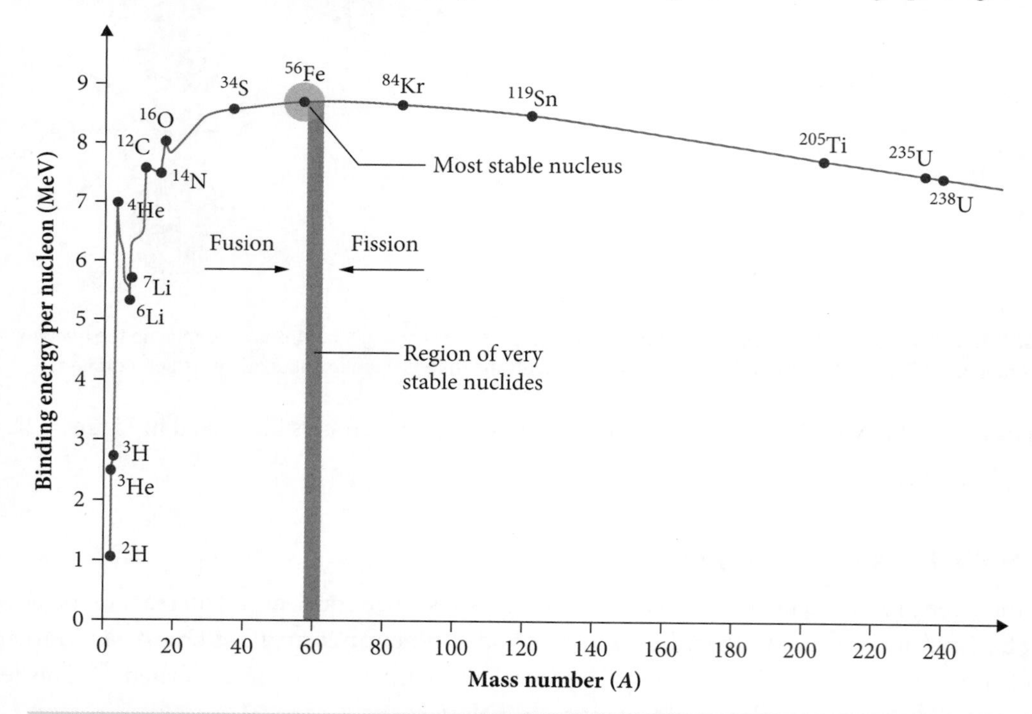

FIGURE 4.23 The graph of binding energy per nucleon against mass number shows the relative stability of various nuclides.

It can be seen that:

- iron-56 (^{56}Fe) is the most stable nucleus (hence the end point for fusion in large stars)
- fission can occur in energy-yielding reactions with parent nuclei larger than ^{56}Fe
- fusion reactions can occur and yield energy with nuclei smaller than ^{56}Fe.

The amount of energy yielded in any nuclear reaction can be determined using data from this type of graph.

For example, in the nuclear fusion reaction in which a deuterium nucleus and a tritium nucleus are fused to create a helium nucleus and a free neutron:

$$^{2}_{1}H + ^{3}_{1}H \xrightarrow{\text{yields}} ^{4}_{2}He + ^{1}_{0}n$$

it can be calculated that the energy to completely separate the nucleons of each of the deuterium (^{2}H) and tritium (^{3}H) nuclei will be given by the number of nucleons times the binding energy per nucleon in each case. (Note the single neutron has no binding energy.)

So, energy used = $2 \times 1.0 + 3 \times 2.8 = 10.4$ MeV.

The energy yielded by the formation of the helium nucleus (^{4}He) will be $4 \times 7.0 = 28$ MeV.

Total energy output from this nuclear fusion reaction is $28 - 10.4 = 17.6$ MeV.

For the nuclear fission reaction given earlier in the chapter (although the level of detail is not available on the graph above):

$$^{1}_{0}n + ^{235}_{92}U \xrightarrow{\text{yields}} ^{92}_{36}Kr + ^{141}_{56}Ba + 3^{1}_{0}n$$

Energy used will be $235 \times 7.6 = 1786$ MeV.

Energy yielded will be $92 \times 8.7 + 141 \times 8.4 = 1984.8$ MeV.

Total energy output from the nuclear fission reaction is $1984.8 - 1786 = 198.8$ MeV.

Note that, although this is a greater energy yield in total than for the fusion reaction, it represents a significantly smaller yield per unit mass of reactants. This is commonly true of fission reactions in comparison to fusion reactions.

This simple graphic in Figure 4.24 might aid understanding.

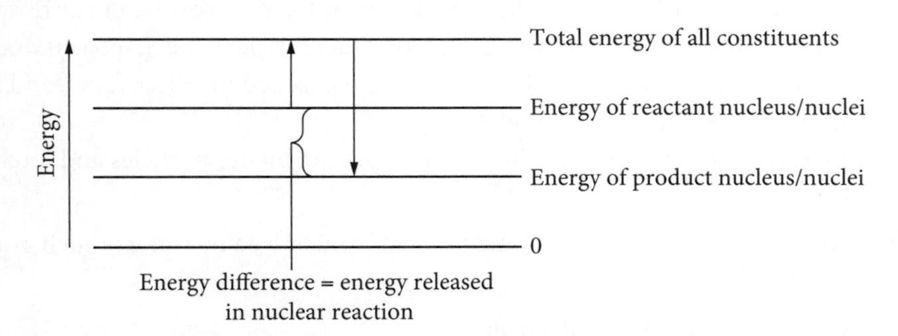

FIGURE 4.24 This is a simple model to explain energy yield in nuclear reactions. The short upwards arrow represents the energy input to separate the nucleus/nuclei into constituents. The longer downwards arrow represents the energy output when new nucleus/nuclei are formed. The difference in arrow length represents the energy released in the reaction.

9780170465304

4.5 Deep inside the atom

By the early part of the 19th century, significant progress had been made in the understanding of the composition of the building blocks of the Universe. Soon, however, experimental evidence highlighted the shortcomings of that model. In reality, the particle model of matter is constantly being updated and our understanding of the nature of matter remains incomplete.

4.5.1 Subatomic particles and the Standard Model of matter

By 1932, with the discovery of the neutron and an understanding of the quantum nature of electron orbits, a simple Universe could be thought of as being constructed from protons, neutrons and electrons. However, from that year onwards studies of high-energy **cosmic ray** collisions began revealing new particles – first the positron, then particles that had a range of masses, half-lives, energy, charge and decay products. These were called muon, **meson**, kaon, lambda, sigma and so forth.

The existence of these new particles did not fit the existing atomic particle model, and physicists wondered if these particles were **fundamental particles** or were **composite particles**.

During the 1950s, increasingly powerful particle accelerators enabled the study of more high-energy collisions and over a hundred more particles were discovered, to the extent that the term 'particle zoo' was coined and scientists were running out of names and symbols for the new particles.

Meanwhile, an increasing body of evidence suggested that protons and neutrons were composite rather than fundamental particles.

- Neutrons had magnetic properties, yet magnetism is associated with charge.
- Beta-minus decay and beta-plus decay processes suggested each involved a transformation of an internal structure.
- Electrons were scattered by protons, in experiments similar to the Geiger–Marsden experiment, in a way that suggested an internally located centre of mass.

Physicists developed a classifications system that organised the existing particles on the basis of their properties and developed a model – the **Standard Model of matter** – to explain the existence of all particles and forces in the Universe on the basis of fundamental particles. The Standard Model of matter made predictions about the existence of particles, which were subsequently discovered, and this is seen as powerful evidence for the correctness of the model.

The Standard Model of matter proposes a Universe comprising matter particles and force particles with the following key points.

- All of the particles of the Standard Model of matter are characterised by features such as mass, charge, half-life, spin and other properties.
- All matter particles have antimatter equivalents.
- The fundamental matter particles can be classified into two groups – **quarks** and leptons.
- Quarks always exist in combination – such combinations are called **hadrons**.
- Composite three-quark combinations – these hadrons are called **baryons** – include the proton and the neutron.
- Composite quark–antiquark combinations are called mesons.
- Leptons always exist alone – both the electron and the neutrino are examples of leptons.
- The four forces are mediated by particles called **bosons**.
- The photon is the boson for the electromagnetic force.
- The strong nuclear force binds nucleons in a nucleus together and bonds quarks together in hadrons. It is mediated by the **gluon**.
- The graviton (the boson through which gravity acts) is yet to be discovered.

9780170465304

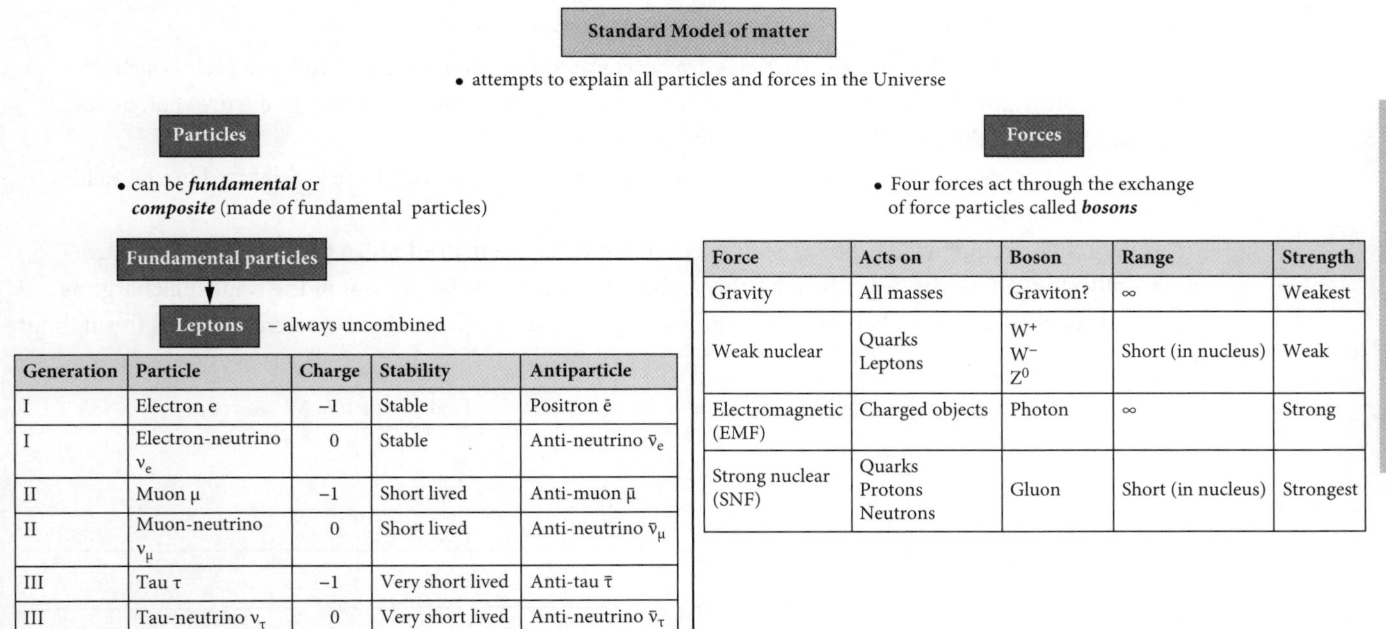

Generation	Particle	Charge	Stability	Antiparticle
I	Electron e	−1	Stable	Positron $\bar{e}$
I	Electron-neutrino ν_e	0	Stable	Anti-neutrino $\bar{\nu}_e$
II	Muon μ	−1	Short lived	Anti-muon $\bar{\mu}$
II	Muon-neutrino ν_μ	0	Short lived	Anti-neutrino $\bar{\nu}_\mu$
III	Tau τ	−1	Very short lived	Anti-tau $\bar{\tau}$
III	Tau-neutrino ν_τ	0	Very short lived	Anti-neutrino $\bar{\nu}_\tau$

Force	Acts on	Boson	Range	Strength
Gravity	All masses	Graviton?	∞	Weakest
Weak nuclear	Quarks Leptons	W^+ W^- Z^0	Short (in nucleus)	Weak
Electromagnetic (EMF)	Charged objects	Photon	∞	Strong
Strong nuclear (SNF)	Quarks Protons Neutrons	Gluon	Short (in nucleus)	Strongest

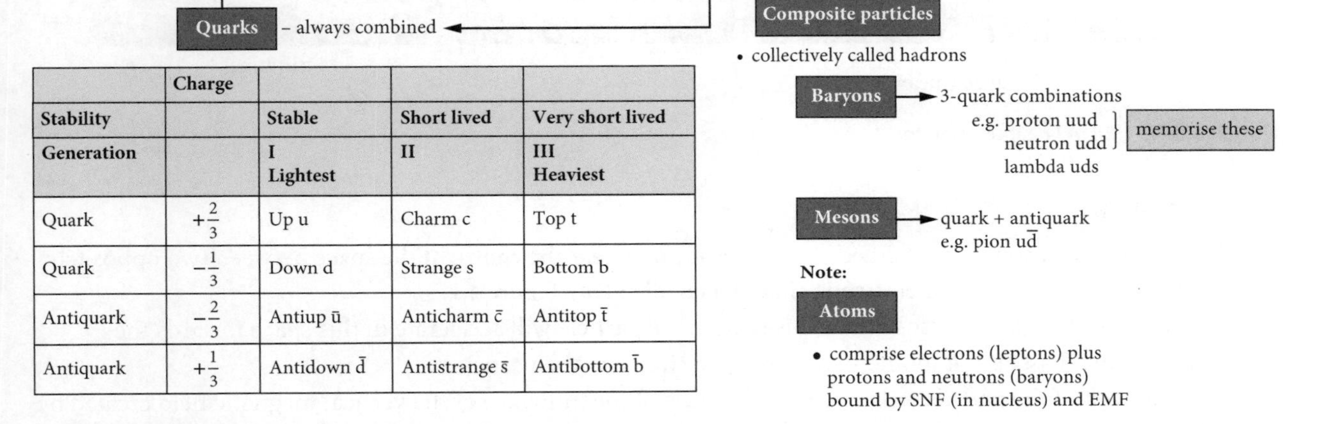

	Charge			
Stability		**Stable**	**Short lived**	**Very short lived**
Generation		**I** **Lightest**	**II**	**III** **Heaviest**
Quark	$+\frac{2}{3}$	Up u	Charm c	Top t
Quark	$-\frac{1}{3}$	Down d	Strange s	Bottom b
Antiquark	$-\frac{2}{3}$	Antiup $\bar{u}$	Anticharm $\bar{c}$	Antitop $\bar{t}$
Antiquark	$+\frac{1}{3}$	Antidown $\bar{d}$	Antistrange $\bar{s}$	Antibottom $\bar{b}$

FIGURE 4.25 A chart of the Standard Model of matter

4.5.2 Evidence from particle accelerators

Particle accelerators have been used to test predictions made by the Standard Model. Particle accelerators rely on large potential differences to accelerate charged particles to extreme velocities to cause sufficiently high particle energy. Because of the mass–energy equivalence, these particles will have much greater mass than their rest mass and, according to de Broglie's equation, very short wavelengths. Sensitive detectors within the particle accelerator are used to collect data that will enable the mass, half-life, charge and path of particles produced by the collisions and the nature of product radiation to be determined.

Particle accelerators can be linear in nature or rely on magnetic fields to change the direction of the moving charged particles so that the path is more circular in nature. Two examples of particle accelerators that reveal the key operation of such devices are the **linear accelerator** and the **cyclotron**.

Linear accelerator

In a linear accelerator, the charged particles are accelerated in a straight line through a series of **drift tubes** (Figure 4.26). The length of the tubes increases over the particle's journey to ensure that, as speed increases, the time the particle spends in each tube is identical.

As there is no electric field inside a hollow charged object, the particles are accelerated by the field between the adjacent tubes.

A very high-frequency AC power source is connected to the drift tubes so that adjacent tubes are oppositely charged. In this way, the particle can be attracted to a tube in front of the opposite charge as itself and repelled by a tube behind of the same charge as itself in the region between the tubes throughout its journey.

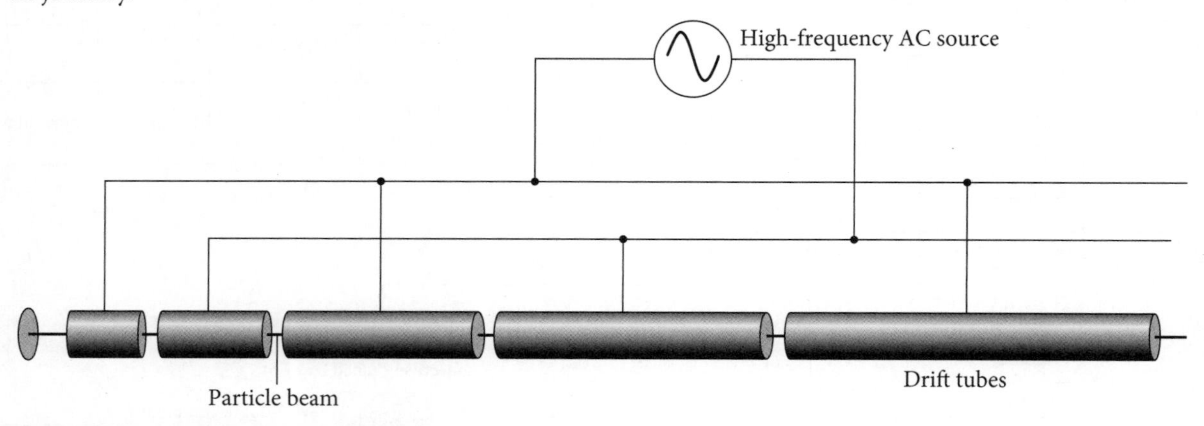

FIGURE 4.26 A linear accelerator

Cyclotron

In a cyclotron, a source of charged particles is located near the centre of the space between two oppositely charged, hollow, D-shaped electrode chambers (called **Ds**)(Figure 4.27).

As a result of the electric field in this region, the particle will accelerate in this space towards the chamber that has, at that moment, the opposite charge to the particle.

The path of the particle will be curved because of the influence of the vertical magnetic field created by the two electromagnets. The particles will move through the circular path within the D at a constant speed because there is no electric field in this region. Upon emerging, the particle will speed up across the gap again because the polarity of the Ds has been changed by the high-frequency AC power source.

The circular path it makes inside the next D is of a greater radius because it is moving with a greater velocity; however, the time taken for each arc is identical. (The frequency can be shown to be independent of the radius.) The particle, therefore, moves in an outwards spiral (as it speeds up each time it moves through the gap) until at some point its radius is large enough to pass through a gate to strike the target.

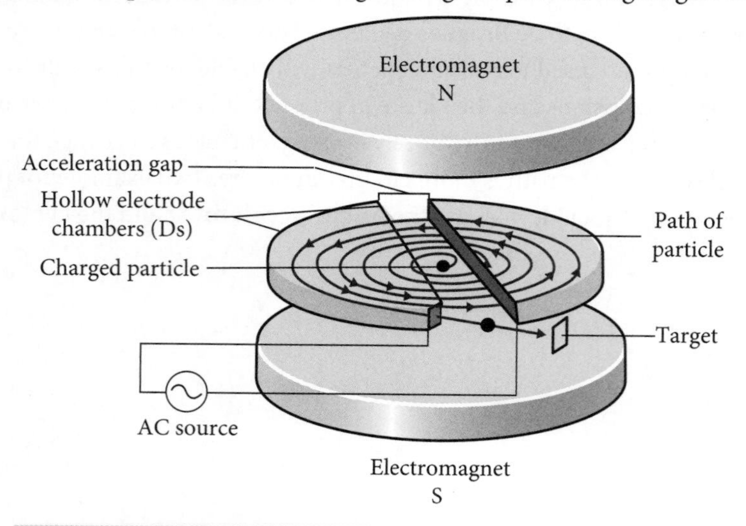

FIGURE 4.27 A cyclotron

Glossary

alpha particle Two protons bound to two neutrons; the nucleus of a helium atom

antineutrino The antimatter equivalent of the neutrino

atomic number The number of protons in the nucleus (Z)

baryon A hadron composed of three quarks

beta particle An electron or positron that is the product of a radioactive decay

Big Bang The initial event that formed the Universe, creating space, time and all matter

binding energy The energy input required to break a nucleus into its constituent parts

black hole A region of space where gravity is so intense that no matter or radiation can escape

boson A particle that mediates a force

carbon–nitrogen–oxygen (CNO) cycle A cycle of nuclear fusion reactions that creates a helium nucleus from four protons in which carbon, nitrogen and oxygen act as catalysts

cathode rays Electron beams emerging from the negative electrode in an evacuated tube

chain reaction A reaction that proceeds in a self-sustaining fashion

composite particle A particle made up of two or more fundamental particles

controlled chain reaction A self-sustaining reaction that proceeds at a constant rate

cosmic rays Highly energetic particles travelling through space at speeds approaching light speed

cyclotron A particle accelerator in which particles move in circles of increasing radius

daughter nuclei Nuclei that result from a transmutation or nuclear decay event

decay constant A measure of the rate of radioactive decay of a radioisotope

deterministic A situation that can be found to have a single solution

drift tubes Hollow, cylindrical, charged conductive tubes that form part of a linear accelerator mechanism

Ds Hollow, charged conductors that form part of a cyclotron mechanism

fission A nuclear reaction in which a large nucleus splits into two smaller nuclei

fluorescence Visible radiation produced by a substance as a consequence of incident radiation

fundamental particle A particle that cannot be broken down into constituent particles

https://get.ga/aplus-hsc-physics-u34

A+ DIGITAL FLASHCARDS
Revise this topic's key terms and concepts by scanning the QR code or typing the URL into your browser.

fusion A nuclear reaction in which two smaller nuclei combine to form a larger nucleus

gamma rays High-frequency electromagnetic radiation

gluon The boson for the strong nuclear force that holds together the quarks that comprise particles such as protons and neutrons; also binds nucleons in the nucleus of an atom

hadron A composite particle made up of quarks

half-life The time taken for half of a sample of radioactive material to decay

Hertzsprung–Russell diagram A plot of luminosity against spectral class/temperature for all stars

ionising ability A measure of the capacity of radiation to ionise material it encounters

isotopes Atoms with the same number of protons but with different numbers of neutrons in their nucleus

lepton An fundamental particle that exists uncombined

linear accelerator A particle accelerator in which particles move in a straight line

luminosity The amount of light energy being emitted by a star

main sequence A region of the Hertzsprung–Russell diagram characterised by stars fusing hydrogen

mass defect The difference between the mass of a nucleus and the sum of the masses of its constituent nucleons; the difference between the mass of reactants and products in a nuclear reaction

matter-wave theory de Broglie's theory that particles with momentum have a wavelength and wave properties

meson A particle made up of a quark and an antiquark

neutrino A very small, nearly massless, fundamental particle with no charge

neutron star A very dense remnant star comprising only neutrons and representing the last stage of the evolution of some high-mass stars

nucleon A particle found in the nucleus

nucleosynthesis The formation of a new nucleus

nuclide A distinct kind of nucleus with a specific number of protons and neutrons

orbital A specific electron energy level according to Bohr's atomic model

parent nucleus A nucleus just before a radioactive decay

penetrating ability A measure of the ability of the products of radioactive decay to pass through various materials

planetary model An atomic model proposed by Rutherford that has electrons orbiting the nucleus like planets orbit the Sun

plum pudding model An atomic model proposed by Thomson that has electrons embedded in the evenly distributed positive mass of an atom

positron The antimatter particle of an electron; positively charged

probabilistic Cannot be predicted with certainty no matter how much is known

proton–proton (p–p) chain A chain of nuclear fusion reactions that creates a helium nucleus from four protons

quark An fundamental particle that always exists in combined form as baryons or mesons

radioactive Describing a substance that undergoes spontaneous nuclear decay

radioisotope An isotope of an element that undergoes spontaneous nuclear decay

singularity An infinitely small point of infinite density

spectral class The group in which a star is placed according to its spectrum

Standard Model of matter The scientific model used to describe all matter and forces in terms of fundamental particles

stationary state An orbit of an electron that is stable and does not result in electromagnetic radiation emission

supernova A massive explosion caused by the gravitational collapse of very large stars after fusion has ceased

transmutation The changing of a nucleus into a different nucleus

uncontrolled chain reaction A self-sustaining nuclear reaction that increases in magnitude over time

Zeeman effect The splitting of single spectral lines into two when the source atoms are subjected to a magnetic field

Exam practice

Multiple-choice questions

Solutions start on page 192.

Origins of the elements

Question 1

The velocity of a variety of galaxies was measured and plotted on a graph as shown.

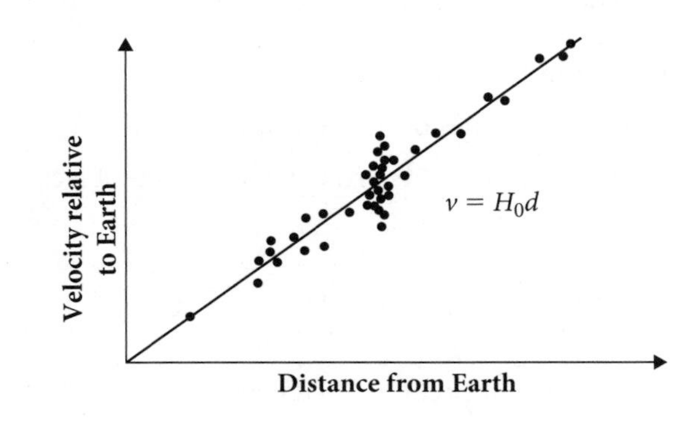

From this graph, which of the following is true about the slope of the graph?

A It is the Hubble constant.

B It is the age of the Universe.

C It is the inverse of Hubble's constant.

D It cannot be used to find the age of the Universe.

Question 2

It is written that $E = mc^2$ applies to stellar processes.

This statement is most likely referring to the fact that

A nuclear fission occurring in stars converts energy into mass according to $E = mc^2$.

B nuclear fission occurring in stars converts mass into energy according to $E = mc^2$.

C nuclear fusion occurring in stars converts mass into energy according to $E = mc^2$.

D nuclear fusion occurring in stars converts energy into mass according to $E = mc^2$.

Question 3

Which of the following statements is true?

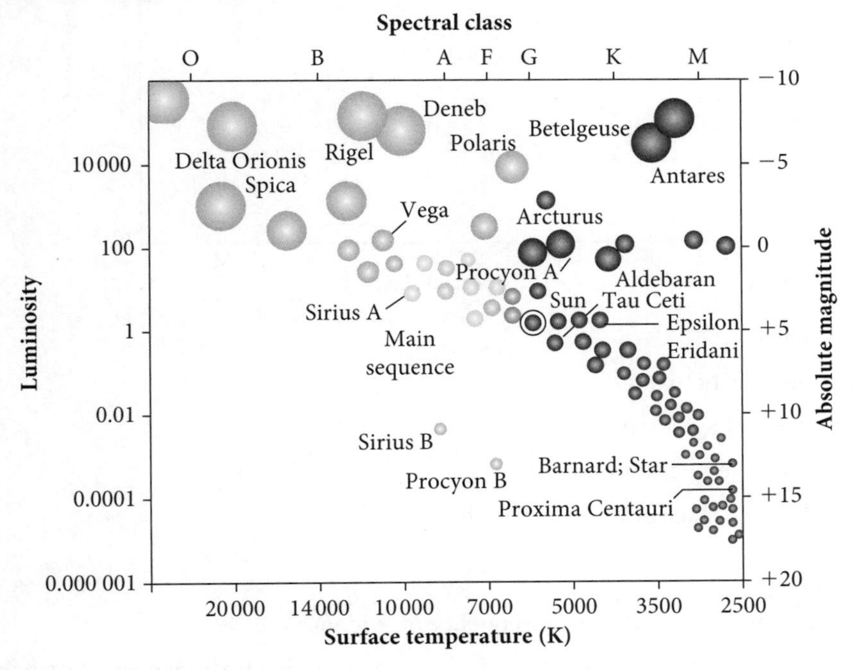

A Delta Orionis (upper left) is orange-red in colour.

B Spica (upper left) is currently capable of fusing heavy elements.

C Betelgeuse (upper right) is still fusing hydrogen into helium.

D Proxima Centauri (lower right) will only have a short life span.

Question 4

Choose the option that best represents the correct sequence of events following the Big Bang.

A Forces become distinct from each other, fundamental particles form from energy, nuclei form, atoms form.

B Nuclei form, atoms form, energy is converted into fundamental matter particles, annihilation leaves behind matter particles only.

C Fundamental particles form from energy, nuclei form, forces separate, atoms form.

D Forces separate, fundamental particles form from energy, atoms form, nuclei form.

Question 5

The spectral lines of two stars are shown.

Which option could correctly identify the stage in the star's life cycle?

	Star P	Star Q
A	Red giant	Red giant
B	Main sequence	Red giant
C	Red giant	Main sequence
D	Main sequence	Main sequence

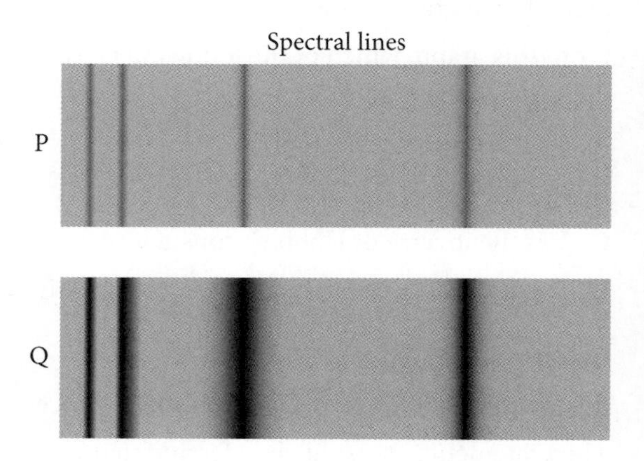

Question 6 ©NESA 2019 SI Q4

Four stars, *P*, *Q*, *R* and *S*, are labelled on the Hertzsprung–Russell diagram.

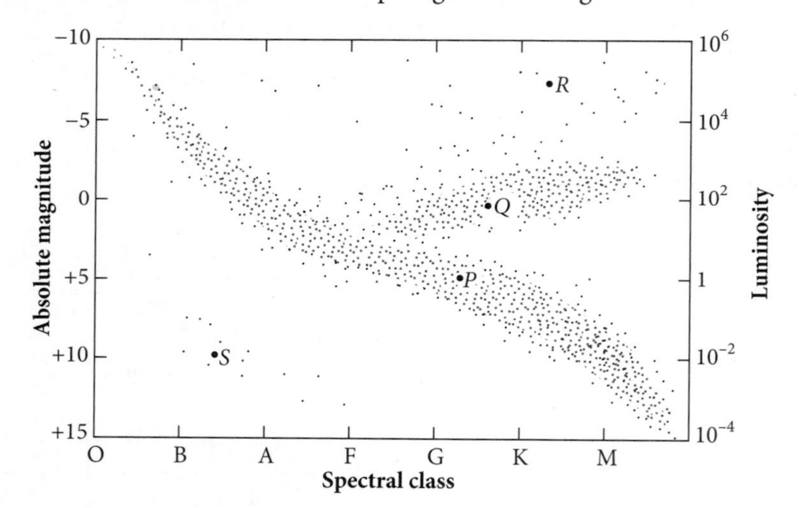

Which statement is correct?

A *S* has a greater luminosity than *Q*.

B *R* is a blue star whereas *S* is a red star.

C *S* has a higher surface temperature than *R*.

D *P* is at a more advanced stage of its evolution than *R*.

Question 7

The Big Bang theory was developed as a consequence of evidence that

A the composition of atoms in the Universe was approximately 75% hydrogen and 25% helium.

B there was leftover radiation, later called cosmic microwave background radiation.

C there was an imbalance between the number of particles and antiparticles in the Universe.

D observed galaxies were moving away from Earth at speeds proportional to their distance from us.

9780170465304

Question 8

The CNO cycle is a nucleosynthesis reaction that occurs predominantly in

A post-main sequence stars that fuse heavier elements.

B large stars with higher core temperatures.

C stars with highest densities, such as white dwarfs and neutron stars.

D small stars with spectral lines indicating the presence of carbon, nitrogen and oxygen.

Structure of the atom

Question 9 ©NESA 2019 SI Q3

Geiger and Marsden carried out an experiment to investigate the structure of the atom.

Which diagram identifies the particles they used and the result that they **initially** expected?

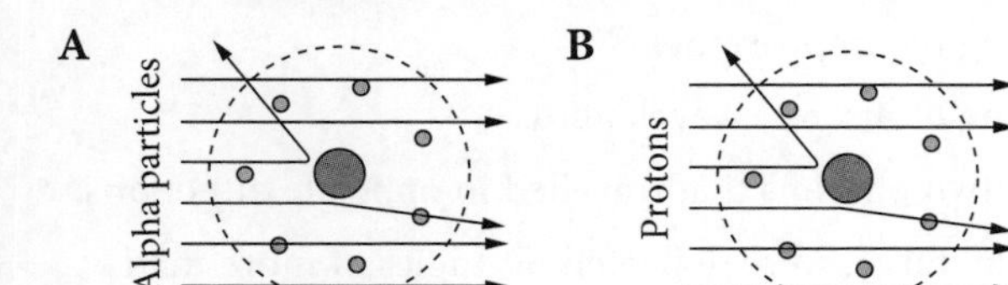

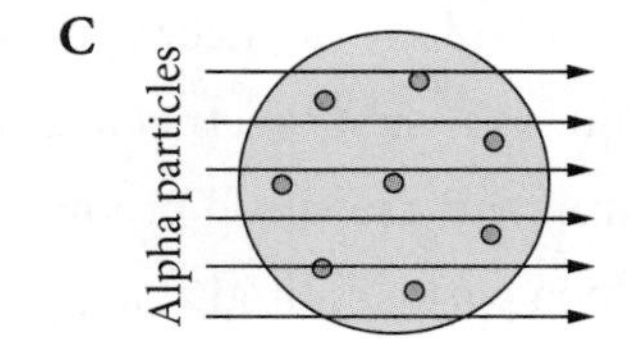

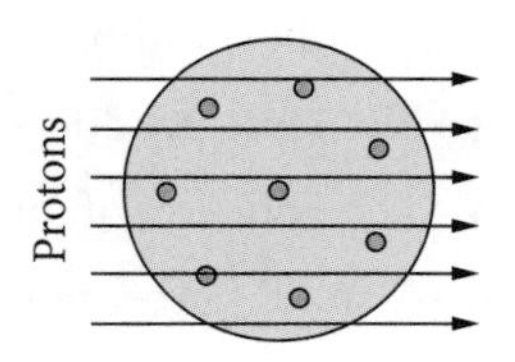

Question 10

Thomson used a magnetic field to make the cathode rays move in a circular arc and was able to use the radius of this arc to calculate his q/m ratio. In one part of the experiment, he exposed the cathode ray simultaneously to an electric field as well as the magnetic field in the same region of the cathode ray tube. The purpose of this was to

A balance the magnetic field so that the cathode ray velocity could be determined.

B accelerate the cathode rays to higher velocities.

C deflect the cathode rays in a direction perpendicular to the magnetic field.

D neutralise the cathode rays so that they travelled in a straight line.

Question 11

Which two forces did Millikan need to balance to collect data for his experiment?

A Strong nuclear force and electromagnetic force

B Magnetic force and gravitational force

C Electrostatic force and magnetic force

D Gravitational force and electrostatic force

Question 12

In an investigation of Chadwick's experiment, university students test the 'unknown' radiation emitted by the beryllium metal after it has been bombarded with alpha particles.

They make three observations.

I The radiation cannot be deflected by magnetic fields.

II The radiation caused protons to be ejected from a paraffin block.

III The radiation cannot cause the photoelectric effect.

Which piece or pieces of evidence are needed to conclude that the radiation is indeed neutrons rather than gamma rays?

A II only **B** I and II **C** I and III **D** III only

Question 13

Debate about the nature of cathode rays centred on whether they were waves or particles.

Evidence that was used to support the wave model of cathode rays included

A rays cast shadows, rays expose a photographic plate, rays are deflected by magnetic field.

B rays travel in straight lines, rays cast shadows, rays expose a photographic plate.

C rays are deflected by magnetic field, rays travel in straight lines, rays cast shadows.

D rays expose a photographic plate, rays are deflected by magnetic field, rays travel in a straight line.

Quantum mechanical nature of the atom

Question 14 ©NESA 2020 SI Q9

Bohr improved on Rutherford's model of the atom.

Which observation by Bohr provided evidence supporting the improvement?

A Elements produced unique emission spectra consisting of discrete wavelengths.

B The collision of an electron and a positron produced two photons that travelled in opposite directions.

C A small percentage of alpha particles fired at a gold foil target were deflected by angles of more than 90 degrees.

D A beam of electrons reflected from a nickel crystal produced a pattern of intensity at different angles, consistent with their wave properties.

Question 15

The diagram shows the energy levels of hydrogen with four different electron transitions labelled A, B, C, D.

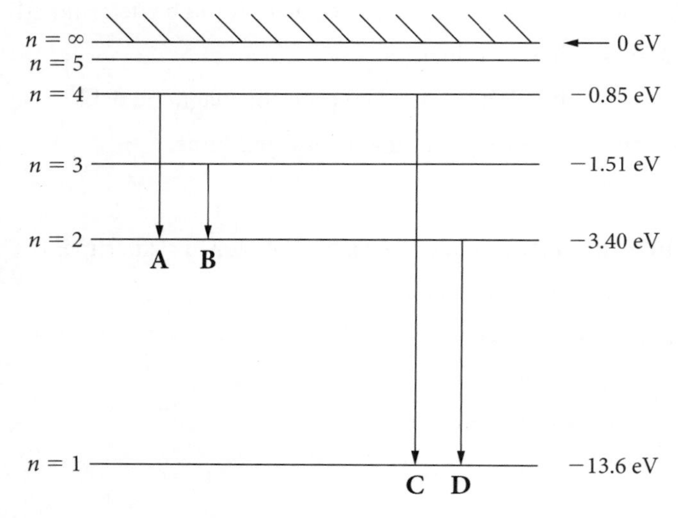

Not to scale

Which of the transitions shown corresponds to an emitted photon of wavelength 486 nm?

Question 16

A cricket ball of mass 152 g, moving at 140 km h^{-1} will have a wavelength of

A 3.11×10^{-38} m **B** 1.12×10^{-37} m **C** 3.11×10^{-35} m **D** 1.12×10^{-34} m

Question 17 ©NESA 2021 SI Q13

The diagram shows electron transitions in a Bohr-model hydrogen atom.

Which transition would produce the shortest wavelength of light?

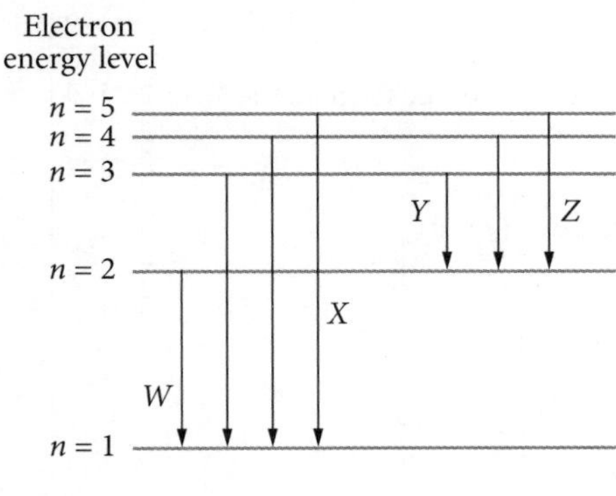

A W **B** X **C** Y **D** Z

Question 18

Identify the spectral line that is the result of the $n = 3$ to $n = 2$ electron transition in hydrogen.

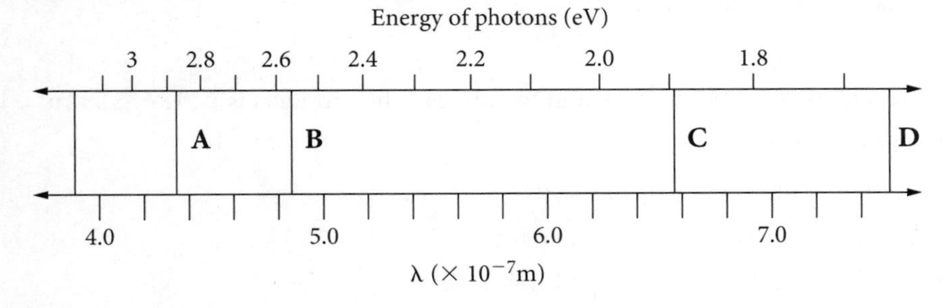

Question 19

The light from a hydrogen discharge tube shines through a diffraction grating and is projected onto a screen. It is possible to observe the four smallest energy transitions of the visible spectrum of hydrogen, as seen in the image. M is the central maximum.

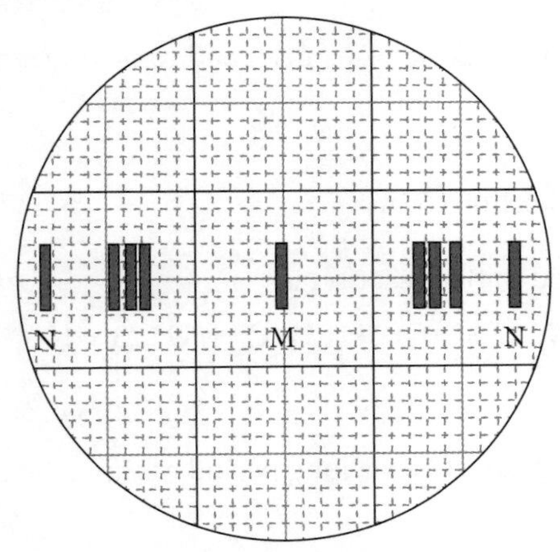

The electron transition within the hydrogen atom that is represented by the band labelled N is from

A energy level 6 to energy level 2.

B energy level 5 to energy level 2.

C energy level 4 to energy level 2.

D energy level 3 to energy level 2.

Properties of the nucleus

Question 20 ©NESA 2020 SI Q8

A uranium isotope, *U*, undergoes four successive decays to produce Q.

Which row of the table correctly shows the decay process *R* and product *Q*?

	Process *R*	Product *Q*
A	α	Pa-230
B	β	Pa-234
C	α	Th-230
D	β	Th-234

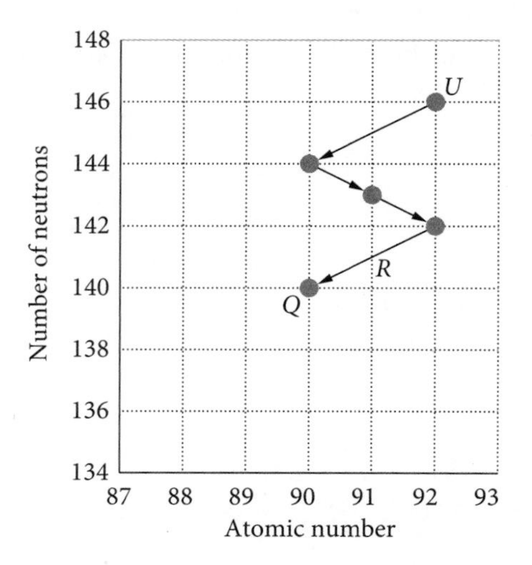

Question 21 ©NESA 2020 SI Q11

Consider the following nuclear reaction.

$${}^{6}_{3}\text{Li} + {}^{1}_{0}\text{n} \rightarrow {}^{4}_{2}\text{He} + {}^{3}_{1}\text{H}$$

The mass of the reactants is 7.023 787 704 u and the mass of the products is 7.018 652 532 u.

What type of reaction is this?

A A fusion reaction in which energy is released

B A fusion reaction that requires an input of energy

C A transmutation reaction in which energy is released

D A transmutation reaction that requires an input of energy

Question 22

Lightning storms can create highly accelerated electrons that, in turn, create gamma photons as a result of the high accelerations.

${}^{13}_{7}\text{N}$ is then produced by the photons striking ${}^{14}_{7}\text{N}$ nuclei and removing a neutron.

$${}^{0}_{0}\gamma + {}^{14}_{7}\text{N} \rightarrow {}^{13}_{7}\text{N} + {}^{1}_{0}\text{n}$$

${}^{14}_{7}\text{N}$ 14.003 07 u, ${}^{13}_{7}\text{N}$ 13.005 74 u, ${}^{1}_{0}\text{n}$ 1.008 67 u

The energy (in MJ) that is absorbed in this reaction is

A $1.692\,23 \times 10^{-6}$ **B** $1.692\,23 \times 10^{-12}$ **C** $1.692\,23 \times 10^{-18}$ **D** $1.692\,23 \times 10^{-27}$

Question 23

All rocks on Earth contain uranium-235. Over a long period of time, the uranium-235 decays into other nuclides.

If a sample of rock contains 5% of the original amount of uranium-235 after a period of 4.5×10^{6} years, calculate the decay constant for uranium-235.

A $6.657 \times 10^{-7}\,\text{s}^{-1}$ **B** $1.1341 \times 10^{-14}\,\text{s}^{-1}$

C $2.111 \times 10^{-14}\,\text{s}^{-1}$ **D** $4.884 \times 10^{-15}\,\text{s}^{-1}$

Question 24

What is the decay constant for a radioactive isotope with a half-life of 172 years?

A $1.28 \times 10^{-10}\,\text{s}^{-1}$ **B** $0.003\,98\,\text{s}^{-1}$ **C** $4.47\,\text{s}^{-1}$ **D** $21.72\,\text{s}^{-1}$

Question 25 ©NESA 2019 SI Q19

Consider the following nuclear reaction.

W + X → Y + Z

Information about W, X and Y is given in the table

Species	Mass defect (u)	Total binding energy (MeV)	Binding energy per nucleon (MeV)
W	0.002 388 17	2.224 566	1.112 283
X	0.009 105 58	8.481 798	2.827 266
Y	0.030 376 64	28.295 66	7.073 915

Which of the following is a correct statement about energy in this reaction?

A The reaction gives out energy because the mass defect of Y is greater than that of either W or X.

B It cannot be deduced whether the reaction releases energy because the properties of Z are not known.

C The reaction requires an input of energy because the mass defect of the products is greater than the sum of the mass defects of the reactants.

D Energy is released by the reaction because the binding energy of the products is greater than the sum of the binding energies of the reactants.

Question 26 ©NESA 2020 SI Q4

The graph shows the mass of a radioactive isotope as a function of time.

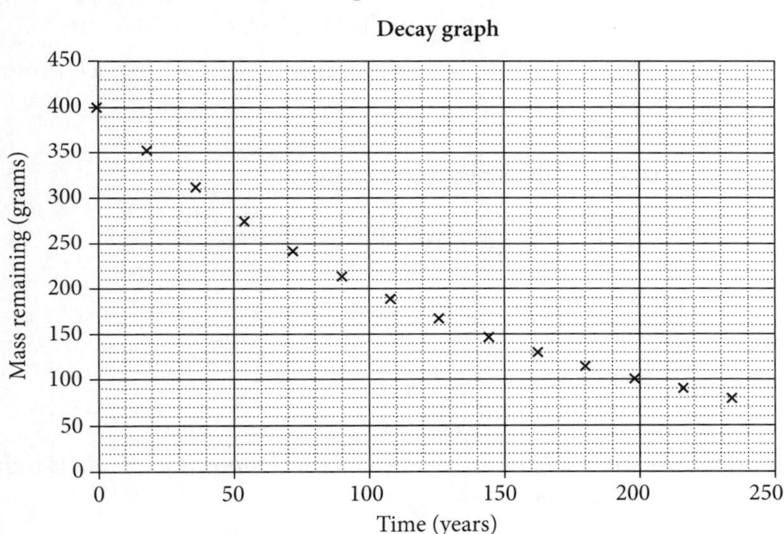

What is the decay constant, in years^{-1}, for this isotope?

A 0.0030 **B** 0.0069 **C** 2.0 **D** 100

Question 27

The difference between a controlled nuclear chain reaction and an uncontrolled nuclear chain reaction is that

A uncontrolled nuclear chain reactions only happen in the Sun, whereas controlled nuclear chain reactions can be created on Earth.

B controlled nuclear chain reactions emit energy at a constant rate, whereas uncontrolled nuclear chain reactions emit energy at an increasing rate.

C uncontrolled nuclear chain reactions only involve fission, whereas controlled nuclear chain reactions only involve fusion.

D controlled nuclear chain reactions emit neutrons at a constant rate, whereas uncontrolled nuclear chain reactions emit neutrons at a decreasing rate.

Question 28

Which line of the table correctly distinguishes features of two of the three main forms of radioactive decay?

	Ionising ability	Penetrating ability
A	Alpha: low Gamma: high	Alpha: low Gamma: high
B	Alpha: high Gamma: low	Alpha: high Gamma: low
C	Alpha: high Gamma: low	Alpha: low Gamma: high
D	Alpha: low Gamma: high	Alpha: high Gamma: low

Deep inside the atom

Question 29

Using the Standard Model of matter to explain beta-plus decay means that

A an up quark changes into a down quark.

B a down quark changes into an up quark.

C a quark changes into a lepton.

D a lepton changes into a quark.

Question 30 ©NESA 2021 SI Q3

Which of the following is **not** a fundamental particle in the Standard Model of matter?

A Electron **B** Gluon **C** Muon **D** Proton

Question 31 ©NESA 2019 SI Q12

The table shows two types of quarks and their respective charges.

Quark	Symbol	Charge
Up	u	$+\frac{2}{3}$
Down	d	$-\frac{1}{3}$

In a particular nuclear transformation, a particle having a quark composition of *udd* is transformed into a particle with a quark composition of *uud*.

What is another product of this transformation?

A Electron **B** Neutron **C** Positron **D** Proton

Question 32

Particle accelerators are used to validate aspects of the Standard Model of matter because they

A give particles sufficient kinetic energy to initiate nuclear reactions.

B give charged particles sufficient kinetic energy to initiate nuclear reactions.

C accelerate bosons to high speeds so that their interactions with particles can be studied.

D give particles sufficient speed to overcome the strong nuclear force to separate gluons.

Short-answer questions

Solutions start on page 196.

Origins of the elements

Question 33 (3 marks)

In Einstein's work on special relativity, published in 1905, he introduced the law of conservation of mass–energy, which featured the derivation of the famous equation for mass–energy equivalence $E = mc^2$. Explain how this equation can account for the vast energy produced by stars.

Question 34 (5 marks)

Sketch a graph showing the results of Hubble's observations, outline how the results provided evidence of an expanding Universe and describe how the graph could be used to estimate the age of the Universe.

Question 35 (7 marks)

The Hertzsprung–Russell diagram shows the position of four stars, P, Q, X and Y.

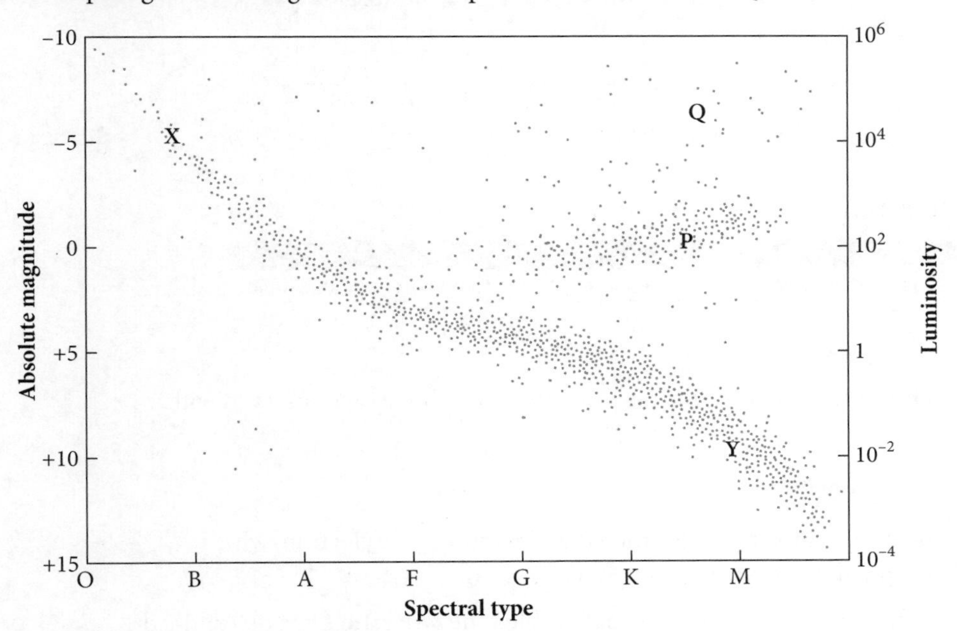

Compare the process by which energy is produced and the remaining stages of the evolutionary path of each of the four stars P, Q, X and Y.

Question 36 (8 marks)

Spica is a large blue-white main sequence star and Barnard's star is a small red main sequence star. Analyse the nucleosynthesis processes that occur in each of the two stars as they are observed now and when they become post-main sequence stars.

Question 37 (4 marks)

Outline the processes involved in the Big Bang.

Question 38 (5 marks)

Compare the physical characteristics of the four stars marked A, B, C and D indicated in the Hertzsprung–Russell diagram.

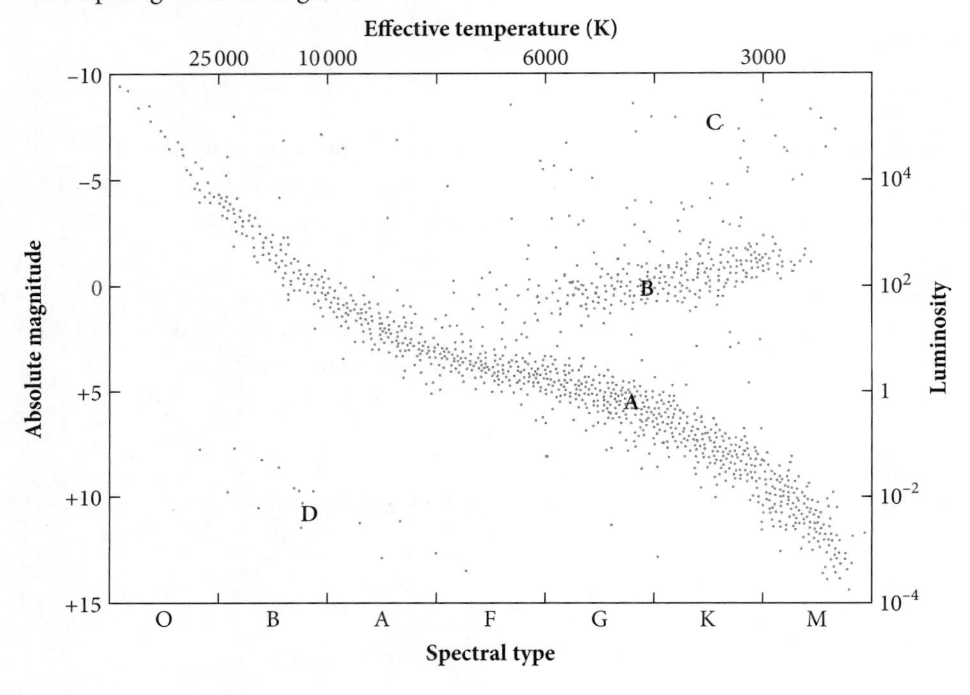

Structure of the atom

Hint
Consider providing a labelled diagram to support your responses to questions of this style.

Question 39 (4 marks)

With the aid of sketch diagrams, describe three properties of cathode rays observed with cathode ray tubes.

Question 40 (5 marks)

Thomson conducted an experiment to determine the q/m ratio of the electron, which eventually helped to confirm the particle nature of the electron.

Describe the steps of the experimental process that enabled the q/m ratio to be determined mathematically from the measured radius of the cathode beam.

Question 41 (7 marks)

Analyse the significance of the work of Millikan and Thomson in the discovery of the electron.

Question 42 (6 marks)

Describe the Geiger–Marsden experiment and the significance of the results.

Question 43 (3 marks)

Recount Chadwick's experiment and describe how his observations enabled him to deduce the existence of the neutron.

Quantum mechanical nature of the atom

Question 44 (5 marks) ©NESA 2020 SII Q21

a Calculate the wavelength of light emitted by an electron moving from energy level 5 to 2 in a Bohr model hydrogen atom. 2 marks

b Describe the behaviour of electrons in the Bohr model of the atom with reference to the law of conservation of energy. 3 marks

Question 45 (9 marks)

The development of the current atomic model has involved the work of multiple scientists and included several specific experiments that led to new discoveries.

Analyse the development of the atomic model through history, with a focus on the work of Rutherford, Bohr and de Broglie.

Question 46 (7 marks)

Bohr's model of the atom was able to explain several features of the atom that previous models could not address. However, its proposal raised other issues that could not be explained.

Explain how Bohr's model of the atom was able to address limitations of previous atomic models and describe limitations of the Bohr model.

Question 47 (5 marks)

The following experimental set-up was used by Davisson and Germer.

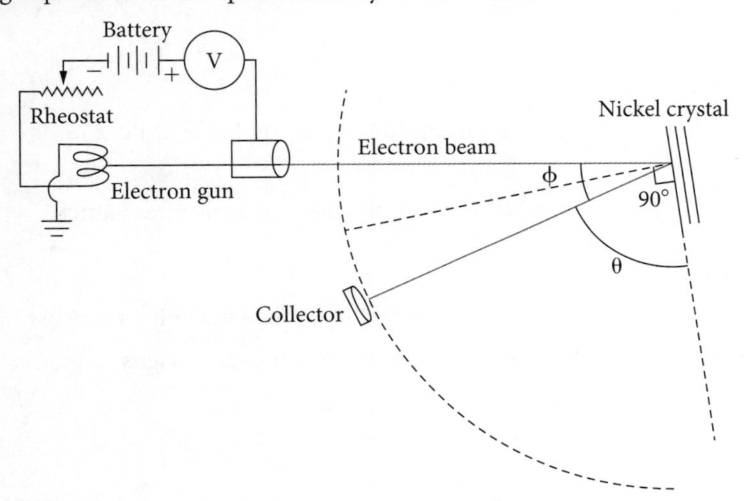

a Identify *one* observation and/or measurement made during this experiment. 2 marks

b Outline how this experimental evidence helped to change the model of the atom by addressing a key limitation of the Bohr and Rutherford atomic models. 3 marks

Question 48 (5 marks)

Outline de Broglie's matter-wave theory and how it addressed some issues of previous models.

Question 49 (5 marks) ©NESA 2021 SII Q29

Bohr, de Broglie and Schrödinger **each** proposed a model for the structure of the atom.

How does the nature of the electron proposed in each of the three models differ?

Properties of the nucleus

Question 50 (4 marks)

The fission of uranium by a neutron is shown in the diagram and equation.

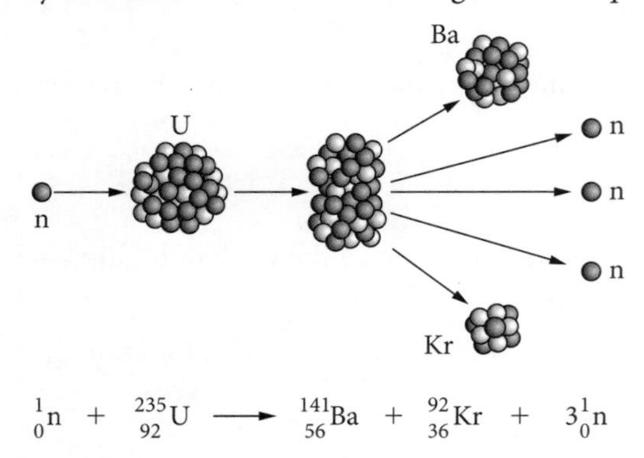

$${}^{1}_{0}n + {}^{235}_{92}U \longrightarrow {}^{141}_{56}Ba + {}^{92}_{36}Kr + 3{}^{1}_{0}n$$

Discuss the stability of the nuclei involved in the nuclear fission reaction above with reference to their binding energy per nucleon. Use an appropriate graph in your response.

Question 51 (6 marks) ©NESA 2021 SII Q35

A spacecraft is powered by a radioisotope generator. Pu-238 in the generator undergoes alpha decay, releasing energy. The decay is shown with the mass of each species in atomic mass units, *u*.

$${}^{238}Pu \rightarrow {}^{234}U + \alpha$$

238.0495u 234.0409u 4.0026u

a Show that the energy released by one decay is 9.0×10^{-13} J. 3 marks

b At launch, the generator contains 9.0×10^{24} atoms of Pu-238. The half-life of Pu-238 is 87.7 years.

Calculate the total energy produced by the generator in the first ten years after launch. 3 marks

Question 52 (4 marks)

One of the reactions that occurs in some red giants is thought to be the triple-alpha process. The process involves steps that fuse three helium-4 nuclei into a carbon-12 nucleus

$$3{}^{4}_{2}He \xrightarrow{\text{yields}} {}^{12}_{6}C$$

and yields two gamma ray photons.

Given the data below, calculate the wavelength of the two gamma ray photons. Assume they share the energy created in the reaction evenly.

Nuclide	Binding energy per nucleon (MeV)
${}^{4}_{2}He$	7.07
${}^{12}_{6}C$	7.68

Question 53 (7 marks) ©NESA 2019 SII Q36

A radon-198 atom, initially at rest, undergoes alpha decay. The masses of the atoms involved are shown in atomic mass units (u).

radon-198	→	polonium-194	+	helium-4
197.999 u		193.988 u		4.002 60 u

The kinetic energy of the polonium atom produced is 2.55×10^{-14} J.

By considering mass defect, calculate the kinetic energy of the alpha particle, and explain why it is significantly greater than that of the polonium atom.

Question 54 (4 marks)

The Sun's mass decreases by approximately 1.34×10^{17} kg each year as it fuses hydrogen to helium in the following net nuclear fusion reaction.

$$4\,{}^{1}_{1}\text{H} \xrightarrow{\text{yields}} {}^{4}_{2}\text{He} + 2e^{+} + 2\nu_{e} + \text{energy}$$

The mass of a helium nucleus is 6.643×10^{-27} kg and a positron has a mass equal to that of an electron.

Assuming the other reaction product has no mass, calculate the power of the Sun and how many nuclear reactions occur per second in the Sun.

Question 55 (4 marks)

The activity of a beta-minus emitting radioactive isotope of phosphorus-32 decreases to 22.64% of its initial activity over the course of 30 days.

a Write a complete nuclear reaction for this beta-minus decay. 2 marks

b Calculate the half-life of phosphorus-32 to the nearest day. 2 marks

Deep inside the atom

Question 56 (4 marks)

Use the Standard Model to describe the structure of the helium-4 atom.

Question 57 (8 marks)

The Standard Model of matter has been extremely successful in making predictions and explaining observations since its development.

a Identify the main features of the Standard Model. 4 marks

b The test of a model is its ability to predict. Identify one prediction made by the Standard Model and where it may be experimentally verified. 2 marks

c Identify two limitations of the Standard Model. 2 marks

Question 58 (3 marks)

Particle accelerators have been an important tool in verifying predictions made by the Standard Model of matter.

Summarise the reasons that justify particle accelerators as an appropriate tool for this purpose.

SOLUTIONS

CHAPTER 1 MODULE 5 ADVANCED MECHANICS

Multiple-choice solutions

1 D

$u = 40\,\text{m s}^{-1}$

$\theta = 20°$

Therefore, $u_y = u \sin\theta = 40 \sin 20°\,\text{m s}^{-1}$.

$v_y = 0$ at maximum height

$a = -9.8\,\text{m s}^{-2}$

$s_y = ?$

Choose equation from data. $v_y^2 = u_y^2 + 2as_y$

Substitute known values. $0 = (40 \sin 20°)^2 + 2 \times -9.8 \times s_y$

Rearrange. $s_y = -\dfrac{(40 \sin 20°)^2}{-19.6}$

Calculate the value. $s_y = 9.549\,\text{m} = 9.5\,\text{m}$ (to two sig. fig.)

2 D

v_x is constant throughout trajectory, hence **C** is incorrect.

v_y must be up at P and Q and down at R and S, hence **A** is incorrect.

v_y must be greatest closer to the ground at P and S and smallest closer to the apex at Q and R, hence **B** is incorrect.

3 B

Expressions for each of the range and maximum height in terms of u and θ only will be required.

Start by deriving an expression for time of flight:

$$s_y = u_y t + \tfrac{1}{2}at^2$$

$$0 = u \sin\theta \times t + \tfrac{1}{2} \times -9.8 \times t^2$$

Solve for t.

$t = 0$ (this is when it was launched) or $\dfrac{u \sin\theta}{4.9}$.

This can be substituted into the range equation.

$$s_x = u_x t$$

$$s_x = u \cos\theta \times t$$

$$s_x = \frac{u^2 \cos\theta \times \sin\theta}{4.9}$$

Now, since u is constant and $\cos 60° \times \sin 60° = \cos 30° \times \sin 30°$, both projectiles will have the same range.

For maximum height $v_y = 0$, so using:

$$v_y^2 = u_y^2 + 2as_y$$

$$0 = (u \sin\theta)^2 + 2 \times -9.8 \times s_y$$

Therefore, $s_y = \dfrac{(u \sin\theta)^2}{-19.6}$.

Since $\sin 30° < \sin 60°$, the maximum height is greater when $\theta = 60°$ than when $\theta = 30°$.

4 B

Using derivations from Question 3:

$$s_x = \frac{u^2 \cos\theta \times \sin\theta}{4.9}$$

$t = 0$ or $\dfrac{u \sin\theta}{4.9}$ and

$$s_y = \frac{(u \sin\theta)^2}{-19.6}$$

As θ increases from 45° to 60°, sin θ increases; therefore, s_y increases and t increases.

Furthermore, since sin 45° × cos 45° > sin 60° × cos 60°, as θ increases from 45° to 60°, s_x decreases.

5 A

Using derivations from Question 3:

$$t = \frac{u \sin\theta}{4.9}$$

It can be seen, therefore, that the factors affecting time of flight are u (launch velocity), θ (launch angle) and acceleration due to gravity (which has contributed the 4.9 value in the denominator).

B and **D** include references to range, which will not be a factor determining time of flight, and **C** omits launch angle.

6 D

$u_x = 10\,\text{m s}^{-1}$

$s_y = -300\,\text{m}$

$\theta = 0$

Therefore, $u_y = 0\,\text{m s}^{-1}$.

$a = -9.8\,\text{m s}^{-2}$

$s_x = ?$

Choose equation from data to determine t.	$s_y = u_y t + \frac{1}{2}at^2$
Substitute known values.	$-300 = 0 + \frac{1}{2} \times -9.8 \times t^2$
Rearrange.	$t = \sqrt{\frac{300}{4.9}}$
Now substitute t into equation to find range.	$s_x = u_x t$
	$s_x = 10 \times \sqrt{\frac{300}{4.9}}$
Calculate the value.	$s_x = 78.246\,\text{m} = 78\,\text{m}$ (to two sig. fig.)

7 A

Kinetic energy, K, is converted to potential energy, U, as the object moves up. At the higher point the object has more U and must, therefore, have less K. As mass is constant, the object must have lower v at P (i.e. it is moving slower).

B is incorrect as it is assumed there is no friction and, therefore, total mechanical energy is constant.

C is incorrect as acceleration due to gravity is constant throughout the trajectory.

D is incorrect as v_x is always constant throughout the trajectory.

8 C

The equation $s_y = u_y t + \frac{1}{2}at^2$, where $u_y = 0$, can be rearranged to give an expression for the time of flight of the object:

$$t = \sqrt{\frac{2s_y}{a}}$$

where s_y and a are the same for both balls P and Q. It can be seen then that time of flight is same for P and Q.

The final speed of the balls is given by the vector sum of the horizontal and vertical components of the velocity in the expression $v = \sqrt{v_x^2 + v_y^2}$. Since ball P has a bigger v_x, it must have a greater v.

9 C

The object will be accelerating uniformly under the influence of gravity so the gradient of the line on the speed–time graph must be constant, i.e. it must be a straight line. Hence **B** and **D** are incorrect. The speed value must be least at the apex, so **A** is incorrect.

10 A

The question states that height is the independent variable and so appears on the horizontal axis, so **C** and **D** are incorrectly arranged.

Using the equation $s_y = u_y t + \frac{1}{2}at^2$, where $u_y = 0$, it can be seen that

$$s_y = \frac{1}{2}at^2 = \frac{1}{2} \times -9.8 \times t^2$$

so s_y is proportional to t^2 and, thus, the graph s_y vs t cannot be linear, so **B** is incorrect.

11 B

The object will be accelerating uniformly under the influence of gravity, so the gradient of the line on the velocity–time graph must be constant, i.e. it must consist of straight lines. Hence **C** and **D** are incorrect. The ball will move both up and down over the time period, so the velocity must be both positive and negative and there must be velocities above and below the t-axis, so **A** is incorrect.

12 A

When the object is moving at 45°, the horizontal component of its velocity must equal the vertical component of its velocity.

So $v_x = v_y = u$.

Velocity is the vector sum of v_x and v_y:

$$v = \sqrt{v_x^2 + v_y^2} = \sqrt{u^2 + u^2} = u\sqrt{2}$$

This value is closest to $1.4u$.

13 C

The lift is slowing down as it approaches the top floor, so it is experiencing negative acceleration. This is mathematically equivalent to the lift speeding up as it moves down.

In those circumstances, the time of flight for the object falling would increase (as the elevator accelerates away from it) and so the range will increase.

Since the object will still have a constant horizontal velocity component and uniform vertical acceleration (although not $9.8\,\text{m s}^{-2}$), it will still move in the recognisable parabolic path.

14 C

From the question it can be concluded that the period of the uniform circular motion of the carriage is 80 seconds. Since $f = \frac{1}{T}$, then $f = \frac{1}{80} = 0.0125\,\text{Hz}$.

15 A

Using $v = \frac{2\pi r}{T} = \frac{2\pi}{80} \times \frac{45}{2} = 1.77\,\text{m s}^{-1}$.

16 A

In uniform circular motion, the centripetal force will be constant in magnitude and always directed to the centre. Tension, as long as it is non-zero, will always be in the same direction as the centripetal force – towards the centre of the circle, because it is providing the centripetal force.

At the top, the force of gravity will 'help' the tension to provide the centripetal force, so the tension force will be less than the centripetal force, i.e. $T = F_c - mg$. **C** is incorrect. However, at the bottom the force of gravity will work against the tension as it provides the centripetal force, so the tension force is larger than the centripetal force, i.e. $T = F_c + mg$. **B** and **D** are incorrect.

17 D

Kinetic energy and potential energy are (usually) constant and no energy leaves the system.

A is incorrect as there is always a net force (centripetal force) acting on objects moving in uniform circular motion.

B is incorrect as objects in uniform circular motion are moving at a constant speed.

C is incorrect since objects in uniform circular motion have both kinetic and potential energy.

18 A

Two objects moving in uniform circular motion on the same rotating system must have the same period. **B** is incorrect.

Since $v = \frac{2\pi r}{T}$, then if T is the same, the object moving with the greater radius must have the greater velocity as v is proportional to r. Therefore N is faster than M. **B** is incorrect.

Combining $F_c = \frac{mv^2}{r}$ and $v = \frac{2\pi r}{T}$, then $F_c = \frac{4\pi^2 mr}{T}$, so F_c is also proportional to r and so the object at the greater radius, N, must have the larger centripetal force, F_c. Therefore **C** and **D** are incorrect.

19 D

The pulley does not rotate when released when the anticlockwise torque created by the force of gravity acting on m_1 is equal to the clockwise torque created by the force of gravity acting on m_2.

Since $\tau = rF\sin\theta$, where $\theta = 90°$, this means that

$rF_{\text{clockwise}} = rF_{\text{anticlockwise}}$, where the force is the force due to gravity.

Therefore, $m_1 g r_1 = m_2 g r_2$.

Cancel g on both sides:

$m_1 r_1 = m_2 r_2$

20 C

Centripetal acceleration, a_c, is a function of v and r as indicated by the equation $a_c = \frac{v^2}{r}$. In contrast, centripetal force F_c is a function of v, r and m ($F_c = \frac{mv^2}{r}$).

In the situation described, v and r are both constant but m increases, so a_c is constant but F_c will increase.

SOLUTIONS

21 C

Following the hint:

If there are n revolutions per second, then the frequency is n. Therefore, since $T = \frac{1}{f}$, $T = \frac{1}{n}$.

$$K = \frac{1}{2}mv^2 \text{ and } v = \frac{2\pi r}{T}$$

Therefore, by substitution

$$K = \frac{1}{2}m\left(\frac{2\pi r}{T}\right)^2 = \frac{2\pi^2 mr^2}{T^2} \text{ and since } T = \frac{1}{n}$$

$$K = 2\pi^2 mr^2 n^2$$

22 A

The tension force acting along the string and the weight force can be used to construct a vector diagram in which the resultant force is the centripetal force. Alternatively, the tension force can be resolved into horizontal and vertical components as seen below.

The vertical component is equal and opposite to the force due to gravity acting on the mass (mg).

The horizontal component provides the centripetal force $\left(\frac{mv^2}{r}\right)$.

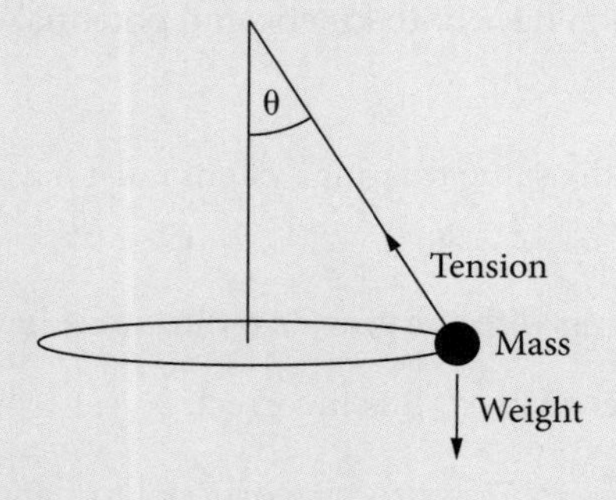

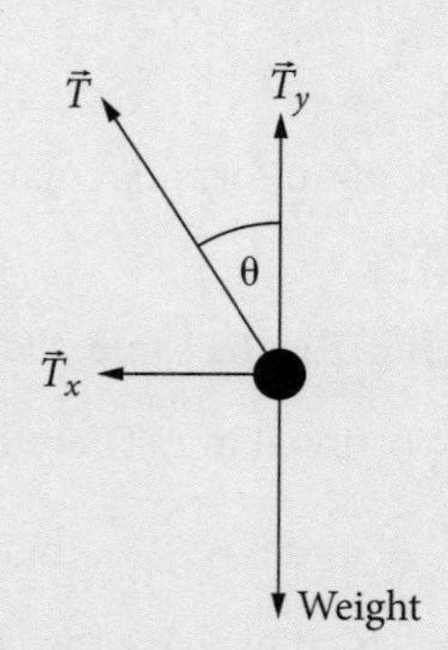

It can be seen that $\tan\theta = \frac{mv^2}{r} \times \frac{1}{mg}$.

So $\tan\theta = \frac{v^2}{rg}$.

$v = r\omega$

Therefore, $\tan\theta = \frac{r^2\omega^2}{rg} = \frac{r\omega^2}{g}$.

23 D

The force due to gravity acting on M provides the centripetal force, relayed through tension in the string, to enable mass m to move in uniform circular motion.

Therefore, $Mg = \frac{mv^2}{r}$

so $M = \frac{mv^2}{rg}$.

24 C

At the minimum speed, the force due to gravity provides all the centripetal force (at greater speed, the normal force will provide some of F_c).

So, at the top, F_g can be equated with F_c.

$$mg = \frac{mv^2}{r}$$

$$v^2 = rg$$

$$\therefore v = \sqrt{rg}$$

$$v = \sqrt{3.6 \times 9.8} = 5.9\,\text{m s}^{-1}$$

25 A

Using Newton's universal law of gravitation:

$$F = \frac{GMm}{r^2} = \frac{6.67 \times 10^{-11} \times 6.39 \times 10^{23} \times 2.0 \times 10^{15}}{(20\,070 \times 10^3)^2} = 2.1 \times 10^{14}\,\text{N}$$

26 A

Using Kepler's third law:

$$\frac{r^3}{T^2} = \text{constant for all objects orbiting Mars}$$

So $$\frac{r_P^3}{T_P^2} = \frac{r_D^3}{T_D^2}$$

Rearranging gives $$\frac{r_P^3 T_D^2}{r_D^3} = T_P^2$$

Substituting gives $$\frac{(0.3 \times 20\,070)^3 \times 30^2}{(20\,070)^3} = T_P^2$$

Therefore, $T_P = 4.9295$ hours.

Hint

When using the equation in this form, there is no need to convert to SI units. The units of the answer will be the same as the units of the unpaired bit of data.

27 A

As a consequence of the attractive force due to gravity, the craft will slow as it moves away from Earth and then, once reaching a point where the Moon's gravitational field is stronger, will speed up approaching the Moon.

Since the rocket, Moon and Earth form an isolated system, the total amount of energy is constant. The rocket's energy will be converted from kinetic energy to gravitational potential energy as it slows moving away from Earth and vice versa as it speeds up approaching the Moon.

28 A

Kepler's third law of periods is a consequence of the conservation of energy. Planets undergoing an elliptical orbit around a star will move slowest (and have the least kinetic energy) when they are furthest from the star (and have the greatest gravitational potential energy).

It must, therefore, move slower than average at the most distant point. **A** is correct.

29 D

Gravitational field strength at a point is a vector sum of gravitational fields due to all objects. This can be achieved by subtracting the smaller right-directed field of Q from the larger left-directed field of P.

So, $$g_E = g_P - g_Q = \frac{GM_E}{r_P^2} - \frac{GM_Q}{r_Q^2} = \frac{2GM}{\left(\frac{R}{2}\right)^2} - \frac{GM}{R^2} = \frac{7GM}{R^2}.$$

30 A

Geostationary satellites are positioned so that their orbital period is 24 hours. It can be determined that a geostationary satellite will have an orbital radius of about 42 000 km (about 6 times that of low-Earth orbit satellites). From this distance, more of Earth's surface will be visible and there is effectively no drag.

Low-Earth orbit satellites have an orbital velocity of approximately 7 km s^{-1} compared to geostationary satellites, which orbit at approximately 3 km s^{-1}.

31 C

From the scale diagram it can be seen that the distance from Y to the centre of Earth is 1.5 times the distance from X to the centre of Earth.

Since the gravitational force is given by $F = \frac{GMm}{r^2}$, then $\frac{F_X}{F_Y}$ will be given by $\frac{GMm}{r^2} \times \frac{(1.5r)^2}{GMm}$.

By cancelling, this is equal to 2.25.

32 D

An object in a stable orbit is undergoing uniform circular motion and is therefore under the influence of a net force directed towards the centre. This is called the centripetal force and is provided by the force due to gravity. Satellites in stable orbits do not require a propulsive force. **A** and **B** are incorrect. The reaction force will act on the other object in the pair (in this case Earth). **C** is incorrect.

33 B

From Kepler's law of periods $\frac{r^3}{T^2} = \frac{GM}{4\pi^2}$.

This can be rearranged to become $T^2 = \frac{4\pi^2 r^3}{GM}$.

It can be seen that since r is common to both planets and other values are constants, T^2 is proportional to $\frac{1}{M}$.

If M_{Xerus} is $4 \times M_{\text{Earth}}$, then T^2_{Xerus} is $\frac{1}{4} \times T^2_{\text{Earth}}$ and thus T_{Xerus} is $\frac{1}{2} \times T_{\text{Earth}}$.

34 B

Escape velocity can be shown to be given by the equation $v_{esc} = \sqrt{\frac{2GM}{r}}$.

Doubling the mass and halving the radius would result in an increase of escape velocity by a factor of $\sqrt{4} = 2$.

Short-answer solutions

35 Data:

$u_y = 20\text{ m s}^{-1}$

$v_y = 0\text{ m s}^{-1}$ at maximum height

$v_y = -20\text{ m s}^{-1}$ (as a consequence of symmetrical nature of flight)

$a = -9.8\text{ m s}^{-2}$

$s_y = 0\text{ m}$

Using equation $v_y = u_y + at$, it can be shown that time of flight is approximately 4 s and time to apex is approximately 2 s.

9780170465304

The graph should be drawn with the following key features.

- Intercept on vertical axis is $+20\,\text{m s}^{-1}$.
- Intercept on horizontal axis is 2 s.
- Graph features a straight line of constant gradient approximately $-10\,\text{m s}^{-2}$.
- Graph finishes at approximately $t = 4$ s and velocity $= -20\,\text{m s}^{-1}$.

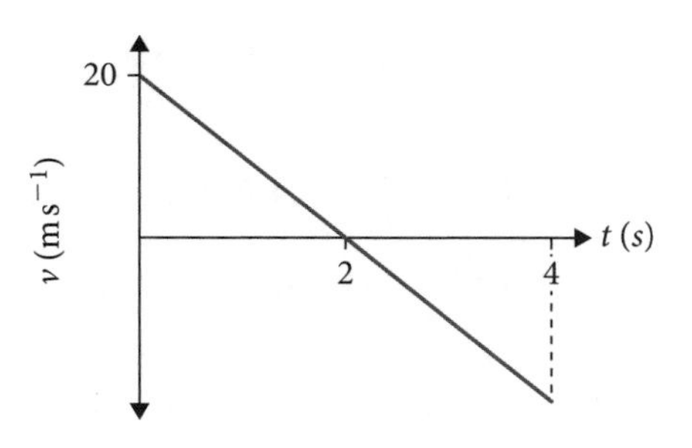

Mark breakdown

- 1 mark: graph sketched with v and t axes labelled and a straight line drawn
- 2 marks: graph sketched with v and t axes labelled and a straight line drawn with at least one quantitative feature correctly indicated
- 3 marks: graph complete with most of listed features
- 4 marks: graph complete with all listed features

36 a The lowest speed of the arrow will occur when the arrow is at the maximum height and v_y is then zero. At that point the total speed will be equal to v_x.

$$v_x = u_x = u\cos\theta$$
$$= 75 \times \cos 20°$$
$$= 70.476\,946\,56$$

Thus $v_x = 70\,\text{m s}^{-1}$ (two sig. fig.).

Mark breakdown

- 1 mark: response either correctly describes the point at which speed is lowest or determines correct value
- 2 marks: response correctly describes the point at which speed is lowest and determines correct value

b $u_x = u\cos\theta = 75 \times \cos 20\,\text{m s}^{-1}$

$s_y = -1.5\,\text{m}$

$\theta = 20°$

Therefore, $u_y = u\sin\theta = 75 \times \sin 20\,\text{m s}^{-1}$

$a = -9.8\,\text{m s}^{-2}$

$s_x = ?$

Hint

In these situations t can be found using $s_y = u_y t + \frac{1}{2}at^2$ and the quadratic formula, or by using $v_y^2 = u_y^2 + 2as_y$ and $v_y = u_y + 2at$ depending on student preference.

Choose equation from data to determine t.

Substitute into $s_y = u_y t + \frac{1}{2}at^2$.

$$-1.5 = 75 \times \sin 20° \times t + \frac{1}{2} \times -9.8 \times t^2$$

Solving gives $t = 5.292\,839\,289$ s.

Now substitute t into equation to find range. $s_x = u_x t$

$s_x = 75 \times \cos 20° \times 5.292\,839\,289$

Calculating $s_x = 373.023\,151\,7$ m

so $s_x = 370$ m (two sig. fig.)

Mark breakdown

- 1 mark: response uses one of the equations of motion to attempt to determine an unknown feature of the arrow's motion
- 2 marks: response attempts to determine time of flight and subsequently calculate range with an error or oversight
- 3 marks: response correctly determines time of flight and subsequently calculates range

37 Gradient $= \frac{\text{rise}}{\text{run}} = \frac{1.4 - 0.56}{2.6 - 1.0} = 0.525$

Gradient $= \frac{s_x}{u_x} = t$ (from $s_x = u_x t$)

So time for object to fall = 0.525 s.

Using $s_y = u_y t + \frac{1}{2}at^2$ for $u_y = 0$:

s_y = height of the table $= \frac{1}{2} \times -9.8 \times 0.525^2 = 1.35\,\text{m}$
$= 1.4\,\text{m}$ (two sig. fig.)

(Responses will vary somewhat around this value depending on subjectivity of line of best fit.)

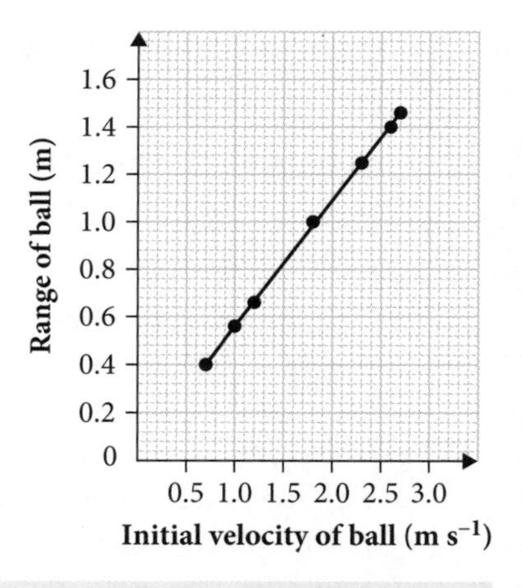

Mark breakdown

- 1 mark: response plots data on graph appropriately
- 2 marks: response plots data on graph appropriately and attempts to use t in an equation of motion to determine s_y
- 3 marks: response correctly determines time of flight and subsequently calculates table height with minor omission or error, such as failing to convert cm to m or inverting gradient
- 4 marks: response correctly uses graph to determine time of flight and subsequently calculates correct table height (within reason)

38 ©NESA 2014 MARKING GUIDELINES SIB Q30B (ADAPTED) For ball P:

As it falls from the maximum height $u = 0$.

Therefore, $s_y = \frac{1}{2}at^2$.

When $t = 1$ s, $s_y = \frac{1}{2}a$ and $s_y = 3$ squares, so $\frac{1}{2}a = 3$ squares.

When $t = 2$ s, $s_y = \frac{1}{2}a \times 2^2$ and $\frac{1}{2}a = 3$ squares, so $s_y = 3 \times 4 = 12$ squares.

When $t = 3$ s, $s_y = \frac{1}{2}a \times 3^2$ and $\frac{1}{2}a = 3$ squares, so $s_y = 3 \times 9 = 27$ squares.

For ball Q:

All vertical displacements will be same as for ball P and horizontal displacement will change 3 squares per second at a constant rate.

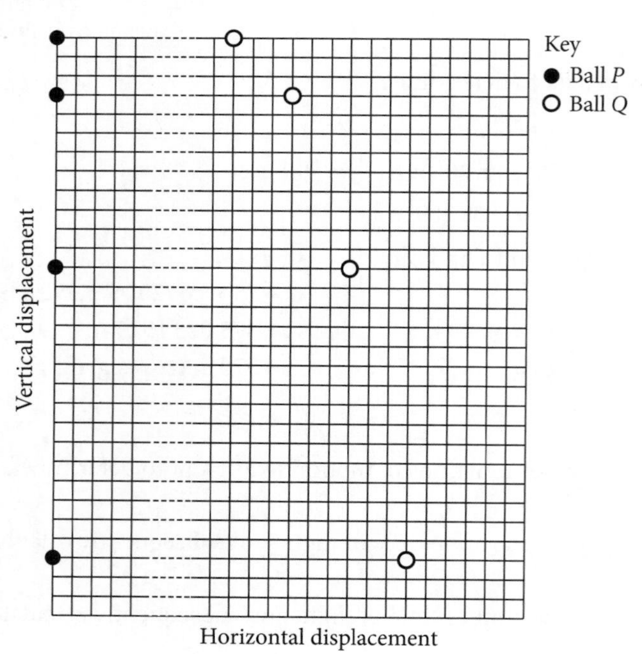

Mark breakdown

- 1 mark: response correctly identifies one feature of the flight of a cannonball
- 2 marks: response shows and/or justifies some features of trajectories
- 3 marks: response shows and/or justifies most features of trajectories
- 4 marks: response correctly shows and justifies with appropriate calculations all features of trajectories

39 $s_y = 34\,\text{m}$

$s_x = 45\,\text{m}$

$a = -9.8\,\text{m}\,\text{s}^{-2}$

$u = ?\,\text{m}\,\text{s}^{-1}$

$\theta = 60°$

Therefore, $u_y = u\sin\theta = u\sin 60°\,\text{m}\,\text{s}^{-1}$ and $u_x = u\cos\theta = u\cos 60°\,\text{m}\,\text{s}^{-1}$.

Choose equation from data. $s_y = u_y t + \frac{1}{2}at^2$

Substitute known values. $34 = u\sin 60t + \frac{1}{2} \times -9.8 \times t^2$ (1)

And $s_x = u_x t$

Substitute known values. $45 = u\cos 60° \times t = \frac{1}{2}ut$

Therefore $t = \frac{90}{u}$ (2)

These two equations need to be solved simultaneously (easiest by substituting t into (1)).

$$34 = u\sin 60° \times \frac{90}{u} - 4.9\left(\frac{90}{u}\right)^2$$

$$34 - 90 \times \sin 60° = -\frac{39\,690}{u^2}$$

$$u^2 = 903.23$$

$u = 30\,\text{m}\,\text{s}^{-1}$ (two sig. fig.)

Mark breakdown

- 1 mark: response correctly identifies one feature of trajectories
- 2 marks: response shows and/or justifies some features of trajectories
- 3 marks: response shows and/or justifies most features of trajectories
- 4 marks: response correctly shows and justifies with appropriate calculations all features of trajectories

40 Any response similar to the one below would be suitable.

Gravitational force: a satellite moves in uniform circular motion (UCM) as a consequence of the force of gravitational attraction due the orbited body providing the centripetal force.

Friction: a bicycle is able to move in a circle around a flat roundabout as friction between the tyres and road provides the centripetal force needed for UCM.

Tension: an ice-skater could move in a circle holding a rope connected to a fixed position at the centre of the circle, where the tension in the rope provides the centripetal force needed for UCM.

Electrostatic force: an electron is able to orbit a nucleus since the electrostatic attraction between oppositely charged objects provides the centripetal force needed for UCM.

Magnetic force: a charged particle moving in a magnetic field experiences a force perpendicular to its direction of motion, which acts as a centripetal force, causing the charged particle to move in a circular path.

Normal force: a person on the 'rotor' at an amusement park is able to move in UCM as the normal force of the wall acting on their back provides the necessary centripetal force.

Mark breakdown

- 1 mark: response correctly describes *one* centripetal force example
- 2 marks: response correctly describes *two* different centripetal force examples

41 $\tau = rF\sin\theta$; therefore, increasing the magnitude of any of the three factors will increase torque.

Angle: the student could arrange the spanner so that the angle between the tension force in the cable and the line joining the point of attachment and the centre of the nut (the spanner) is closer to 90°. This is because the maximum value of $\sin\theta$ occurs when $\theta = 90°$ and τ is proportional to $\sin\theta$.

Force: by adding mass, the tension force will increase as force due to gravity increases, since $T = mg$ and τ is proportional to T.

Radius: by moving the attachment point further from the centre of the nut, the radius will increase and τ is proportional to r.

Mark breakdown

- 1 mark: response correctly describes and justifies *one* factor increasing torque, using the equation and/or proportionality, *or* response correctly describes *three* factors increasing torque
- 2 marks: response correctly describes and justifies *two* factors increasing torque, using the equation and/or proportionality
- 3 marks: response correctly describes and justifies *three* factors increasing torque, using the equation and/or proportionality

42 The greatest speed will be the speed at which the force due to gravity acting on the skier is sufficient to provide the centripetal force.

$$F_c = F_g$$

$$\frac{mv^2}{r} = mg$$

$$v^2 = rg$$

Therefore, $v = \sqrt{10 \times 9.8} = 9.89949 = 10\,\text{m s}^{-1}$ (two sig. fig.).

Mark breakdown

- 1 mark: response attempts to equate F_c with F_g
- 2 marks: response correctly equates F_c with F_g to get correct answer

43 a The force of friction is given by $F_f = \mu N$, where N is the normal force, which is equal and opposite to the weight force (mg).

Consequently, the force of friction that provides the centripetal force is $F_f = F_c = \frac{mv^2}{r}$.

So $\mu mg = \frac{mv^2}{r}$.

Cancelling m from each side gives $\mu gr = v^2$.

$$0.9 \times 9.8 \times 200 = v^2$$

$$v_{max} = 42\,\text{m s}^{-1}$$

This is a maximum speed, since at greater speeds more force would be required – which cannot be provided by friction.

Mark breakdown

- 1 mark: response attempts to determine speed, using frictional coefficient and centripetal force
- 2 marks: response uses equations for frictional force and centripetal force to arrive at speed value with one minor error or omission
- 3 marks: response correctly uses equations for frictional force and centripetal force to arrive at correct speed value

b On a frictionless surface, the centripetal force must be provided by the normal force. As shown in the solution to Question 22, the equation governing a banked corner is $\tan\theta = \frac{v^2}{rg}$.

So $\tan\theta = \frac{20^2}{200 \times 9.8}$

Thus $\theta = 12°$ to the nearest degree.

Mark breakdown
- 1 mark: response attempts to use normal force to determine a value for the angle
- 2 marks: response arrives at correct value without derivation *or* uses derivation with a minor error or omission
- 3 marks: response correctly derives equation relating angle to speed and uses the equation to arrive at correct value

44 a

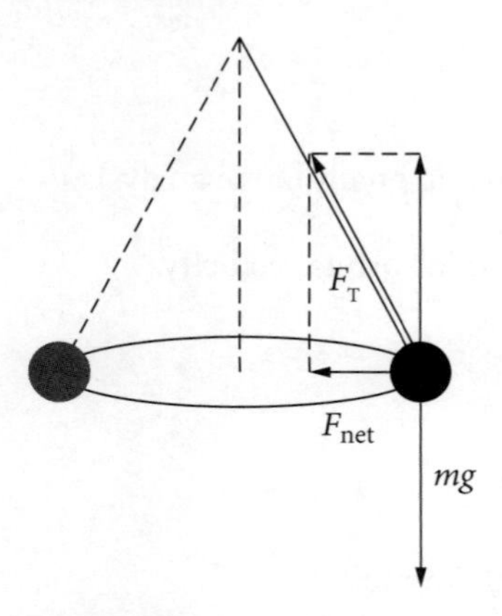

Mark breakdown
- 1 mark: diagram features some relevant force vector arrow
- 2 marks: diagram is constructed with each of three forces – tension, weight force (mg) and net force – with some errors or omissions, such as labelling or arrow direction
- 3 marks: diagram is correctly constructed and appropriately labelled with features of each of three forces – tension, weight force (mg) and net force

b Use the forces annotated above to create a vector diagram that can be used to solve the problem.

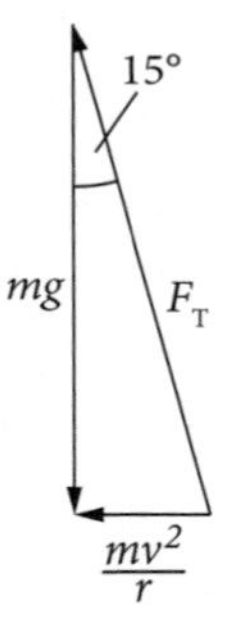

Since $\cos 15° = \frac{mg}{F_T}$, then $F_T = \frac{mg}{\cos 15°}$.

Hence $F_T = \frac{490}{\cos 15°} = 507.285\,328\,4$

$F_T = 510\,\text{N}$ (two sig. fig.)

Now $\tan 15° = \frac{mv^2}{r \times mg} = \frac{v^2}{rg} = \frac{v^2}{58.8}$

So $v = \sqrt{58.8 \tan 15°}$

Since $v = \frac{2\pi r}{T}$, period $T = \frac{2\pi r}{v} = \frac{2\pi r}{\sqrt{58.8 \tan 15°}}$.

$T = 9.497\,651\,492$

$T = 9.5\,\text{s}$ (two sig. fig.)

Mark breakdown
- 1 mark: response attempts to calculate tension *or* period with significant errors or omissions
- 2 marks: response determines values for tension *or* period, using working from relevant vector diagram *or* attempts to calculate tension and period with significant errors or omissions
- 3 marks: response determines values for tension and period, using working from relevant vector diagram with some minor error or omission
- 4 marks: response correctly determines values for tension and period, using full working from relevant vector diagram

45 a Weight is a function of gravitational field strength. When the HST is far from Earth, where the gravitational field is weaker, its weight will be less.

From the equation $g = \frac{GM}{r^2}$, it can be seen that increasing the orbital radius from 6371 km (radius of Earth) to 6911 km (orbital radius of HST) will result in a proportional reduction of gravitational field strength (equivalent to a factor of $\frac{6371^2}{6911^2}$).

Mark breakdown

- 1 mark: response attempts to justify difference in weight
- 2 marks: response is able to validly justify the difference in weight, using quantitative analysis

b The derivation starts with equating F_c with F_g to yield an equation for orbital velocity.

$$\frac{mv^2}{r} = \frac{GMm}{r^2}$$

$$v = \frac{\sqrt{GM}}{r}$$

Substitute known values. $v = \sqrt{\frac{6.67 \times 10^{-11} \times 6.0 \times 10^{24}}{6911 \times 10^3}} = 7609.709811$

Thus $v = 7.6 \times 10^3\,\text{m s}^{-1}$ (two sig. fig.)

Mark breakdown

- 1 mark: response attempts to calculate orbital velocity with some errors or omissions
- 2 marks: response correctly determines orbital velocity

c At a lower orbit, it will have a greater orbital velocity (from the equation above) but will experience more negative acceleration (due to frictional forces experienced as a result of atmospheric drag).

Mark breakdown

- 1 mark: response describes change to one feature of motion
- 2 marks: response describes changes to two features of motion *or* explains change to one feature of motion
- 3 marks: response explains change to two features of motion

46 Since it can be shown that the kinetic energy of an object in a stable orbit is given by $K = \frac{GmM}{2r}$, then rearranging gives $r = \frac{GmM}{2K} = \frac{6.67 \times 10^{-11} \times 420 \times 10^3 \times 6.0 \times 10^{24}}{2(1.23 \times 10^7 \times 10^6)} = 6\,832\,682.927\,\text{m}$.

So, by subtracting the radius of Earth, the altitude becomes 460 km (two sig. fig.).

Mark breakdown

- 1 mark: response attempts to calculate orbital radius with some valid equation
- 2 marks: response determines orbital radius, using working from relevant rearranged equation
- 3 marks: response determines altitude, using working from relevant rearranged equation to determine orbital radius

47 $v_{esc} = \sqrt{\frac{2GM}{r}}$ so if r is decreased, the escape velocity will increase.

Mark breakdown

- 1 mark: response attempts to relate escape velocity and planet radius with errors or omissions.
- 2 marks: response makes clear relationship between escape velocity and planet radius.

48 a Model X assumes that the strength of Earth's gravitational field is constant, whereas model Y assumes that it varies as distance from centre of Earth changes.

Mark breakdown
- 1 mark: response includes some relevant information
- 2 marks: response correctly identifies assumptions made for both models

b Since the radius of orbit is very close to the radius of Earth, the gravitational field strength given by $g = \frac{GM}{r^2}$ is only marginally different in the two cases.

Mark breakdown
- 1 mark: response correctly identifies that the two radii are very similar in magnitude

c $$v = \sqrt{\frac{GM}{r}}$$

Substitute known values. $$v = \sqrt{\frac{6.67 \times 10^{-11} \times 6.0 \times 10^{24}}{6.58 \times 10^{6}}}$$

Thus $v = 7797\ \text{m s}^{-1} = 7.8 \times 10^{3}\ \text{m s}^{-1}$ (two sig. fig.)

Mark breakdown
- 1 mark: response attempts to calculate some relevant feature of orbital motion
- 2 marks: response determines value for orbital velocity with minor errors or omissions
- 3 marks: response determines value for orbital velocity

49 The mass that is released from Tower A above the pole will accelerate towards the centre of Earth under the influence of Earth's gravitational field. As the field gets stronger closer to the surface, the rate of acceleration will gradually increase. At some point the mass will start to encounter atmospheric gases, which will exert a frictional force, with the effect of limiting the acceleration towards Earth.

Before the mass is released from Tower B above the equator, it is in a position and moving with an orbital period that corresponds effectively with the characteristics of an object in a geostationary orbit, as it has an orbital period equal to 24 hours. When it is released, it will remain in orbit at the same point as released relative to the tower.

Mark breakdown
- 1 mark: response makes some valid statement about motion of one mass
- 2 marks: response describes motion of both masses correctly or explains motion of one mass correctly
- 3 marks: response explains motion of both masses with some minor error or omission
- 4 marks: response correctly explains motion of both masses

50 Since planet A and planet B both orbit Regulus, Kepler's third law can be used to state that:

$$\frac{r_A^3}{T_A^2} = \frac{r_B^3}{T_B^2}$$

Rearranging, $\frac{r_A^3}{r_B^3} = \frac{T_A^2}{T_B^2}$ and from information in question, $r_B = 4r_A$.

Substituting: $$\frac{r_A^3}{(4r_A)^3} = \frac{T_A^2}{T_B^2}$$

$$\therefore \frac{r_A^3}{64r_A^3} = \frac{T_A^2}{T_B^2}$$

$$\therefore \frac{1}{64} = \frac{T_A^2}{T_B^2}$$

Taking square root of both sides: $$\frac{1}{8} = \frac{T_A}{T_B}$$

Ratio of period $T_A : T_B = 1:8$.

SOLUTIONS

Mark breakdown

- 1 mark: attempts to use Kepler's law to solve problem
- 2 marks: uses Kepler's law in a valid fashion with calculation error
- 3 marks: uses Kepler's law correctly to calculate a correct value

CHAPTER 2 MODULE 6 ELECTROMAGNETISM

Multiple-choice solutions

1 B

Electric field lines indicate the direction of a force on a positively charged particle; hence, an electron in the field will move in a direction against the field lines. **A** and **C** are incorrect.

The strength of the electric field will determine the force and thus acceleration of the electron ($F = qE$ and $F = ma$, so $a = \frac{qE}{m}$). In an electric field diagram, closer field lines indicate a stronger field. As the electron is moving from a region of more widely spaced lines to one of less widely spaced lines, the field experienced by the electron over time is stronger and its acceleration will be increasing.

2 A

The particle is at rest if electrostatic and gravitational forces acting on it are balanced ($F_g = F_E$ so $mg = qE$).

The negative particle will accelerate downwards if the upwards force decreases since the downwards force of gravity on the particle cannot be changed.

In order to decrease the upwards force on the particle, the electric field strength or the charge on the particle must be reduced.

The length of the plates plays no role in electric field strength. **C** is incorrect.

Both increasing the potential difference between the plates and decreasing the separation will increase the field strength since $E = \frac{V}{d}$. Hence **B** and **D** are incorrect.

3 C

Electric field strength due to two charges is a vector sum of the fields due to the individual charges. The electric field strength due to an individual charge is given by $E = \frac{kq}{r^2}$ ($k = \frac{1}{4\pi\varepsilon_0}$). The units of cm can be ignored as relative strength is required.

At X: $E = \frac{kq}{2^2}$ left due to $+q$ plus $E = \frac{kq}{6^2}$ right due to $-q$. $E_{total} = \frac{2kq}{9}$ left

At Y: $E = \frac{kq}{1^2}$ left due to $+q$ plus $E = \frac{kq}{5^2}$ right due to $-q$. $E_{total} = \frac{24kq}{25}$ left

At Z: $E = \frac{kq}{1^2}$ right due to $+q$ plus $E = \frac{kq}{3^2}$ right due to $-q$. $E_{total} = \frac{10kq}{9}$ left

4 D

By equating the magnetic force acting on the proton with the centripetal force that it provides, $F_c = F_B$ and thus $\frac{mv^2}{r} = qvB$, it can be shown that $r = \frac{mv}{qB}$.

Doubling the strength of the magnetic field will halve the radius; halving the velocity will halve the radius again.

Consequently, the radius will be a quarter of r.

5 B

With currents travelling in the same direction and using the right-hand grip rule, it can be seen that the region between the wires will feature fields in opposite directions. This will result in a weakening of fields, which is represented by field lines that are more widely spaced, so **C** is incorrect.

Magnetic field lines cannot cross. **A** is an incorrect representation.

Magnetic field lines surround a current-carrying conductor but do not emerge from one. **D** is an incorrect representation.

6 C

From the derivation in the solution to Question 4, it can be seen that radius is proportional to velocity and, hence, the spiralling inwards indicates a slowing of the charges so **B** and **D** are incorrect.

Using the right-hand push rule, the initial upwards force on the positive charge is created by a field directed into the page. **A** and **B** are incorrect.

7 D

Using the right-hand grip rule, the magnetic field surrounding the current-carrying wire can be shown to be into the page on the left side of the wire and out of the page on the right side. Using the right-hand push rule, the initial force on each charge created by the field surrounding the wire can be determined.

Both charges on the left side will experience a force down the page.

The positive charge on the right side of the wire will be forced away from the wire initially and thus the negative charge on the right side will be forced towards the wire.

8 D

Using the right-hand push rule, it can be seen force is not in the plane with B and I and that it is possible to change the direction of B and I with no effect on the direction of the force. **A**, **B** and **C** are incorrect.

9 A

Using the right-hand push rule, it can be shown that the bar will move to the right.

10 B

Using the right-hand push rule, it can be shown that the arrangement that will result in a force up the page is with current coming out of the page. Therefore **A** is incorrect.

There is no force if the current flows parallel to the magnetic field. **C** and **D** are incorrect.

11 D

The relationship for the force between wires carrying parallel currents is $\frac{F}{l} = \frac{\mu_0 I_1 I_2}{2\pi r}$, so increasing the force by a factor of 8 by altering the values of I and r can be achieved by doubling both currents and halving r.

In order to make the wires attract, the currents must flow in the same direction. Since the wires repel, the currents must be flowing in opposite directions. Reversing the direction of current in one wire will cause them to attract. **A** is incorrect.

Halving the length of the wire cannot result in an increase in force. **C** is incorrect.

9780170465304

12 A

In each of the four 'leaves' of the wire loop, the current on one side is in the opposite direction to the current on the other side. Since unlike currents repel, the two sides of each 'leaf' will repel. Since the wire is flexible, it will be likely to change shape (**C** is incorrect) and that change will be due to the repulsion. **B** is incorrect.

D is not feasible from the situation illustrated and consequently a circle is the most likely shape.

13 A

The force experienced by the electron will be described by the equation $F = qvB\sin\theta$.

Substituting values of $\theta = 90°$, $v = 5.4\,\text{m}\,\text{s}^{-1}$, $B = 5.2 \times 10^{-3}\,\text{T}$ and $q = 1.602 \times 10^{-19}\,\text{C}$ yields a force value of $4.5 \times 10^{-21}\,\text{N}$. Using the right-hand push rule, it can be seen that a positive charge would be forced from X towards Y and so an electron will experience a force towards X.

14 A

The force experienced by the wire will be described by the equation $F = BlI\sin\theta$.

Substituting values of $\theta = 90°$, $F = 40 \times 10^{-3}\,\text{N}$, $B = 200 \times 10^{-3}\,\text{T}$ and $l = 0.1\,\text{m}$ yields a current value of 2.0 A. Using the right-hand push rule, it can be seen that an upwards force will result from a current that flows from P to Q.

15 D

There are two ways to consider this question. First, as the field is uniform and magnetic flux is represented by field lines, more field lines thread the coil in position D than in any other position shown. Alternatively, magnetic flux is a function of the area of the coil (A), the magnetic field strength (B) and the angle (θ) made by the normal to the plane of the coil with the magnetic field, as seen in the equation $\Phi = BA\cos\theta$. Hence the closer the normal to the plane of the coil is to the magnetic field lines (θ approaches 0), the greater the flux.

16 A

Using the equation $\Phi = BA\cos\theta$, where $B = 0.015\,\text{T}$, $A = (0.04)^2$ and $\theta = 30°$ (the normal is at right angles to the plane of the coil and θ will be the complementary angle to the angle marked 60°), yields a value of $2.078\,46 \times 10^{-5}\,\text{Wb}$. Accounting for the three loops means multiplying that value by three to give $6.235\,382 \times 10^{-5}\,\text{Wb} = 6.2 \times 10^{-1}\,\text{Wb}$ (to two sig. fig.).

17 D

Change in flux ($\Delta\Phi$) will be given by final flux (Φ_{final}) – initial flux (Φ_{initial}). Since the flux in the initial position is zero (no field passes through the coil and $\theta = 90°$), it can be said that $\Delta\Phi = \Phi_{\text{final}}$.

In the position after a rotation of 90°, $\theta = 0°$.

Using the equation $\Phi = BA\cos\theta$, where $B = 0.2\,\text{T}$, $A = (0.36)^2$ and $\theta = 0°$, a value of $\Phi = 0.025\,92\,\text{Wb}$ is obtained.

18 D

A step-up transformer is designed to give a greater output voltage than input voltage by having more secondary coils than primary coils. Conversely a step-down transformer is designed to give a smaller output voltage than input voltage by having fewer secondary coils than primary coils.

The diagram shows a step-down transformer with approximately half as many secondary coils as primary coils. The output voltage would be approximately half the input voltage as a consequence, according to the equation $\frac{N_p}{N_s} = \frac{V_p}{V_s}$.

19 D

According to the equation $\frac{N_\text{p}}{N_\text{s}} = \frac{V_\text{p}}{V_\text{s}}$, the secondary voltage will be 5 times the primary voltage, 50 V. **A** and **C** are incorrect.

The emf in the secondary coil will be related to the rate of change of flux in the primary coil, so the emf in the secondary coil is greatest when the flux of the primary coil is changing at the greatest rate.

The flux of the primary coil will change at the greatest rate when the primary voltage is changing at the greatest rate; that is, when the primary V vs t graph has the greatest gradient. This means that the secondary emf will be 90° out of phase with the primary emf.

20 A

By considering a single positive charge in the conducting rod and using the right-hand push rule, it can be seen that positive charges will experience a force towards the top end of the rod. **B** and **D** are incorrect.

There will be a very brief movement of charge when the rod starts moving, but very soon a build-up of positive charge at the top and negative charge at the bottom will prevent any further movement of charge. Since there is no further movement of charge, there cannot be any sustained force on the conductor in the magnetic field. **C** is incorrect.

21 C

There are two ways to solve this question.

Method 1 is to consider a positive charge at both the top and bottom of the ring during the motion.

Position 1: The right-hand push rule shows that a positive charge at the bottom of the ring experiences a force to the left while the positive charge at the top of the ring experiences no force as it has yet to enter the field. This results in a clockwise current through the ring.

Position 2: The right-hand push rule shows that positive charges at the bottom and the top of the ring experience a force to the left. These opposing forces result in no current through the ring.

Position 3: The right-hand push rule shows that a positive charge at the top of the ring experiences a force to the left while the positive charge at the bottom of the ring experiences no force as it has now left the field. This results in an anticlockwise current through the ring.

Method 2 is to use Lenz's law to determine the direction of the induced magnetic field and thus the induced current in the ring.

Position 1: The ring experiences an increase in magnetic flux out of the page. A current flows in a direction to create a magnetic field that opposes this change in flux. The right-hand grip rule shows that this requires a clockwise current to flow around the conducting ring.

Position 2: There is no change in flux in the area enclosed by the ring and so no emf and no current.

Position 3: The ring experiences a decrease in magnetic flux out of the page. A current flows in a direction to create a magnetic field that opposes this change in flux. The right-hand grip rule shows that this requires an anticlockwise current to flow around the conducting ring.

SOLUTIONS

22 C

There are two ways to solve this question.

Method 1 is to consider a positive charge at each of the side of the ring during the motion as the magnet moves from above (call it position 1) to below (call it position 2) the ring. In this method, it is best to consider the magnetic field to be radial in shape (like the spokes of a bicycle wheel) as it intersects with the conducting ring.

Position 1: A positive charge on the right side of the ring is moving upwards relative to the magnet in a magnetic field pointing right. The right-hand push rule shows that this charge experiences a force towards the back of the ring. A positive charge on the left side of the ring is moving upwards relative to the magnet in a magnetic field pointing left. The right-hand push rule shows that this charge experiences a force towards the front of the ring. Hence, an anticlockwise current flows in the ring.

Position 2: A positive charge on the right side of the ring is moving upwards relative to the magnet in a magnetic field pointing left. The right-hand push rule shows that this charge experiences a force towards the front of the ring. A positive charge on the left side of the ring is moving upwards relative to the magnet in a magnetic field pointing right. The right-hand push rule shows that this charge experiences a force towards the back of the ring. Hence, a clockwise current flows in the ring.

The current changes direction, so **A** and **B** are both incorrect.

The current that opposes the change in flux will be minimal when the magnet is furthest and will be greatest when the magnet is closest and the flux is greatest (**C**).

Method 2 is to use Lenz's law to determine the direction of the induced magnetic field and thus the induced current in the ring as the magnet moves from above (call it position 1) to below (call it position 2) the ring.

Position 1: The ring experiences an increase in magnetic flux as a north pole approaches. A current flows in a direction to create a magnetic field that opposes this change in flux. The right-hand grip rule shows that this requires an anticlockwise current to flow around the conducting ring.

Position 2: The ring experiences a decrease in magnetic flux as the south pole moves away. A current flows in a direction to create a magnetic field that opposes this change in flux. The right-hand grip rule shows that this requires a clockwise current to flow around the conducting ring.

The current changes direction, so **A** and **B** are both incorrect.

The current that opposes the change in flux will be minimal when the magnet is furthest and will be greatest when the magnet is closest and the flux is greatest.

23 C

The magnitude of torque in a motor is determined by the equation $\tau = nIAB\sin\theta$, where θ is defined as the angle between the normal to the plane of the coil and the magnetic field lines. Since $\sin\theta$ is a maximum when $\theta = 90°$, this will occur when the plane of the coil is parallel to the field lines.

24 B

Doubling the number of coils will double the output amplitude but not change the curve in any other way.

Rotating the generator at double the speed will double the amplitude of the curve but would also halve the length of each complete cycle. **A** is incorrect.

Adding a set of slip rings without any other changes will be unlikely to cause any change. **C** is incorrect.

Including another set of magnets perpendicular to the original magnets will change the shape of the output as well as increasing the output. **D** is incorrect.

9780170465304

25 B

The split-ring commutator is a ring that is split (not surprisingly), which corresponds with the part labelled P.

The brush is the part that maintains electrical contact with the split-ring commutator as it spins. This part is labelled Q.

Any part that is stationary comprises the stator – of the labelled parts R is the appropriate choice.

The armature is the rotating structure of a motor or generator, including wire coils. It is labelled T.

26 A

It has a handle and no power source, so it must be a generator.

The slip rings rather than split-ring commutator are indicative of the AC nature of the device.

27 A

The force F_m will be determined by the relationship described by the equation $F = BIl \sin\theta$. In this motor, and motors generally, B, I and l are constant. Since θ remains 90° (the wire is always perpendicular to the magnet), then $F = BIl$ and the magnitude of F_m is unchanged throughout its rotation.

28 B

The torque experienced by the motor coil as a consequence of the magnetic field is exactly equal and opposite to the torque the coil experiences as a consequence of the force of gravity acting on the suspended mass.

From the equation $\tau = rF \sin\theta$, where θ is the angle between the pivot arm and the applied force, it can be determined that the torque is given by $\tau = 0.125 \times 1.5 \times 9.8 \times \sin 140° = 1.18\,\text{N m}$.

29 B

By focusing on the current through just one part of the wire in the end view at a time, a direction of the motor effect force can be determined using the right-hand push rule.

For example, start with the uppermost section of wire with a current going left in a magnetic field going down the page. The resulting force is out of the page.

This process can be repeated with the same result on the bottom, right- and left-hand sides.

So the direction of the force on the coil in the side view when the current flows would be towards the right.

Short-answer solutions

30 a Just as masses have energy as a consequence of their position in a gravitational field, charged particles have energy as a consequence of their position in an electric field. This energy is electrical potential energy (analogous to gravitational potential energy). Because the particle is free to move, the potential energy is converted to kinetic energy.

Mark breakdown

- 1 mark: response identifies the energy as electrical potential energy or kinetic energy
- 2 marks: response clearly connects the force exerted by the electric field and the object's position within the field to the electrical potential energy it has or the kinetic energy it gains (or both)

b Because the proton is midway between the two charged objects, the potential difference through which it will move will be half of the total potential difference, thus $V = 6.25 \times 10^3$ V. The energy gained by the proton will be equal to the work done on it by the field:

$$\Delta E = W = qV = 1.602 \times 10^{-19} \times 6.25 \times 10^3 = 1.00 \times 10^{-15}\text{ J (two sig. fig.)}$$

Mark breakdown

- 1 mark: response attempts to use appropriate work equation to determine energy gained
- 2 marks: response effectively uses appropriate work equation and correct data to determine energy gained

c $\Delta E = K_{\text{gained}} = \frac{1}{2}mv^2 = 1.00 \times 10^{-15}$ ($K_{\text{initial}} = 0$)

$$\text{Then } v^2 = \frac{2 \times 1.00 \times 10^{-15}}{1.673 \times 10^{-27}}$$
$$= 1.195\,472\,62 \times 10^{12}$$
$$v = 1.1 \times 10^6\text{ m s}^{-1}\text{ (two sig. fig.)}$$

Mark breakdown

- 1 mark: response attempts to use appropriate equation to determine speed gained
- 2 marks: response effectively uses appropriate equation and correct data to determine energy gained

31 a The electric field between a pair of parallel plates is given by

$$E = \frac{V}{d}$$

The force on a charged particle between charged parallel plates is given by

$$F = qE = \frac{qV}{d}$$

And the work done is given by

$$W = F_{||}s$$

Because the force on each particle is upwards (as it hits the top plate at P) and only the displacement in this direction is caused by the force, the work done on particle 1 will be

$$W = F_{||}s = \left(\frac{q_1 V}{d}\right)\left(\frac{d}{2}\right) = \frac{q_1 V}{2}$$

The work done on particle 2 will be

$$W = F_{||}s = \left(\frac{q_2 V}{2d}\right)\left(\frac{3d}{2}\right) = \frac{3q_2 V}{4}$$

As the charges are identical, the field does 1.5 times more work on q_2 than it does on q_1.

Mark breakdown

- 1 mark: response uses some correct equations to make minor progress towards comparative values
- 2 marks: response correctly manipulates data and equations without a valid comparison made *or* response manipulates data and equations and presents a comparison with a minor error or omission
- 3 marks: response correctly manipulates data and equations and presents a valid comparison

b This question can be answered quantitatively (preferable since quantitative data is provided) or qualitatively as follows.

The horizontal component of the velocity is equal and constant in both cases. Given the upwards acceleration will be $a = \frac{F}{m} = \frac{\left(\frac{qV}{d}\right)}{m} = \frac{qV}{dm}$ and the initial vertical component of the velocity is zero, the time of flight for the first particle can be determined using

$s = ut + \frac{1}{2}at^2$ and hence $t = \sqrt{\frac{2s}{a}} = \sqrt{\frac{2\left(\frac{d}{2}\right)dm}{qV}} = \sqrt{\frac{md^2}{qV}}$.

The horizontal distance travelled by q_1 in this time will be

$$s = vt = v\sqrt{\frac{md^2}{qV}}$$

Similarly, for q_2,

$$a = \frac{F}{m} = \frac{\left(\frac{qV}{2d}\right)}{m} = \frac{qV}{2dm}$$

$$t = \sqrt{\frac{2s}{a}} = \sqrt{\frac{2\left(\frac{3d}{2}\right)2dm}{qV}} = \sqrt{\frac{6md^2}{qV}}$$

$$s = vt = v\sqrt{\frac{6md^2}{qV}}$$

Therefore, particle q_2 will travel a horizontal distance of $(\sqrt{6})s$, where s is the horizontal distance travelled by q_1.

or

The vertical force ($F = \frac{qV}{d}$) and, hence, vertical acceleration of q_2 will be half as great as the vertical acceleration of q_1. As the vertical distance to the top plate is 3 times further for q_2 than for q_1 and as q_2 is accelerating less rapidly, q_2 will be in flight for a longer time than q_1 before it hits the top plate. Because both charges have the same horizontal velocity, q_2 will therefore travel a greater horizontal distance before it hits the top plate than q_1.

Mark breakdown

- 1 mark: response uses some correct equations to make minor progress towards comparative values or expression
- 2 marks: response correctly manipulates data and equations without a valid comparison made *or* response manipulates data and equations and presents a comparison with a minor error or omission
- 3 marks: response correctly manipulates data and equations and presents a valid comparison

32 By equating the magnetic force acting on the proton with the centripetal force that it provides $F_c = F_B$ and thus $\frac{mv^2}{r} = qvB$, it can be shown that $r = \frac{mv}{qB}$.

Since v, q and B are the same in each case:

$$r_1 = \frac{m_1 v}{qB} \text{ and } r_2 = \frac{m_2 v}{qB}$$

The distance d will be the diameter of semicircle 1 – diameter of semicircle 2. So

$$d = 2(r_1 - r_2)$$

thus $d = 2\left(\frac{m_1 v}{qB} - \frac{m_2 v}{qB}\right) = 2\frac{v(m_1 - m_2)}{qB}$.

Mark breakdown

- 1 mark: response equates F_c with F_B but is unable to determine an equation for r that can be used to derive d expression
- 2 marks: response equates F_c with F_B to determine an equation for r that is used to derive d expression with minor error or omission
- 3 marks: response equates F_c with F_B to determine an equation for r that is used correctly to derive d expression

33 As seen in Question 32, by equating the magnetic force acting on the proton with the centripetal force that it provides, $F_c = F_B$ and thus $\frac{mv^2}{r} = qvB$. On this occasion it can be shown that $v = \frac{rqB}{m}$.

As it is in uniform circular motion, $v = \frac{2\pi r}{T}$.

Then $\frac{2\pi r}{T} = \frac{rqB}{m}$

and rearranging yields $T = \frac{2\pi m}{qB}$. The period of the uniform circular motion undergone by the particle is independent of its velocity.

Substituting for the electron:

$$T = \frac{2\pi m}{qB}$$

$$T = \frac{2\pi \times 9.11 \times 10^{-31}}{1.602 \times 10^{-19} \times 1.45 \times 10^{-3}}$$

$T = 2.464 \times 10^{-8}$ s for a completed circle.

$t = 1.23 \times 10^{-8}$ s (three sig. fig.) to complete a semicircle.

Mark breakdown

- 1 mark: response attempts to rearrange from $F_c = F_B$ and $v = \frac{2\pi r}{T}$, but is unable to derive expression or calculate values successfully
- 2 marks: response eliminates velocity from the equation for period to show T independent of r *or* substitutes to get correct time value with minor error or omission
- 3 marks: response can eliminate velocity from the equation for period to show T independent of r *or* successfully substitutes to get correct time value
- 4 marks: response eliminates velocity from the equation for period to show T independent of r and successfully substitutes to get correct time value

34 a Using the right-hand push rule, it can be seen that a force to the right, required to deflect the wire in the direction shown, will result from a current that flows from the back of the page towards the front.

Mark breakdown

- 1 mark: response identifies direction successfully

b An object will be at rest if the forces acting on it are balanced (or are in equilibrium). In this situation, the forces are the magnetic force to the right, the gravitational force down and the tension force in the section of wire connecting the main wire to the top anchor point. The vector sum of these three forces is zero.

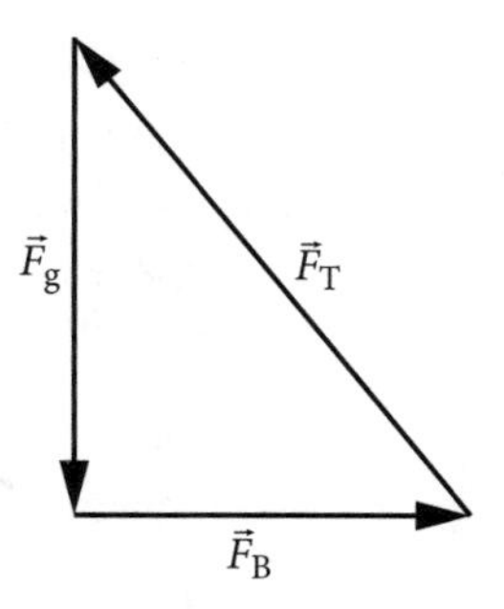

Mark breakdown
- 1 mark: response attempts to describe a situation of balanced forces, including two relevant forces
- 2 marks: response correctly identifies the equilibrium of the specific three forces

c The reduction of the current will result in a smaller magnetic force according to the equation $F = BIl \sin\theta$. The new equilibrium position will be decreasingly deflected from the vertical until it hangs in a vertical position.

The reversal of current direction will result in deflection to the left instead of the right (seen by applying the right-hand push rule) and as the current increases this deflection will increase.

Mark breakdown
- 1 mark: response identifies both changes *or* explains one change successfully
- 2 marks: response correctly identifies and explains both changes that occur

35 a Since the current in adjacent loops will flow in the same direction, there will be a constant force of attraction between adjacent loops as a consequence of the direct current.

Mark breakdown
- 1 mark: response identifies that the force is attractive *or* constant
- 2 marks: response describes that the force is both attractive and constant

b Whenever the alternating current was flowing, the adjacent loops would again be carrying current in the same direction and, hence, a force of attraction would again occur. However, the magnitude of the force would vary between a maximum and zero 100 times per second.

Mark breakdown
- 1 mark: response identifies that force is not constant or is still attractive
- 2 marks: response correctly identifies that force oscillates between maximum attraction and zero

36 a Wire 1 will experience a downwards force due to wire 2 (attraction) and an upwards force due to wire 3 (repulsion). The sum of these force will reveal the total force.

Using the equation $\frac{F}{l} = \frac{\mu_0}{2\pi}\frac{I_1 I_2}{r}$ for wires 1 and 2, then $\frac{F}{l} = \frac{\mu_0}{2\pi}\frac{40 \times 30}{0.1} = 0.0024\,\text{N m}^{-1}$ downwards.

Using the equation $\frac{F}{l} = \frac{\mu_0}{2\pi}\frac{I_1 I_3}{r}$ for wires 1 and 3, then $\frac{F}{l} = \frac{\mu_0}{2\pi}\frac{40 \times 50}{0.2} = 0.0020\,\text{N m}^{-1}$ upwards.

The total force per metre is $0.0024\,\text{N m}^{-1}$ downwards + $0.0020\,\text{N m}^{-1}$ upwards = $0.0004\,\text{N m}^{-1}$ downwards or $4 \times 10^{-4}\,\text{N m}^{-1}$ downwards.

Mark breakdown
- 1 mark: response attempts to determine force on wire 1 due to one of wire 2 or 3
- 2 marks: response determines forces acting on wire 1 due to each of wire 2 and 3 with minor errors or omissions
- 3 marks: response uses forces due to both wires 1 and 2 to determine correct force value and direction

b Consider the wire to be 1 m long. The total upwards force acting on wire 2 due to wire 1 (equal to the value of the force acting down on wire 2 due to wire 1 calculated in part **a**) is 0.0024 N.

In order to be suspended, the downwards force due to gravity must be equal to the upwards force.

Since $F_g = m \times 9.8$, then $m = \frac{0.0024}{9.8} = 0.000\,244\,897\,95\,\text{kg}$.

Therefore, the mass per metre would need to be $0.24\,\text{g}\,\text{m}^{-1}$ in order for the wire to be suspended by the magnetic force.

Mark breakdown

- 1 mark: response attempts to determine force on wire 2 due to wire 1
- 2 marks: response uses force due to wire 1 on wire 2 to determine mass per metre with minor errors or omissions
- 3 marks: response uses force due to wire 1 on wire 2 to determine mass per metre

37 When the current flows in the direction shown in the diagram, the magnetic force is upwards according to the right-hand push rule. This magnetic force is equal and opposite to the force exerted downwards by gravity on the wire rectangle; hence, it is balanced.

When the current is reversed, the magnetic force will now be downwards but of the same magnitude as before. Now there needs to be a 20 g mass on the left side to bring the balance to equilibrium.

Since it is known from the first paragraph that the downwards force due to gravity is equal to the upwards force due to the magnetic field, then each must be equivalent to the force exerted down by a 10 g mass; that is, 0.098 N.

Hence, using $F = nBIl \sin\theta$, where $\theta = 90°$, $F = 0.098\,\text{N}$, $l = 0.40\,\text{m}$, $n = 20$ turns and a current value of 0.50 A, we get a value of $B = 0.0245\,\text{T} = 0.025\,\text{T}$ (two sig. fig.).

Mark breakdown

- 1 mark: response attempts to determine force, using an appropriate relationship
- 2 marks: response uses valid logic and physics to determine unknowns but is unable to complete process to magnetic field strength
- 3 marks: response uses force generated in both directions to calculate a value for electric field strength with minor errors or omissions
- 4 marks: response successfully uses force generated in both directions to calculate the electric field strength

38 The total magnetic field strength at P will be given by summing the magnetic field strength for each wire given by $B = \frac{\mu_0 I}{2\pi r}$.

When both wires are included, $B_{total} = c \times I_M + d \times I_N$ (where c and $d \times I_M$ are constants).

This is of the form $y = mx + b$ and hence B against I_N is the appropriate plot.

Data should be plotted and line of best fit created to give a graph similar to that shown.

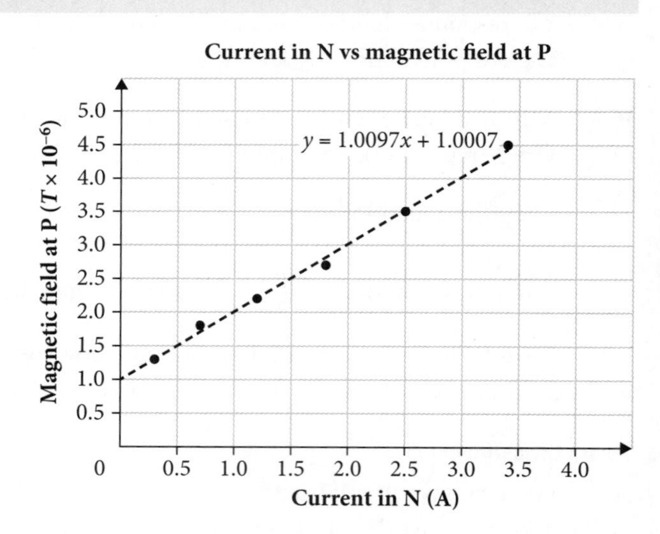

By extrapolating the line of best fit back to the intercept on the vertical axis, it can be shown that when the current in N is zero and the magnetic field is entirely due to the current in M, the field strength is approximately $1 \times 10^{-6}\,\text{T}$. From this it can be determined that the $\frac{I_N}{d}$ must equal 5.

The field gets stronger as the current in N is increased, so both wires are creating a field out of the page at point P and therefore the current in M must be flowing right to left.

The straight line plotted above will fit the linear equation $y = mx + b$, where y is the total magnetic field strength at P, x is the current in wire N and b is the magnetic field strength due to wire M.

By equating $y = mx + b$ with the equation for the total field strength at P, the gradient can be used to find d.

$y = mx + b$ is equivalent to $B_{total} = mI_N + B_M$ and $B = \frac{\mu_0 I}{2\pi r}$

Therefore, $mI_N = \frac{\mu_0 I_N}{2\pi d}$

Hence, gradient $= \frac{\mu_0}{2\pi d}$

Since the gradient of the graph is approximately 1×10^{-6}, a value of $d = 0.2$ m is obtained from the graph. Now since the $\frac{I_N}{d} = 5$ and $d = 0.2$ m, the current in N must be 1 A.

Mark breakdown

- 1 mark: response attains one piece of relevant unknown information
- 2 marks: response uses a feature of the graph to determine distance d *or* magnitude of current in M within reasonable bounds *or* the direction of current in M
- 3 marks: response uses features of the graph to determine two of distance d, magnitude of current in M and direction of current in M reasonably
- 4 marks: response uses gradient and intercept of graph to determine distance d and the direction and magnitude of current in M with minor errors and/or omissions
- 5 marks: response successfully uses gradient and intercept of graph to determine distance d and the direction and magnitude of current in M within reasonable bounds

39 a The secondary voltages are less than the primary voltages, so this is a step-down transformer.

Mark breakdown

- 1 mark: response correctly identifies step-down transformer

b

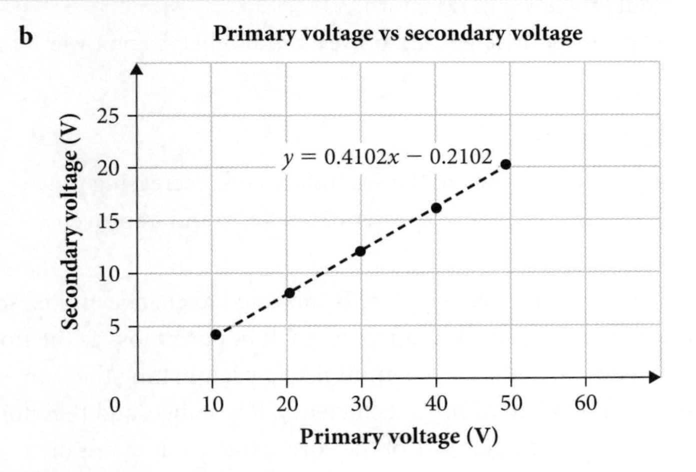

Mark breakdown

- 1 mark: graph has some correct aspect
- 2 marks: graph has key features for a correct answer with minor errors or omissions
- 3 marks: graph has axes correctly labelled and scaled, with primary voltage on x-axis, plotted correctly and with an appropriate straight line of best fit

c Since $\frac{N_p}{N_s} = \frac{V_p}{V_s}$, this rearranges to become $N_s = \frac{N_p V_s}{V_p}$.

From the graph, the gradient of the line of best fit = $\frac{\text{rise}}{\text{run}} = \frac{V_s}{V_p}$.

A value for N_s can be determined by multiplying the gradient by N_p.

$N_s = 0.4102 \times 100$ = approximately 40 turns.

Mark breakdown

- 1 mark: response attempts to determine unknown by some valid step(s)
- 2 marks: response is mostly correct with minor errors or omissions
- 3 marks: response correctly determines a value for gradient, rearranges transformer equation to enable substitution of gradient value and multiplies by 100 to gain approximately 40 turns in secondary coil

d Efficiency = $100 \times \frac{P_{\text{output}}}{P_{\text{input}}} = 100 \times \frac{V_s I_s}{V_p I_p}$

For the first data set, efficiency = $\frac{100 \times 4.2 \times 3.2}{10.5 \times 1.6} = 80\%$.

For the last data set, efficiency = $\frac{100 \times 20.2 \times 13.9}{49.4 \times 8.9} = 64\%$.

Mark breakdown

- 1 mark: response attempts to calculate efficiencies
- 2 marks: response successfully calculates both efficiencies

e It can be seen that the transformer is most efficient when current and voltages are lower and increasingly inefficient as current and voltages increase. In a transformer, energy is transformed into unwanted forms. One of these forms is resistive heating. At higher current values, power loss is greater according to the equation $P_{\text{loss}} = I^2R$.

Mark breakdown

- 1 mark: response makes valid statement about efficiency in transformers
- 2 marks: response shows clear relationship between efficiency and power loss and between power loss and increasing current

f Improving efficiency means decreasing power loss.

Since this step-down transformer carries higher currents in the secondary coil, decreasing the resistance of the secondary coil by making it from wire of a greater diameter would improve efficiency as $P_{\text{loss}} = I^2R$.

The iron core of this transformer is solid and conductive. The core is subjected to change of flux, so eddy currents will be generated in the iron core. These eddy currents result in power loss as the iron core heats due to resistive heating. These eddy currents can be minimised by laminating the iron core. Smaller eddy currents mean less power loss and improved efficiency. Resistance, and therefore resistive heating, can be reduced by lowering the temperature of the core, using cooling fins or circulating cold oil or water over or through the iron core.

Mark breakdown

- 1 mark: response identifies one change that could be made to improve efficiency
- 2 marks: response identifies two changes that could be made to improve efficiency of this transformer or justifies one valid change
- 3 marks: response correctly identifies two changes that could be made to improve efficiency of this transformer and justifies the improvement to efficiency

40 a

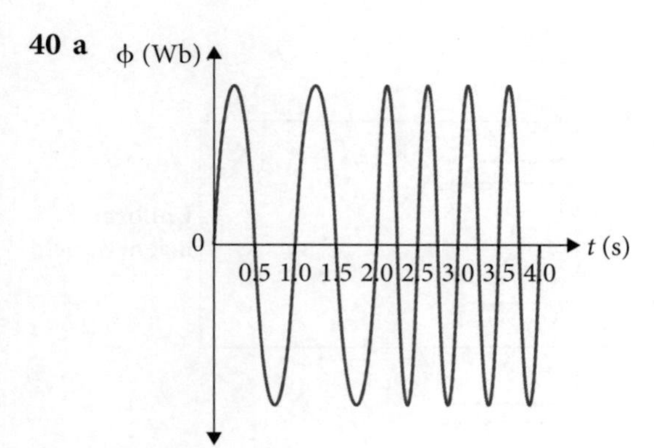

Mark breakdown

- 1 mark: graph features some correct aspect
- 2 marks: response features a graph with correct characteristics but with minor errors or omissions
- 3 marks: graph features correct sinusoidal shape, starting point (at either + or – half maximum amplitude) for first 2 seconds then a continuation of the sinusoidal shape with the same amplitude and double the frequency for the next 2 seconds

b

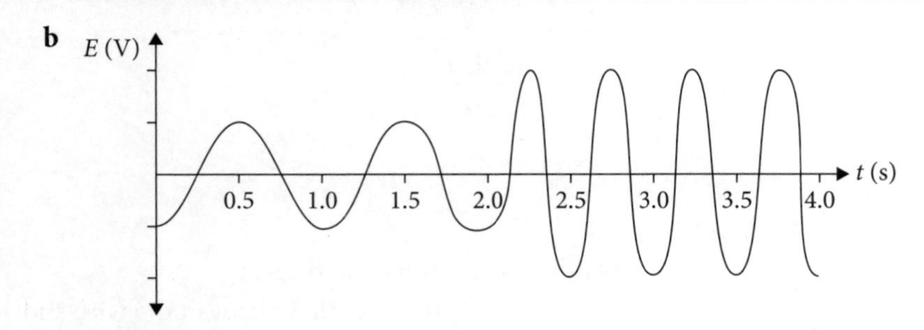

Mark breakdown

- 1 mark: graph features some correct aspect
- 2 marks: response features a graph with correct characteristics but with minor errors or omissions
- 3 marks: graph features correct sinusoidal shape, starting point (at either + or – half maximum amplitude) for first 2 seconds then a continuation of the sinusoidal shape with double amplitude and double frequency for the next 2 seconds

41 I: Entering the field

Similarities: Both the loop and the sheet will experience a change in flux and have an emf induced (Faraday's law). Since both are able to provide complete circuits, currents will flow in each and the direction will be the same. The consequence of this current/these currents will be a magnetic field that will interact with the external field, resulting in a force that opposes the change in flux (Lenz's law). Both will therefore experience a retarding force.

Difference: The currents in the wire loop will flow through the wires but the currents in the sheet will be eddy currents.

II: Within the field

Similarities: Since neither the sheet nor the loop will experience a change in flux when wholly within the field, there will be no emf and no current.

III: Leaving the field (same as entering the field (although current will flow in opposite direction) and may thus be omitted)

Similarities: Both the loop and the sheet will experience a change in flux and have an emf induced (Faraday's law). Since both are able to provide complete circuits, currents will flow in each and the direction will be the same. The consequence of this current/these currents will be a magnetic field that will interact with the external field, resulting in a force that opposes the change in flux (Lenz's law). Both will therefore experience a retarding force.

Difference: The currents in the wire loop will flow through the wire, whereas the currents in the sheet will be eddy currents.

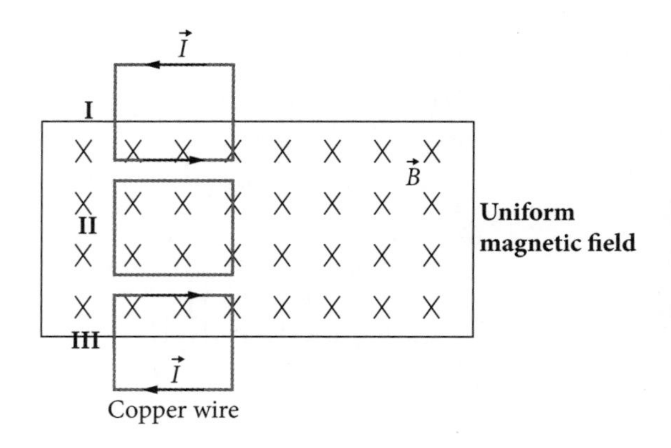

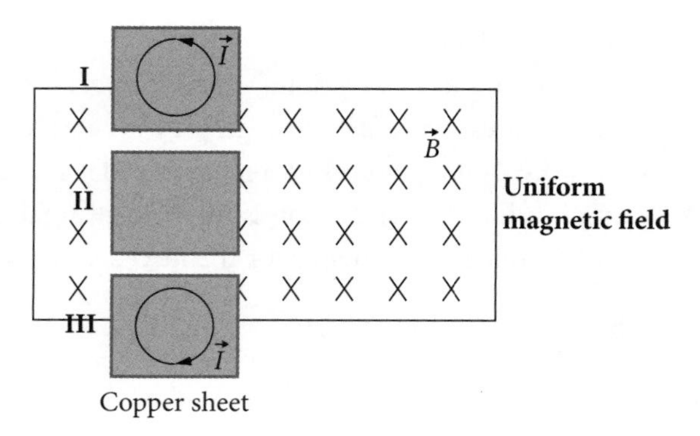

Mark breakdown

- 1 mark: response makes some relevant points
- 2 marks: response gives sound comparison of processes that occur in wire and sheet
- 3 marks: response gives thorough comparison of electromagnetic processes that occur in the wire and sheet supported by an effective diagram, or gives extensive comparison of electromagnetic processes without an effective diagram
- 4 marks: response gives extensive comparison that features both similarities and at least one difference of electromagnetic processes supported by an effective diagram

42 a Using Faraday's law, $\varepsilon = -N\frac{\Delta\Phi}{\Delta t}$, $\Phi = BA$ and $N = 1$ for this situation, where the change in flux is due to the area enclosed by the conducting loop changing, then $\varepsilon = -\frac{B\Delta A}{\Delta t}$.

The area is changing by $0.25 \times 5\,\text{m}^2$ each second.

Therefore, $\varepsilon = -0.04 \times 0.25 \times \frac{5}{1} = 0.05\,\text{V}$.

In a circuit with a $0.25\,\Omega$ resistor, $I = \frac{V}{R} = \frac{\varepsilon}{R} = \frac{0.05}{0.25} = 0.20\,\text{A}$.

From Lenz's law, it is known that the current will flow in a direction so as to create a magnetic field that will oppose the change in flux. This means it will flow in a direction that will result in a force in the opposite direction to the motion. Using the right-hand push rule indicates that the current will flow up the conducting rod and through the circuit in a clockwise direction.

Mark breakdown

- 1 mark: response reasonably attempts to determine current magnitude or direction
- 2 marks: response calculates a value for current and determines current direction with several minor errors or omissions
- 3 marks: response shows full working to calculate a value for current and determines current direction with an error or omission
- 4 marks: response shows full working to correctly calculate a value for current and determines current direction correctly

b Current is determined by factors seen in the equations $\varepsilon = -N\frac{\Delta\Phi}{\Delta t}$, $\Phi = BA$ and $I = \frac{V}{R} = \frac{\varepsilon}{R}$, so an increase in the magnitude of the current can be achieved by any of the following: increasing the magnetic field strength (B); increasing the velocity of the bar, which will increase the rate at which the area enclosed by the conducting loop changes; increasing the area (A) by separating the rails by a greater distance; decreasing the resistance (R) or increasing the number of rods that are used (N).

To reverse the direction of the current, the field direction could be reversed (then the right-hand push rule would show charge moves in the opposite direction through the rod) or move the rod left instead of right (then the right-hand push rule would show charge moves in the opposite direction through the rod).

Mark breakdown

- 1 mark: response states one way the current can be increased or reversed
- 2 marks: response states at least three ways in which the magnitude of current can be increased and both ways in which the current direction can be reversed without an explanation *or* gives reason for one way the current can be increased and one way it can be reversed
- 3 marks: response give reasons for at least two ways in which the magnitude of current can be increased and one way in which the current direction can be reversed
- 4 marks: response give reasons for at least three ways in which the magnitude of current can be increased and both ways in which the current direction can be reversed

43 When the electromagnet is on and the wheel spins past, sections of the wheel that are entering and leaving the magnetic field will experience a change in flux. This change in flux will result in an emf (Faraday's law) and, consequently, eddy currents will flow in the wheel.

These eddy currents will create magnetic fields with directions that will result in forces against the motion of the wheel. As seen in the diagram, this will mean that a magnetic field upwards out of the wheel will be created in the area approaching the external field and a magnetic field downwards into the wheel will be created in the area leaving the external field.

Alternatively: The eddy currents will circulate in a direction such that charges within the field will move towards the axle. It can be considered that these currents create a motor effect force in the opposite direction to the direction of motion. The eddy current is completed by charges moving towards the edge of the disc outside the field, as seen in the diagram.

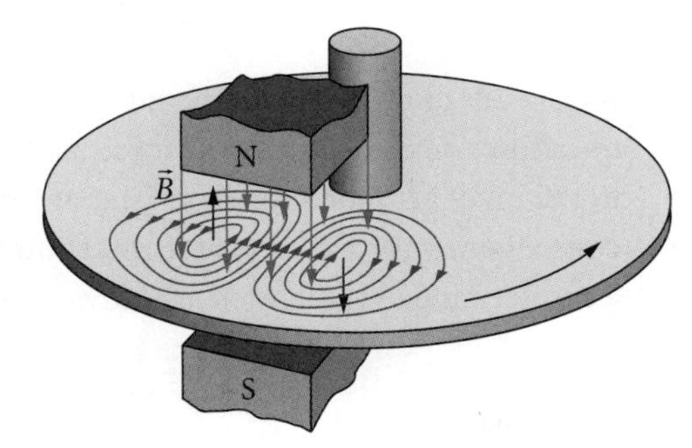

Mark breakdown

- 1 mark: response makes some relevant points
- 2 marks: response gives sound explanation of electromagnetic braking
- 3 marks: response gives thorough explanation of electromagnetic braking supported by an effective diagram or gives extensive explanation of electromagnetic braking without an effective diagram
- 4 marks: response gives extensive explanation of electromagnetic braking supported by an effective diagram

44 The magnitude of torque in a motor is determined by the equation $\tau = nIAB\sin\theta$, where θ is defined as the angle between the normal to the plane of the coil and the magnetic field lines. Since $\sin\theta$ is a maximum when $\theta = 90°$, the maximum torque will be given by $\tau = nIAB$.

The ratio of the torque at an angle to the maximum torque then becomes $nIAB\sin\theta : nIAB$. This simplifies to $\sin\theta : 1$.

From the definition of θ in the first paragraph, in this situation $\theta = 65°$.

Therefore, the ratio is $\sin 65° : 1$ or $0.91 : 1$.

Mark breakdown

- 1 mark: response features some relevant information
- 2 marks: response correctly identifies torque equation to yield ratio with some error or omission
- 3 marks: response effectively uses torque equation to validly arrive at correct value

45 As long as the conductor through which constant direct current flows is perpendicular to the magnetic field, then the force is constant. This is true of both motor designs.

From the equation $\tau = rF\sin\theta$, where θ is the angle between the pivot arm and the applied force, it can be determined that the torque is a maximum when θ is 90°.

In a motor with flat-ended magnets, which features a uniform magnetic field, this results in the torque maximum at only one position per half rotation, as seen in the graph below.

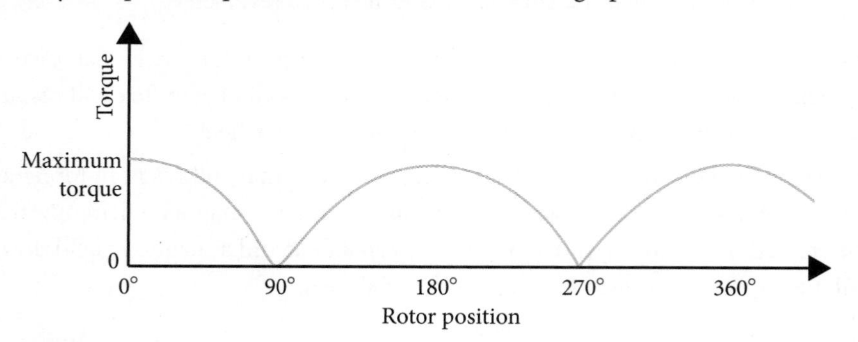

As a result of the shape of the field in the motor in the question, the direction of the force stays perpendicular to the pivot arm for a much greater array of angles and consequently torque stays at its maximum value over a greater range, which is clearly advantageous.

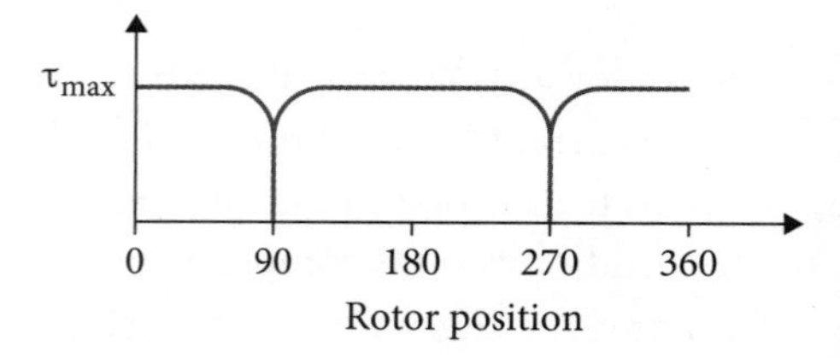

Mark breakdown

- 1 mark: response makes some relevant points
- 2 marks: response gives sound explanation of change that occurs to torque
- 3 marks: response gives thorough explanation of the advantageous change that occurs to torque as a result of the changed magnetic field shape
- 4 marks: response gives extensive explanation of the advantageous change that occurs to torque, with force unchanged, as a result of the changed magnetic field shape

46 The Faraday motor is a simple application of the motor effect. With the mercury and its container acting as conductors to complete the circuit, the hanging wire is a current-carrying conductor in a magnetic field, consequently experiencing a force.

A top view considering the shape of the magnetic field emerging from the north pole is helpful in this case. The field will be a collection of arrows radiating away from the pole like the spokes of a wheel before they bend back towards the south pole further down. The current will then be going up the wire and in this field the direction of the force can be shown to be perpendicular to the current, using the right-hand push rule.

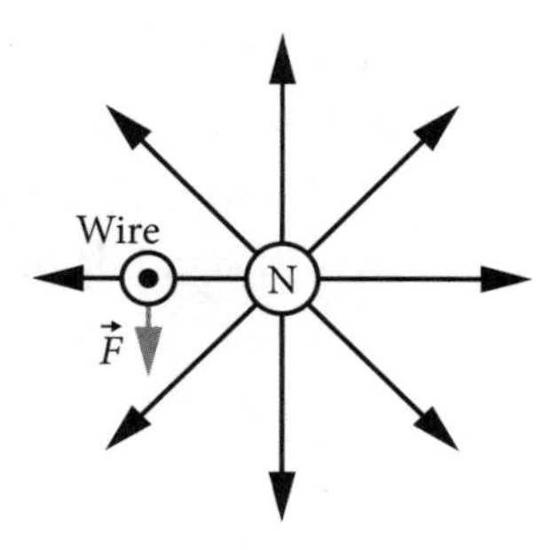

Since the wire is free to pivot about the loop at the top, it will swing around in the field, and it will then experience a force in a new direction. At all times, this force will act perpendicularly to the field lines and current and so the wire will move in an anticlockwise circle.

Mark breakdown

- 1 mark: response makes some relevant points
- 2 marks: response gives sound explanation of the process that enables the wire to move
- 3 marks: response gives thorough explanation of the process that enables the wire to move and of the direction of that motion
- 4 marks: response gives extensive explanation of the process that enables the wire to move and of the direction of that motion

47 Phased AC current is provided to pairs of electromagnets positioned on opposite sides of the stator, as seen in the diagram.

As a consequence, a change of flux is experienced by the bars of the squirrel cage rotor.

According to Faraday's law, an emf is induced in the squirrel cage rotor and thus a current flows in a circuit along the bars and connecting end plates.

The direction of this current establishes a magnetic field that interacts with the magnetic field of the stator coil.

This interaction results in a magnetic force that results in a torque on the squirrel cage rotor.

Mark breakdown

- 1 mark: response features an aspect of the process
- 2 marks: response outlines some relevant processes for the functioning of the AC induction motor
- 3 marks: response outlines a substantially correct process with significant errors or omissions
- 4 marks: cause and effect response effectively incorporates aspects of the diagram to follow a chain of logic that links the AC supply current through all key steps to the delivered torque, with a significant error or omission or without reference to the diagram
- 5 marks: cause and effect response effectively incorporates aspects of the diagram to follow a chain of logic that links the AC supply current through all key steps to the delivered torque

48 a As soon as a motor coil begins turning in the stator magnetic field, the coil experiences a change of flux. This change of flux will induce an emf (Faraday's law) that will oppose the cause of the change of flux (Lenz's law).

This means that the emf will act against the supply emf and result in a decrease in the total emf according to the relationship $\varepsilon_{total} = \varepsilon_{supply} - \varepsilon_{back}$.

The motor will continue to turn faster as a result of the torque caused by this total emf notwithstanding the influence of friction and/or a load.

As the motor turns faster, the back emf increases and total emf decreases; thus, the rate at which the motor is increasing its rate of rotation decreases.

At some point, the total emf is just enough to match friction and/or the load and the motor will not be able to get any faster. Thus, back emf has set a maximum operating speed for the motor.

Mark breakdown

- 1 mark: response makes some relevant points
- 2 marks: response gives sound explanation of effect of back emf on motor speed with some key points made
- 3 marks: response gives thorough explanation for the process of back emf capping the motor's maximum speed, including reference to Lenz's law, the impact on total emf and on motor rotation rate in which there are some omissions or weak points
- 4 marks: response gives extensive explanation for the process of back emf capping the motor's maximum speed, including reference to Lenz's law, the impact on total emf and on motor rotation rate

b

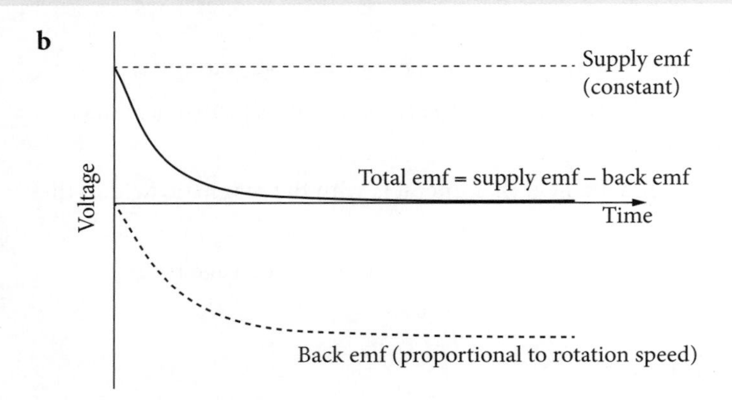

Note only the total emf line is necessary for the correct response.

Mark breakdown

- 1 mark: graph has one correct feature
- 2 marks: graph has most of features correct
- 3 marks: graph starts at a maximum value and decreases steadily over time until reaching zero

c If the rate of rotation is low, then there is a minimal amount of change of flux for the coil. As a result, the back emf is small. Since the $\varepsilon_{total} = \varepsilon_{supply} - \varepsilon_{back}$, when ε_{back} is low then $\varepsilon_{total} = \varepsilon_{supply}$ approximately. Since $I = \frac{V}{R} = \frac{\varepsilon_{supply}}{R}$, the resistance of copper wire is very low and, hence, currents can be very high. These currents cause heating of the wire, which can burn through the insulation and, hence, burn out the motor.

Mark breakdown

- 1 mark: response makes some relevant points
- 2 marks: response gives sound explanation of reason for motor burn out with most of the key points
- 3 marks: response gives thorough explanation of reason for motor burn out with complete logic sequence, including back emf, total emf and resistance

d Conservation of energy states that the total energy of a system is conserved. It would seem that when an ideal DC motor is connected to a power supply, the input electrical energy should result in increasing (rotational) kinetic energy of the motor.

However, the motor reaches a maximum speed.

Since back emf reduces the total emf to zero at the maximum speed, the power source is no longer providing any energy to the motor and thus its constant (rotational) kinetic energy is consistent with conservation of energy.

Mark breakdown

- 1 mark: response makes some relevant points
- 2 marks: response gives reasoned explanation of the role of back emf in reducing total emf to zero for an ideal motor at top speed and linking this to conservation of energy

CHAPTER 3 MODULE 7 THE NATURE OF LIGHT

Multiple-choice solutions

1 D

Maxwell was a theoretical physicist rather than an experimental physicist, so **A** and **B** are misdirections. **C** seems to refer to the work of Planck, culminating in his equation $E = hf$. Maxwell is known for his equations that link electricity and magnetism. Going through the syllabus and knowing the work and contribution made by all named scientists is vital.

2 D

It is important to use the scale in this instance rather than just attempt to vertically align the spectral lines. This can be approached with a process of elimination most easily where the lines are sparse. Lithium features a spectral line well above the 650 nm mark but none appears in that region on the spectrum for the new star, so lithium cannot be present. Sodium features a pair of lines around 590 nm, but only one appears in the star's absorption spectrum.

3 D

The apparatus featured in this diagram relies on the mirror turning through an eighth of a revolution, and thus presenting another face of the mirror in the identical position, during the time the light makes its journey from the mirror wheel to the distant mirror and back again; therefore, light will be observed. If the rate of rotation is known, then the time for this journey can be calculated. In any other position, the rotating mirror will prevent light reaching the observer. The apparatus cannot create an interference pattern or a continuous spectrum.

4 B

The degree to which a spectral pattern is shifted (either red or blue) is determined by the rate of motion relative to Earth. In this instance, the pattern of spectral lines is shifted to a different extent and so the rate of motion must be different. The pattern of spectral lines features the same spacing in each case, which indicates the presence of the same elements.

5 B

Light takes time to travel from Jupiter to Earth. At position Q, Earth moves further from Jupiter while the orbit of Jupiter's moon is being observed. The time that the light takes to travel the additional distance through which Earth has moved would be time added to the orbital period as measured by the observer on Earth. This will cause the orbital period measured at Q to be longer than orbital periods measured in other positions.

6 A

Shorter wavelengths (approaching 400 nm) represent the blue end of the spectrum and longer wavelengths (approaching 700 nm) represent the red end of the spectrum. Star A has had its spectrum shifted towards the red end and it is therefore moving away from Earth. Conversely, star B has had its spectrum shifted towards the blue end and it can be concluded that it is moving towards Earth. Since the pattern of star A has been shifted further, it can be assumed it is moving faster.

7 A

At *Y*, the spectrum shows a decrease in the intensity of specific wavelengths; therefore, it is an absorption spectrum. Discharge tubes produce emission spectra. The general shape of the spectrum is similar to the shape of a black body radiation curve for a star. It can be concluded that this is indicating an absorption spectrum for a star.

8 B

The spectral lines of Arcatares are positioned in line with those of Zenophym but are broader and darker and feature blurred edges. The spectra are aligned, so the speeds of the stars relative to Earth are similar. Generally, broader spectral lines are indicative of greater rate of rotation and/or of a more dense atmosphere. A spectrum alone cannot be used to compare star age.

9 D

This representation of refraction assumes light is a wave and uses Huygens' principle, where each point on a wavefront acts as a source of secondary wavelets.

10 D

25°C is equivalent to 298 K. Using the equation $\lambda_{max} = \frac{b}{T}$, where $b = 2.898 \times 10^{-3}$ m K, yields a value of $\lambda_{max} = 9.72 \times 10^{-6}$ m.

11 D

Refraction occurs when light moves from one medium into another medium. Polarisation involves the passage of light through a polarising filter. Since neither of these features occur, **A**, **B** and **C** are not correct. An interference pattern is produced on the screen by light diffracting around the edges of the disc and meeting at the screen.

12 A

Polarisation is a phenomenon that could not be explained satisfactorily by the particle model of light, but a transverse wave model of light can be used to explain polarisation. Malus' law states that $I = I_0 \cos^2 \theta$ and so in the context of this question $\frac{I_B}{I_0} = \cos^2 30° = 0.75$.

13 A

The equation $d \sin \theta = m\lambda$ describes the relationship between variables in the diffraction grating situation shown here along with double-slit situations. More maxima will strike the screen if either of two changes are made: reducing θ or moving the screen closer to the diffraction grating (reducing L) so that it intercepts the maxima before they have spread past the edge of the screen (this can be shown by trigonometry using $\sin \theta = \frac{mL}{d}$). **D** will not change anything since the same equation with the same variables will yield the same outcome: θ and the distance from grating to screen are unchanged. **C** will result in fewer maxima reaching the screen. **B** effectively means decreasing the distance between slits (d), which will result in θ increasing if wavelength is constant. **A** means a decrease in wavelength, which will result in a decrease in θ for a constant value of d.

14 C

It can be seen from the stimulus that the slit separation $d = 1\ \mu m$. The angle θ is between the line drawn to the central maximum and the line drawn to the first maximum. In this instance, trigonometry shows that angle to be $\theta = \tan^{-1}\left(\frac{0.08}{0.3}\right)$.

15 D

When white light is incident on a diffraction grating, it is separated into visible continuous spectra spreading on both sides of the central maximum according to the equation $d \sin\theta = m\lambda$, whereby light with the shortest wavelength is deflected at the smallest angle. This will result in the spectra all featuring violet nearest the centre and red farthest from the centre.

16 C

When unpolarised light is incident on an ideal polariser, the emerging polarised light has half the intensity of the incident unpolarised light. The position of the transmission axis has no effect on the intensity of the emerging polarised light. Consequently, rotating the transmission axis of the first polariser will not change the intensity I_1. However, the light intensity emitted from the second polariser is related to the angle between the transmission axes of the two polarisers by Malus' law, which states that $I = I_0 \cos^2\theta$. As a consequence, rotating the first polariser will only change the intensity I_2.

17 B

Each photon of light with a wavelength of 550 nm will carry energy given by the equation $E = hf$.

Since $c = f\lambda$, this can be written as $E = \frac{hc}{\lambda}$. So it can be determined that each photon carries energy $E = \frac{6.626 \times 10^{-34} \times 3 \times 10^{8}}{550 \times 10^{-9}}$ J. A laser of power of 500 mW used for one minute will have an output of $500 \times 10^{-3} \times 60$ J. The number of photons will be given by the total energy divided by the energy per photon: $\frac{500 \times 10^{-3} \times 60}{(6.626 \times 10^{-34} \times 3 \times 10^{8}) / 550 \times 10^{-9}} = 8.3 \times 10^{19}$

18 D

Curves such as those represented in the question were part of the experimental data that contradicted theoretical expectations for black body radiation. Max Planck was ultimately able to provide a solution to the discrepancy by assuming that energy exchanged within the black body could only occur in a discrete or quantised way. This gave birth to quantum physics.

19 A

The incident light frequency is above the threshold frequency for both metals and, hence, photoelectrons will be emitted from both metals. Reading from the graph, it can be seen that the maximum kinetic energy of the photoelectrons emitted from metal Y (and, hence, the maximum velocity of the photoelectrons) is less that the maximum kinetic energy of the photoelectrons emitted from metal X.

20 A

The energy of the photons is given by the equation $E = hf$. Thus the energy will be 7.45×10^{-19} J. Watts are units of power and amps units of current, so **B** and **D** respectively are not valid responses. **C** is the energy removed from photoelectrons by the potential difference between the emitter plate and the collector and is not related to the photons.

21 C

A 2.56 V stopping voltage acts to prevent electrons reaching the collector. This potential difference will do work on the electrons given by the equation $W = qV$. In order to reach the collector, the electrons must have kinetic energy in excess of the amount of work done in the opposite direction. Hence $K = \frac{1}{2}mv^2 = qV = 1.602 \times 10^{-19} \times 2.56$. Solving for velocity gives $v = 948\,923\ \text{m s}^{-1}$.

22 D

As can be seen in the stimulus, calcium has a greater work function than caesium and so photoelectrons will be emitted with even less energy. This will not increase the current and so **A** is incorrect. Increasing intensity will result in more photons incident upon the emitter plate, but as none of the electrons will have sufficient energy to reach the collector, the current will not increase. **B** is incorrect. If the frequency is increased to 11.26×10^{14} Hz, the photons will now have energy given by $E = hf = 6.626 \times 10^{-34} \times 11.26 \times 10^{14} = 7.460\,876 \times 10^{-19}$ J. The work function of the metal is 2.1 eV, which is $2.1 \times 1.602 \times 10^{-19}$ J. The stopping voltage does work equivalent to $W = qV = 1.602 \times 10^{-19} \times 2.56$ J against the electrons. Since the sum of the work function plus the work done by the stopping voltage exceeds the energy of the photons, no electrons will reach the collector. Hence **C** is incorrect. Decreasing the stopping voltage to 2.5 V will result in the sum of the work function and the work done by the stopping voltage being less than the energy of the photons and, hence, photoelectrons will reach the collector and a current will flow.

23 D

Although there are a few different ways of phrasing the postulates upon which Einstein founded his special theory of relativity, it is imperative to understand that the first postulate relates to inertial frames of reference and that the second postulate informs of the constancy of the speed of light for all inertial observers and does not depend on the speed of either the source or the observer.

24 B

The regular event of the flashing is occurring on the spaceship and the astronaut is in that inertial frame of reference. The astronaut will record the proper time for the event of 6 minutes. The observer on Earth will record a dilated (greater) time for the event due to the effects of special relativity. The value of the time can be calculated using the time dilation equation. (Note it is unnecessary to do any calculations to determine the correct option because option B is the only option longer than the proper time.) The flashes will reach Earth at the speed of light according to the second postulate.

25 C

The power of the Sun is 3.83×10^{26} W, which means that each second it generates 3.83×10^{26} J of energy. So, in one year it must generate $3.83 \times 10^{26} \times 60 \times 60 \times 24 \times 365$ J of energy. According to the equation $E = mc^2$, 1 kg of matter will convert to 9×10^{16} J of energy. Hence the mass decrease of the Sun in one year will be given by the energy radiated by the Sun in one year divided by the energy created per kg of mass converted, which is $\frac{3.83 \times 10^{26} \times 60 \times 60 \times 24 \times 365\ \text{J}}{9 \times 10^{16}\ \text{J kg}^{-1}} = 1.34 \times 10^{17}$ kg.

26 A

The multiplying factor for this situation, often called the γ factor, is given by the equation

$$\gamma = \frac{1}{\sqrt{\left(1 - \frac{v^2}{c^2}\right)}}$$

$$= \frac{1}{\sqrt{(1 - 0.999998^2}}$$

$$= 500$$

27 C

Both observers will see the light pulse travelling at the same speed, but *Y* sees the light pulse travelling a shorter distance because the sensor moved to the right as the pulse was travelling towards it. Hence observer *X* will measure the time to be longer than that measured by observer *Y*. In terms of time dilation, observer *Y* will see observer *X*'s clock running slow relative to their own clock and, therefore, *Y* will see *X* measure a longer time for the pulse to reach the sensor because *X*'s clock is running slower than *Y*'s clock.

28 B

Since the carriage is moving relative to observer B, they will measure a shorter length of carriage than that measured by observer A due to length contraction. This will only occur in the dimension parallel to the direction of relative motion. Thus, the carriage height will remain 2.4 m. Without any need for calculation, it can be determined that the carriage length, measured by observer B will be 5.3 m (since it is the only option with $x < 8$ m).

Short-answer solutions

29 **i** This is a continuous spectrum or the visible part of a black body spectrum that is produced by hot objects. Most substances will in fact produce a continuous spectrum featuring electromagnetic radiation across a range of frequencies. However, to produce significant radiation in the visible part of the spectrum, the object needs to be sufficiently hot (like the filament of an incandescent light globe) since the intensity of the radiation at shorter wavelengths is related to the temperature of the object by Wien's law ($\lambda = \frac{b}{T}$).

ii This is an emission spectrum, which features specific wavelengths of light only. These wavelengths are emitted by electrons releasing energy as they move back down atomic energy levels in an excited gas sample. This excitation is usually achieved by an electrical current or by heat.

iii This is an absorption spectrum, which features a continuous spectrum with some wavelengths missing. It is produced when a continuous spectrum is passed through a gas and certain wavelengths are absorbed by the electrons of the gaseous atoms as they move to higher atomic energy levels.

Mark breakdown

- 1 mark: response relates one correct piece of information to an appropriate spectrum
- 2 marks: response correctly accounts for the production of two spectra only, *or* superficially accounts for all three
- 3 marks: response accounts for the production of all three spectra with some detail lacking or an omission or error
- 4 marks: response correctly accounts for the production of all three spectra in detail

30 Red shift occurs as a consequence of an optical Doppler effect and results in an increase in wavelength seen by an observer when the distance between the source of light and the observer increases over time.

Comparison of spectra of known elements on Earth with spectra of stars and the amount by which wavelengths are shifted towards the red end of the spectrum can be used to determine the relative velocity of the star to us.

Mark breakdown

- 1 mark: response features some relevant detail regarding red shift
- 2 marks: response effectively explains the process that results in red shift *or* correctly describes how data from stellar spectra can be used to determine relative velocity
- 3 marks: response effectively explains the process that results in red shift and describes how data from stellar spectra can be used to determine relative velocity

31 Maxwell used existing equations and understanding of electricity and magnetism to develop a single theoretical model for the production and propagation of light. Maxwell's theory made significant and bold predictions in regards to light that dramatically improved our understanding. His theory:

- predicted a value for the speed of light (electromagnetic waves) in a vacuum that corresponded to experimentally determined values for the speed of light
- proposed a process for the production of light by oscillating charged particles
- developed a model for light propagation, which was not dependent on a medium, and which consisted of mutually inducing perpendicular oscillating electric and magnetic fields
- predicted the existence of electromagnetic radiation with frequencies greater than and less than the visible spectrum
- predicted that an electromagnetic wave could cause a charged particle to oscillate.

Maxwell's predictions were proved correct in the years after his death. After centuries of experimental and theoretical work making gradual progress towards our understanding the nature of light, Maxwell's theory heralded significant progress and led to the development of many subsequent theories relating to light and electromagnetic radiation. As such, Maxwell made a colossal contribution to our understanding of light.

Mark breakdown

- 1 mark: response relates one correct piece of information to Maxwell's contribution
- 2 marks: response describes two features of Maxwell's work *or* describes one piece of work and attempts to discuss his contribution
- 3 marks: response describes several features of Maxwell's work *or* describes two pieces of work and attempts to discuss his contribution
- 4 marks: response correctly describes most key features of Maxwell's contribution in detail without a discussion *or* response correctly describes some features of Maxwell's contribution in detail and discusses his contribution to our understanding of light
- 5 marks: response correctly describes most key features of Maxwell's contribution in detail and effectively discusses his contribution to our understanding of light *or* response correctly describes all key features of Maxwell's contribution in detail but does not discuss his contribution to our understanding of light
- 6 marks: response correctly describes all key features of Maxwell's contribution in detail and effectively discusses his contribution to our understanding of light

32 Features of the spectrum of a star can be used to determine various properties.

- Chemical composition of stellar atmosphere: Analysis of the star's absorption spectrum, examining the proportion and the wavelengths of absorbed light, can be used to identify the relative abundance of elements in the star's atmosphere.
- Surface temperature: Analysis of the wavelength of maximum intensity in the continuous spectrum emitted by the star (apart from some absorbed wavelengths) can be used to predict the temperature of a black body, using Wien's law ($\lambda = \frac{b}{T}$).
- Translational velocity: The optical Doppler effect will cause red shift or blue shift of the star's spectrum. The extent to which the absorption spectral lines are shifted to the red or blue end of the spectrum can be used to determine the translational velocity of the star.
- Rotational velocity: Depending on the axis of rotation of a star, when it is rotating, one side of the star may be approaching Earth while the opposite side is moving away. Consequently, there will be simultaneous red shift and blue shift occurring, which results in broader spectral lines. The extent of the broadening can be used to determine rotational velocity.

- Density: The density of gases of the outer (atmospheric) layers of the star will play a role in determining the appearance of the spectral lines. Stars with a low-density atmosphere will have sharper, finer spectral lines than those with a high-density atmosphere.

Mark breakdown

- 1 mark: response names at least two uses of spectra and describes at least one way in which spectra can be used to determine star properties
- 2 marks: response names at least three uses of spectra and describes at least two ways in which spectra can be used to determine star properties
- 3 marks: response names and describes at least three ways in which spectra can be used to determine star properties
- 4 marks: response names and describes at least four ways in which spectra can be used to determine star properties
- 5 marks: response correctly describes five ways in which spectra can be used to determine star properties in detail

33 In order for light from the source to reach the mirror, the light must pass through a gap in the rotating toothed wheel. While the light travels to the reflecting mirror and back, the toothed wheel must rotate through an angle such that a gap is again in the path of the light. The light can then travel through the partially reflecting mirror and reach the observer. Since this will happen when any gap reaches the appropriate position, the experiment will need to find the lowest rate of rotation that enables light to reach the observer.

If the rate of rotation of the wheel, at this lowest speed, and the number of gaps in the wheel are both known quantities, then the time for the next gap in the wheel to move into position to allow light passage can be determined.

The distance the light travels from the toothed wheel to the reflecting mirror and back will need to be measured.

With values of distance and time, velocity can be calculated using $v = \frac{d}{t}$.

Mark breakdown

- 1 mark: response provides some relevant detail regarding the apparatus
- 2 marks: response describes how the apparatus is used and how the measurements of distance and time are obtained with multiple major omissions and/or errors *or* response describes one of how the apparatus is used or how the measurements of distance and time are obtained with minor omission or error
- 3 marks: response describes how the apparatus is used and how the measurements of distance and time are obtained with several omissions and/or errors *or* response correctly describes one of how the apparatus is used or how the measurements of distance and time are obtained in detail
- 4 marks: response describes how the apparatus is used and how the measurements of distance and time are obtained with a few minor omissions or errors
- 5 marks: response correctly describes how the apparatus is used and how the measurements of distance and time are obtained in detail

9780170465304

34 When monochromatic light is incident on a double slit, it will diffract through each of the openings. The resulting diffracted waves will interfere with each other as a consequence. This will result in both constructive and destructive interference occurring on both sides of the central maximum (a region of constructive interference located directly ahead of the two slits).

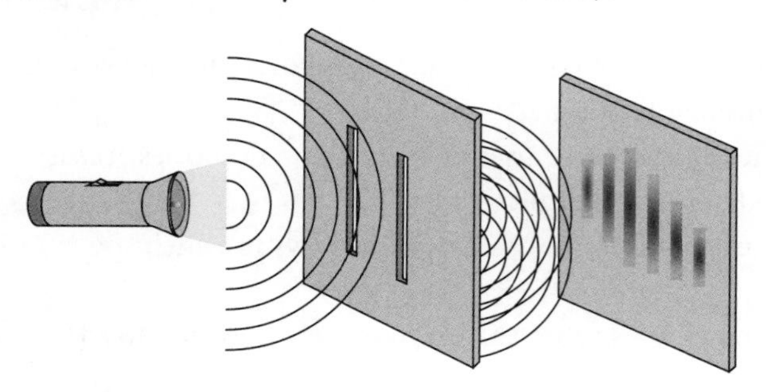

This diffraction and interference pattern is projected onto the screen and can be analysed according to the equation $d\sin\theta = m\lambda$, where d is the separation of the slits, θ is the angle made between the central line and a line drawn to any maximum and m is the maxima number from the central maximum ($m = 1, 2, 3, \ldots$). Substituting into this equation data from any of the maxima will yield an experimental value for wavelength. Best experimental technique would suggest collecting a range of data and averaging or plotting data on a graph and using a line of best fit to determine wavelength. The value of θ can be found by using trigonometry by measuring the distance from slit to screen and from central maximum to maximum of interest.

Mark breakdown

- 1 mark: response features some relevant information regarding the use of Young's double-slit apparatus to find wavelength
- 2 marks: response correctly describes some of a process in which data can be collected from a double-slit experiment and then substituted appropriately into the correct equation/plotted on a graph to be used alongside the equation to determine wavelength
- 3 marks: response describes most of a valid process in which data can be collected from a double-slit experiment and then substituted appropriately into the correct equation/plotted on a graph to be used alongside the equation to determine wavelength
- 4 marks: response correctly describes how data can be collected from a double-slit experiment and then substituted appropriately into the correct equation/plotted on a graph to be used alongside the equation to determine wavelength in detail

35 According to Malus' law $I = I_0\cos^2\theta$, so when polarised light is passed through a polarising filter, the intensity of light transmitted is proportional to $\cos^2\theta$. When the plane of polarisation of the incident light is parallel to the polarising axis of the filter, $\theta = 0$ and so $\cos^2\theta = 1$ and maximum transmission occurs. When the plane of polarisation of the incident light is perpendicular to the polarising axis of the filter, $\theta = 90$ and so $\cos^2\theta = 0$ and no transmission occurs. The graph will be a $\cos^2\theta$ curve. Since the starting point of the experiment is not specified, the curve can start anywhere between 0 and maximum.

In contrast, unpolarised light is reduced in intensity to 50% and this will occur irrespective of the angle at which the polarising filter axis is arranged; hence, a straight line represents the intensity vs filter angle.

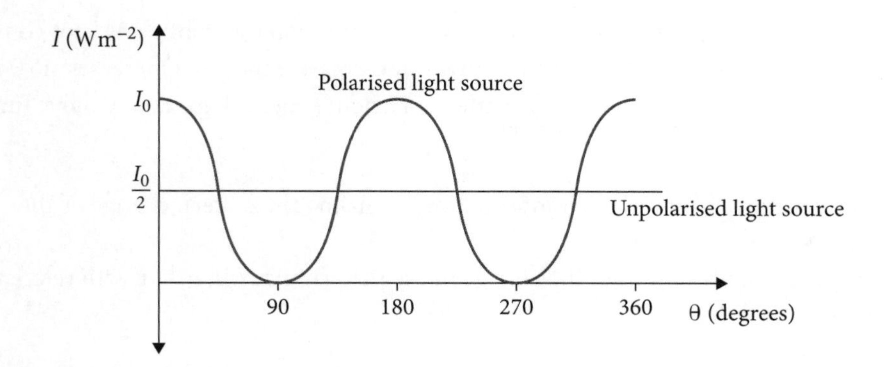

Mark breakdown

- 1 mark: response features some relevant information regarding the curve or Malus' law
- 2 marks: response features some aspects of the curves and justification that are correct *or* both curves drawn correctly on labelled and scaled axes *or* a detailed justification for each curve, including appropriate reference to Malus' law equation
- 3 marks: response features most of both curves drawn correctly on labelled and scaled axes, with a detailed justification for each curve, including appropriate reference to Malus' law equation
- 4 marks: response features both curves drawn correctly on labelled and scaled axes with a detailed justification for each curve, including appropriate reference to Malus' law equation

36 a Maxima are created when waves from each of the two slits arrive in phase and constructively interfere. This will occur when the difference in the length of the path travelled by the waves from each slit differ by 0λ, 1λ, 2λ etc. The third-order maximum is created when the path difference is 3λ.

Mark breakdown

- 1 mark: response provides some relevant detail to creation of maxima
- 2 marks: response correctly explains how maxima are created and the specific detail of the creation of a third-order maximum

b The angle θ must first be determined by using trigonometry where $\tan\theta = \frac{3.62}{6.10}$. Substituting into the equation $d\sin\theta = m\lambda$ yields

$$d \times \sin\left(\tan^{-1}\left(\frac{3.62}{6.10}\right)\right) = 3 \times 585 \times 10^{-9}$$

$$d = 3.44 \times 10^{-6}\,\text{m}$$

Thus 1 m will have $\frac{1}{3.44 \times 10^{-6}}$ slits; 290 794 slits.

This equates to 3000 slits per cm to the nearest thousand.

Mark breakdown

- 1 mark: response features some relevant information regarding the effect of changing the wavelength of incident light
- 2 marks: response uses the appropriate equation to determine a value related to *d* with some errors or omissions
- 3 marks: response successfully uses the appropriate equation to determine the correct value of slits per cm to the nearest thousand

c As can be seen in the equation $d \sin\theta = m\lambda$, if wavelength increases while d is kept constant, then for any maxima number (i.e. any m value) $\sin\theta$ will increase. Since as θ increases $\sin\theta$ increases, the angular separation is increasing. The pattern of bright spots will gradually move further apart.

Mark breakdown

- 1 mark: response features some relevant information regarding the pattern change or the relationships governing the pattern
- 2 marks: response correctly describes the change in the pattern that will occur with reference to relevant physics relationships

37 Newton and Huygens had very different explanations for the phenomenon of refraction by using their respective models of light.

Newton proposed that particles of light (corpuscles) changed direction towards the normal on entering a more dense medium because they accelerate as a consequence of an attractive force.

In contrast, Huygens proposed that secondary wavelets would move more slowly in a more dense medium. He theorised that the new wavefront formed by joining the secondary wavelets will be at an angle to the original wavefront. As a consequence, the slower progress of wavelets arriving at the interface first will bend the direction of travel towards the normal.

Mark breakdown

- 1 mark: response features some relevant information regarding either Newton's or Huygens' explanation of refraction
- 2 marks: response features some relevant information regarding both Newton's and Huygens' explanation of refraction *or* clear and detailed explanation of either Newton's or Huygens' explanation of refraction
- 3 marks: response features largely correct and complete description of both Newton's and Huygens' explanation of refraction
- 4 marks: response features clear, correct and detailed comparison of both Newton's and Huygens' explanation of refraction.

38 The quantum model of light suggests that light exists as tiny packets of energy called photons, and that these packets behaved like particles that have an amount, or quantum, of energy that is related to the light frequency. This model was first proposed by Planck to solve the disparity between the predictions of classical physics and the experimental data for black body radiation experiments.

Planck had theorised that the energy exchanged between the atomic oscillators of a black body could only do so in discrete quanta and that this quantum related to the oscillating frequency according to the equation $E = hf$, where h = Planck's constant.

The success of the quantum model of light in explaining the puzzling results of the black body experiments was extended when Einstein applied the quantum idea successfully to explain the photoelectric effect. This led to the development of quantum theory and quantum mechanics, which revolutionised the way the smallest aspects of the Universe are understood.

Therefore, Planck was vital in development of the quantum model of light, which irrevocably changed the course of science.

Mark breakdown

- 1 mark: response features some relevant information regarding Planck and/or the quantum model of light
- 2 marks: response features some analysis of the work of Planck solving issues with black body radiation experiments with a model that enabled Einstein to propose a model of light as a particle with some major omissions or errors
- 3 marks: response features predominantly clear, correct and detailed analysis of the work of Planck resolving incongruous results for black body radiation experiments with a model that enabled Einstein to propose a model of light as a particle with some minor omissions or errors
- 4 marks: response features clear, correct and detailed analysis of the work of Planck resolving incongruous results for black body radiation experiments with a model that enabled Einstein to propose a model of light as a particle

39 Photoelectric effect experiments tested the effect of changing independent variables of frequency and intensity of incident light on the maximum kinetic energy and number of photoelectrons emitted from a metal surface.

Experimental observation	Wave model of light prediction
Electrons were only emitted at frequencies that exceeded a threshold frequency that was characteristic of the metal.	Electrons should be emitted at any frequency of light provided the intensity is sufficiently high.
Above the threshold frequency the number of electrons was determined by light intensity only.	The number of electrons emitted would depend on frequency of light.
The kinetic energy of emitted electrons was dependent on frequency of light.	The maximum kinetic energy of the emitted electrons would be a function of intensity not frequency.
Photoemission of electrons was immediate above the threshold frequency regardless of light intensity.	At low levels of light intensity, the electrons would be emitted with a delay related to the accumulation of absorbed energy.

This contrast between experimental evidence and the predictions made by the wave model of light was significant evidence that the wave model of light was inadequate to explain all observations of light phenomena.

Mark breakdown

- 1 mark: response features some relevant information regarding the experimental observations of the photoelectric effect or the predictions made by the wave model of light
- 2 marks: response features an analysis of a key experimental observation of the photoelectric effect alongside a prediction made by the wave model of light that exemplifies its inadequacy *or* an attempt to analyse multiple aspects of evidence and prediction with significant errors or omissions
- 3 marks: response features an analysis of key experimental observations of the photoelectric effect alongside predictions made by the wave model of light that exemplifies its inadequacy with errors and/or omissions
- 4 marks: response features a clear and complete analysis two of the four key experimental observations of the photoelectric effect and two of the four incorrect predictions made by the wave model of light that exemplify its inadequacy
- 5 marks: response features a clear and complete analysis of most of the four key experimental observations of the photoelectric effect and the four incorrect predictions made by the wave model of light that exemplify its inadequacy
- 6 marks: response features a clear and complete analysis of the four key experimental observations of the photoelectric effect and the four incorrect predictions made by the wave model of light that exemplify its inadequacy

40 a Values of K_{max} are given by the product of the stopping voltage and the charge of an electron by using $W = q_eV = K_{max}$.

The K_{max} values for caesium, down the column, will be 9.612×10^{-20} J, 1.5219×10^{-19} J, 2.0826×10^{-19} J, 2.4831×10^{-19} J, 3.0438×10^{-19} J.

The K_{max} values for the unknown metal, down the column, will be 2.403×10^{-20} J, 8.01×10^{-20} J, 1.3617×10^{-19} J, 1.7622×10^{-19} J, 2.3229×10^{-19} J.

Mark breakdown

- 1 mark: values are mostly correct
- 2 marks: values are correctly calculated

b

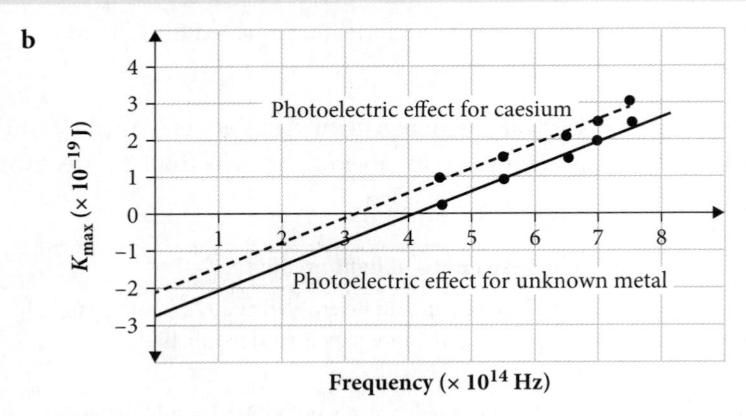

Mark breakdown

- 1 mark: response attempts to plot points and/or draw a line of best fit by using the data.
- 2 marks: points are plotted for both data sets and a line of best fit drawn for each with omissions or errors.
- 3 marks: points are plotted correctly for both data sets and an appropriate line of best fit drawn for each.

c The threshold frequency of each metal will be given by the intercept on the horizontal axis. Hence for caesium $f_0 = 3.1 \times 10^{14}$ Hz and for the unknown metal $f_0 = 4.2 \times 10^{14}$ Hz. There will be some variation around these values due to the subjectivity of the lines of best fit.

Mark breakdown

- 1 mark: response attempts to use the graph to determine threshold frequency
- 2 marks: response correctly determines threshold frequency for each metal

d The work function is indicated by the intercept on the vertical axis. It is clear from the graphed lines that the unknown metal has a greater work function (approximately 2.9×10^{-19} J) than caesium (approximately 2.1×10^{-19} J).

Mark breakdown

- 1 mark: response attempts to use the graph to determine which work function is greater
- 2 marks: response identifies that the unknown metal has a greater work function and justifies the choice with information from the graph

e Planck's constant is given by the average gradient of lines on this style of graph, according to $K = hf - \phi$. These should each be analysed by using two clear points on the respective lines of best fit. Taking care to account for the powers of 10 on each axis, the values are approximately 6.69×10^{-34} and 6.67×10^{-34} respectively, which would yield an average value for Planck's constant of 6.68×10^{-34}.

Mark breakdown

- 1 mark: response attempts to find a gradient of a graphed line
- 2 marks: response determines two gradient values from respective graphs by using two points on each graph and gets a value for Planck's constant with an error of working or missing working
- 3 marks: response shows all working to determine two gradient values from respective graphs by using two points on each graph and how the two values are used to get a value for Planck's constant

f A straight line ($y = mx + b$) plotted on a graph with a vertical axis of K_{max} and a horizontal axis of frequency will have the form $K_{max} = hf - \phi$, as indicated by Einstein. Conservation of energy indicates that the total energy before an interaction is equal to the total energy after an interaction. The energy before the interaction is provided by the photon (hf). After the interaction, the metal has gained energy equal to the work function (ϕ) and the remaining energy is taken by the electron (K_{max}).

Mark breakdown

- 1 mark: response features some relevant information regarding conservation of energy or aspects of the graph
- 2 marks: response explains the relationship between the energies expressed in Einstein's equation and the variables and constants in the formula for a straight line and shows how energy is conserved in this expression with some errors and/or omissions
- 3 marks: response effectively explains the relationship between the energies expressed in Einstein's equation and the variables and constants in the formula for a straight line and shows how energy is conserved in this expression

41 This question is best answered with the aid of a labelled diagram similar to the following.

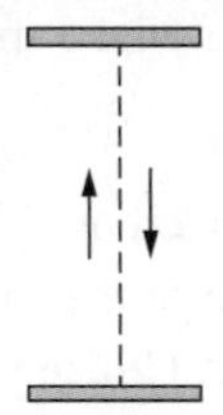

Path of light seen by observer on train

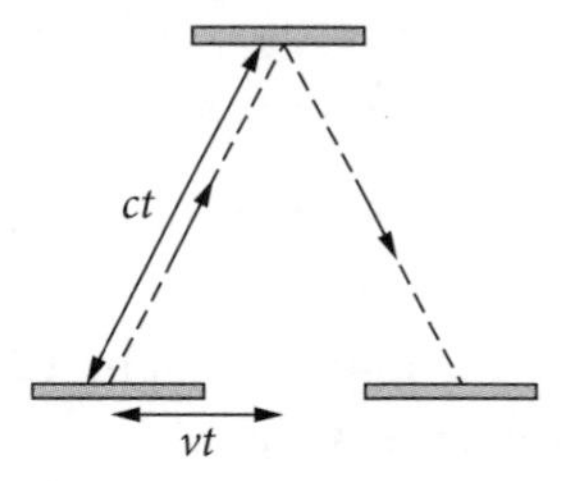

Path of light seen by outside observer

Einstein described a situation in which a train carriage is moving at a constant velocity along a track. There are two observers – one in the carriage, the other observer outside the carriage.

Within the carriage a light clock emits light from a point on the floor. The light travels directly upwards to the roof and reflects back down to the same point from which it is emitted.

The observer in the carriage is in the same inertial frame of reference as the light clock and sees the light move straight up and down in the carriage.

However, the observer outside the carriage, also in an inertial frame of reference, will see the carriage move during the motion of the light and, hence, the light moves through a path diagonally up to the mirror and diagonally back down to the source. This observer knows the light must, therefore, travel a greater distance.

Since the speed of light is a constant for all inertial observers, if it appears to the outside observer that the light travels further, it must also appear to take more time to make the journey. This is time dilation.

Numerous experiments subsequently confirmed this theoretical work. For example, the Hafele–Keating experiment compared the times recorded by several very accurate atomic clocks. Clocks were started simultaneously. One was flown east and one was flown west twice around the world. After landing, both clocks were stopped simultaneously and compared with a clock that had remained at the airport. Time values recorded by the clocks corresponded with the theoretical values predicted by time dilation.

Mark breakdown

- 1 mark: response features some relevant information of the thought experiment or evidence of time dilation
- 2–3 marks: response soundly describes the thought experiment with some relevant details of the event and the two observers' locations in different inertial frames of reference and the relative motion of the carriage *and* outlines a piece of experimental evidence for time dilation
- 4–5 marks: response thoroughly describes the thought experiment, including most of the relevant details of the event and the two observers' locations in different inertial frames of reference and the relative motion of the carriage *and* describes a piece of experimental evidence for time dilation
- 6–7 marks: response extensively describes the thought experiment, including all relevant details of the event and the locations of the two observers in different inertial frames of reference as well as the relative motion of the carriage and how the constant speed of light and the different distances of observed journeys of light leads to a logical conclusion of time dilation, *and* extensively describes a piece of experimental evidence for time dilation

42 a The relativistic momentum is given by the equation $p_v = \dfrac{m_0 v}{\sqrt{\left(1 - \frac{v^2}{c^2}\right)}}$.

Substituting gives $p_v = \dfrac{1.673 \times 10^{-27} \times 2.66 \times 10^8}{\sqrt{\left(1 - \frac{\left(2.66 \times 10^8\right)^2}{\left(3.0 \times 10^8\right)^2}\right)}} = 9.624 \times 10^{-19}\,\text{kg m s}^{-1}$.

Mark breakdown

- 1 mark: response makes some minor errors in calculating the relativistic momentum of the proton
- 2 marks: response shows full working to obtain a correct value

b It can be seen that as the value of v is increased towards the speed of light, the denominator of the fraction will approach zero and, consequently, the momentum of the proton will approach infinity. Since infinite momentum cannot be attained, v cannot reach c. Consequently the idea that v can exceed c is clearly unreasonable.

Mark breakdown

- 1 mark: response features some relevant information regarding speed approaching the speed of light
- 2 marks: response explains the improbability of the theoretical outcome as speed approaches the speed of light

43 The mass of both an electron and a positron is 9.109×10^{-31} kg, so the total mass that is converted to energy in this process is 1.8218×10^{-30} kg. Using $E = mc^2$, a value of energy released can be determined.

$E = 1.8218 \times 10^{-30} \times (3 \times 10^8)^2 = 1.639\,62 \times 10^{-13}$ J

Each photon will carry half this energy: 8.1981×10^{-14} J.

Using $E = hf$, the frequency can be found.

$E = 6.626 \times 10^{-34} \times f$

$f = 1.237 \times 10^{20}$ Hz.

Mark breakdown

- 1 mark: response attempts some steps towards a frequency calculation
- 2 marks: response makes some minor errors in calculating the frequency of the emitted gamma ray photons
- 3 marks: response shows full working to obtain a correct value

44 a When the train is in motion relative to the outside observer, the effects of length contraction will be observed. The train is 60 m and must be contracted to a length of 56 m. The length contraction equation is $l = l_0\sqrt{\left(1 - \frac{v^2}{c^2}\right)}$.

Substituting yields $56 = 60\sqrt{\left(1 - \frac{v^2}{(3 \times 10^8)^2}\right)}$, which rearranges to become

$$\frac{56}{60} = \sqrt{\left(1 - \frac{v^2}{(3 \times 10^8)^2}\right)}$$

Thus, $\left(\frac{56}{60}\right)^2 = 1 - \frac{v^2}{(3 \times 10^8)^2}$

So $\frac{v^2}{(3 \times 10^8)^2} = 1 - \left(\frac{56}{60}\right)^2$

So $v^2 = (1 - \left(\frac{56}{60}\right)^2) \times (3 \times 10^8)^2$.

Thus $v = 1.077 \times 10^8\,\text{m s}^{-1}$.

Mark breakdown

- 1 mark: response attempts some steps towards a length contraction calculation
- 2 marks: response makes some minor errors in calculating the velocity required to cause the desired length contraction
- 3 marks: response shows full working to obtain a correct value

b An observer inside the train would notice that the tunnel has become shorter by the same factor as the train became shorter to the outside observer $\left(\frac{56}{60}\right)$. Hence the tunnel would now be measured to be 52.27 m in length by the onboard observer while the train's length would be 60 m. Hence the observer inside the train would observe that more of the train extends outside the tunnel than previously!

Mark breakdown

- 1 mark: response features some relevant information regarding the train passenger's observations, qualitative or quantitative
- 2 marks: response qualitatively and quantitatively identifies observations that would be made by a train passenger

45 Several choices are possible in response to this question.

Einstein's first postulate states that the laws of physics hold for all inertial frames of reference. The second postulate states that the speed of light is a constant for all inertial observers.

One option would be to discuss evidence supporting the postulates by using subatomic particles.

Subatomic particles that move at close to the speed of light sometimes undergo a spontaneous decay and some of their mass is converted to energy in the form of a gamma ray. Measurements made of the gamma rays emitted during these decays have shown that no matter what the direction of the original motion of the subatomic particle, the speed of the gamma ray relative to the laboratory observer is constant.

Mark breakdown

- 1 mark: response features some relevant information regarding a piece of evidence for one postulate
- 2 marks: response discusses the evidence for one postulate with some lack of detail or an error or omission
- 3 marks: response effectively discusses the evidence for one postulate in appropriate detail

CHAPTER 4 MODULE 8 FROM THE UNIVERSE TO THE ATOM

Multiple-choice solutions

1 A

The gradient of the graph is known as Hubble's constant. The inverse of the graph can be used to estimate the age of the Universe. **B**, **C** and **D** are related distractors.

2 C

Fusion is the process that occurs in stars and all fusion reactions convert mass into energy.

3 C

Betelgeuse is a giant and has, therefore, finished fusing hydrogen in its core but will be fusing hydrogen in its outer shell.

The left-hand side of the HR diagram features blue-white and white stars, so **A** is incorrect.

Spica is a main sequence star and is, therefore, only fusing hydrogen to helium in nucleosynthesis reactions predominated by the CNO cycle, so **B** is incorrect.

Proxima Centauri is a very low mass star and will have a very long life as it consumes hydrogen very slowly, so **D** is incorrect.

4 A

In simple, abbreviated form, forces separated first, then fundamental matter and antimatter particles formed from energy. An unknown event left mostly matter particles after annihilation. Atoms were gradually built from these matter particles as the Universe cooled. First, quarks combined to make hadrons, including neutrons and protons (relying on the strong nuclear force), then they combined to make nuclei (relying on the strong nuclear force), and finally they captured electrons to make atoms (relying on the electromagnetic force). From there, gravity collected atoms into gas clouds and thence stars, and gathered stars into galaxies.

5 C

The spectra appear to show the same wavelengths, therefore each star must be fusing the same element(s). Broader spectral lines are associated with stars with stellar atmospheres of greater density. Since red giants are created by an enormous expansion of a main sequence star, it stands to reason that red giant stars are much less dense than main sequence stars. So, star Q is likely to be main sequence and star P to be red giant.

6 C

The surface temperature of a star is represented by the spectral class of the star. A star with spectral class of A or B has a much higher surface temperature than a star with spectral class K or M.

The luminosity scale (compared to our Sun) is from most luminous at the top to least luminous at the bottom of the HR diagram. **A** is, therefore, incorrect.

Blue stars will belong in spectral class O and red stars in spectral class M, so **B** is incorrect.

Star *R* is a giant and has advanced past the main sequence stage of its evolution, so **D** is incorrect.

7 D

The Big Bang theoretical model suggested that there would be leftover cosmic microwave background radiation and that the Universe would be predominantly made of hydrogen and helium. The timeline of the Big Bang notes that at some point an imbalance of matter and antimatter particles led to a matter-built Universe after an annihilation event. So, **A**, **B** and **C** are predictions of the Big Bang theory rather than providing evidence that enabled the theory to be developed. The Big Bang theory was developed after observations of galaxies moving away from Earth at speeds proportional to their distance from Earth.

8 B

The CNO cycle is one of the two main nucleosynthesis reactions in which hydrogen is fused into helium (the other is the p–p chain). The CNO cycle is the predominant reaction for stars of larger mass on the main sequence and the p–p chain is the predominant reaction in stars of lower mass on the main sequence. Larger stars have higher core temperatures, which are able to have particle energy of sufficient magnitude to overcome the greater electrostatic repulsion between a proton (^1_1H) and the carbon, nitrogen and oxygen nuclei that catalyse the nucleosynthesis.

White dwarfs and neutron stars are not carrying out any nucleosynthesis reactions, so **C** is incorrect.

9 C

Geiger and Marsden bombarded gold foil with alpha particles so **B** and **D** are incorrect as they represent proton bombardment.

Under Rutherford's direction the experiment was to confirm Thomson's model, which featured an atom that was positive in nature and of relatively low density. It was, therefore, expected that the dense and fast-moving alpha particles would pass straight through as in **C**. **A** shows the result they actually obtained.

10 A

The cathode rays were accelerated by an electric field in a different region of the cathode ray tube to cause them to pass across the tube. Hence, **B** is incorrect.

The electric and magnetic fields are perpendicular in orientation, but the forces act in opposite directions, so **C** is incorrect.

Neutralising the cathode rays would not be desirable for the experiment, nor would it be achievable by the process described and so **D** is incorrect.

The electric field was established perpendicular to the magnetic field and adjusted so that the sum of the magnetic force and electric force on the cathode ray was zero, as indicated by the undeflected path of the ray. This was necessary as it enabled a mathematical calculation of the velocity of the ray.

11 D

Millikan's experiment involved oil droplets becoming charged as a consequence of exposure to X-rays and then being suspended by an electric field. This relied on balancing the force on the oil droplet due to gravity with the force on the droplet due to the electric field.

12 D

Observation I: neither neutrons nor gamma rays would be deflected by a magnetic field, so this does not distinguish the two possibilities.

Observation II: both gamma rays and neutron impact could dislodge protons from paraffin, so this was not a distinguishing observation.

Observation III: the photoelectric effect is a result of photon energy causing electrons to be ejected from metals. Gamma ray photons are the most energetic of photons and will cause a photoelectric effect. The radiation must be neutrons.

13 B

The wave model was supported by several observations; however, deflection by magnetic fields was not one of them. **A**, **C** and **D** are all incorrect.

14 A

The positron was first observed in 1933, more than 20 years after Bohr published his work, so **B** is incorrect.

Observations of alpha particle deflection through gold foil provided the foundation Rutherford's model, which Bohr subsequently improved upon, so **C** is incorrect.

The beam of electrons directed at a nickel crystal is referring to the experiment Davisson and Germer used to confirm de Broglie's hypothesis; de Broglie's model was after Bohr's model, so **D** is incorrect.

Observations of emission spectra were yet to be explained when Bohr proposed an atomic model that improved Rutherford's planetary model.

15 A

A photon of wavelength 486 nm has energy given by the equation:

$$E = hf = \frac{hc}{\lambda} = \frac{6.626 \times 10^{-34} \times 3 \times 10^{8}}{486 \times 10^{-9}} = 4.090\,123\,457 \times 10^{-9}\,\text{J}$$

Converting this value to eV is $\dfrac{4.090\,123\,457 \times 10^{-9}}{1.602 \times 10^{-19}} = 2.553\,135\,74\,\text{eV}$

The only transition that could yield this amount of energy is the transition from $n = 4$ (−0.85 eV) to $n = 2$ (−3.40 eV).

16 D

$m_{\text{ball}} = 0.152\,\text{kg}$

$v_{\text{ball}} = \dfrac{140}{3.6} = 38.8889\,\text{m s}^{-1}$

$$\lambda = \frac{h}{mv} = \frac{6.626 \times 10^{-34}}{0.152 \times 38.8889} = 1.121 \times 10^{-34}\,\text{m}$$

17 B

The shortest wavelength corresponds to the highest frequency and, therefore, the greatest energy released. This will occur when the energy difference between the electron energy levels is greatest, which is transition X.

18 C

The spectral line associated with a transition of an electron from energy level 3 to energy level 2 will be associated with light released of a wavelength that can be determined by the equation

$$\frac{1}{\lambda} = R\left(\frac{1}{n_f^2} - \frac{1}{n_i^2}\right) = 1.097 \times 10^{7} \times \left(\frac{1}{2^2} - \frac{1}{3^2}\right) = 1\,523\,611.111$$

$\lambda = 6.5633 \times 10^{-7}\,\text{m}$

19 D

The key to unpacking this question is to understand that the four spectral lines represented on either side of the central maximum are the four visible hydrogen spectral lines of the first-order diffraction maximum, that is $m = 1$.

From the equation $m\lambda = d\sin\theta$, it can be seen that light of greater wavelengths will diffract at greater angles, as seen with the spectral line labelled N.

Longer wavelengths of spectral lines are associated with the smallest energy transitions, since $E = hf = \dfrac{hc}{\lambda}$.

So, spectral line N must be associated with the transition from energy level 3 to energy level 2.

20 C

Process R involves a decrease in atomic number of two and in neutron number of two, which means it has 'lost' two neutrons and two protons. This characterises an alpha decay. The product Q has 90 protons and 140 neutrons and so it has an atomic number of 90 and a mass number of 230. From the periodic table, it can be seen that this is thorium.

21 C

This reaction does not involve two or more nuclei combining and is, therefore, a transmutation rather than a fusion. **A** and **B** are incorrect.

The total mass of products is less than total mass of reactants, so energy will be released in the reaction according to $E = mc^2$.

22 C

The energy absorbed in this reaction is calculated from the mass defect: the products have a greater mass than the reactants. The mass defect is determined by subtraction:

$1.008\,67 + 13.005\,74 - 14.003\,07 = 0.011\,34\,\text{u}$.

Multiplying this value by the constant from the data and formula sheet yields

$0.011\,34 \times 931.5\,\text{MeV}/c^2 = 10.563\,21\,\text{MeV} = 10.563\,21 \times 10^6\,\text{eV}$

Converting to J by multiplying by 1.602×10^{-19} yields

$10.563\,21 \times 10^6 \times 1.602 \times 10^{-19} = 1.\,692\,226\,242 \times 10^{-12}\,\text{J} = 1.692\,226\,242 \times 10^{-18}\,\text{MJ}$

23 C

$N_t = N_0 e^{-\lambda t}$ and 4.5×10^6 years is $1.419\,12 \times 10^{14}$ seconds (converting because the units of the answers are in s^{-1})

$$5 = 100e^{-\lambda \times 1.41912 \times 10^{14}}$$

$$0.05 = e^{-\lambda \times 1.41912 \times 10^{14}}$$

$$\ln(0.05) = -\lambda \times 1.419\,12 \times 10^{14}$$

$$\lambda = -\frac{\ln(0.05)}{1.41912 \times 10^{14}} = 2.110\,978\,827 \times 10^{-14}\,\text{s}^{-1}$$

24 A

Converting 172 years into seconds since the units of the decay constant can be seen to be in s^{-1} gives a half-life of $5.424\,192 \times 10^9$ seconds.

$$\lambda = \frac{\ln 2}{t_{\frac{1}{2}}} = \frac{\ln 2}{5.424192 \times 10^9} = 1.277\,880\,983 \times 10^{-10}\,\text{s}^{-1}$$

25 D

The binding energy (and, therefore, mass defect associated with the formation of the nucleus) of Y greatly exceeds the sum of the binding energies (and, therefore, mass defect associated with the separation of the nuclei into constituents) of W and X and, consequently, energy must be released in this reaction. Particle Z will either have zero binding energy or increase the difference if its binding energy is not zero.

A is incorrect only because it does not consider the sum of the reactants. **B** is incorrect as Z's role is redundant in finding an answer. **C** is mathematically incorrect.

26 B

Using the data plotted on the graph, the time for half of the radioisotopes to decay is 100 years.

$$\lambda = \frac{\ln 2}{t_{\frac{1}{2}}} = \frac{\ln 2}{100} = 0.0069\,\text{year}^{-1}$$

27 B

The terms controlled and uncontrolled chain reaction are used in reference to fission reactions that are either controlled to proceed at a constant rate or allowed to proceed unchecked.

These chain reactions are possible since one neutron can initiate a fission and 2 or 3 neutrons can be produced by a single fission.

Control is achieved by ensuring that, on average, only one product neutron causes a fission to occur and in this way the energy output is steady, as in a nuclear reactor. Uncontrolled nuclear fission reactions occur in explosions and accidental nuclear power plant events such as meltdowns.

28 C

Alpha particles are large and charged and as such cannot travel far without interacting with other atoms, but on contact cause ionisation.

Gamma rays are uncharged but highly energetic and consequently tend to pass through most materials with little ionisation occurring.

29 A

Leptons cannot change into quarks nor quarks into leptons, so **C** and **D** are incorrect.

Beta-plus decay involves a proton changing into a neutron and emitting a positron. Since the proton is made up of two up quarks and a down quark, while a neutron is one up quark and two down quarks, then the transition has involved one of the up quarks changing into a down quark.

30 D

An electron and a muon are both leptons. which are fundamental particles in the Standard Model.

Similarly, a gluon is a boson, which is a fundamental particle in the Standard Model.

A proton is a baryon – a composite particle made up of three quarks.

31 A

It can be determined from the data, if not from memory, that the udd baryon is a neutron and the uud baryon is a proton. Hence the transformation described is a beta-minus decay. This decay has a decay product of an electron, which provides electrical neutrality across the nuclear reaction. None of the other three options, neutron, positron and proton, could do this.

32 B

Bosons cannot be accelerated nor separated by particle accelerators and, hence, **C** and **D** are incorrect. The distinguishing feature between **A** and **B** is that particle accelerators can only accelerate charged particles.

Short-answer solutions

33 In nuclear fusion reactions in the Sun (and other stars) that convert four hydrogen nuclei into a helium nucleus, a small proportion of mass in each reaction is 'lost'. This mass is converted to energy according to Einstein's equation $E = mc^2$. Thus the stars produce energy as a consequence of a reduction in their mass. The huge energy output of stars can be considered to be the result of both the enormous amount of mass that they are able to 'lose' and of the c^2 multiplying factor, which sees each kilogram of mass loss equating to 9×10^{16} J of energy released.

Mark breakdown

- 1 mark: response makes some appropriate substitution into a relevant equation *or* features some relevant information about the equation, mass loss or nuclear fusion
- 2 marks: response uses the equation to explain how mass is converted to energy by stars
- 3 marks: response clearly and thoroughly links features of the equation to the conversion of mass to energy, including reference to hydrogen as fuel and to c^2 as a feature that helps us to understand the magnitude of energy released

34 Hubble's observations yielded spectral data that could be used to measure the speed of galaxies and used Cepheid variables to ascertain the distance to those galaxies. The velocity of the galaxies was plotted against their distance from Earth to give the graph below.

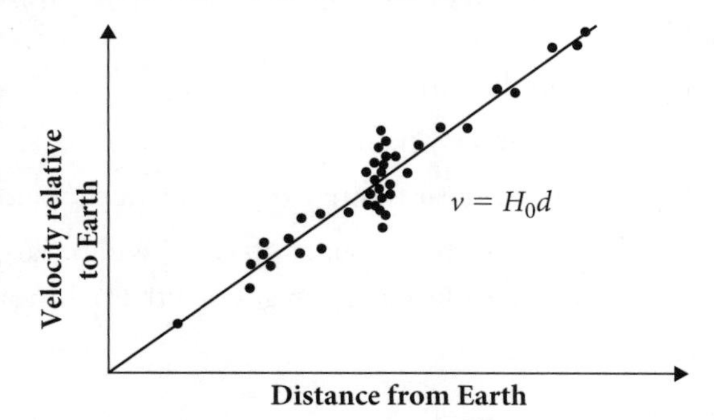

The results showed a linear relationship between the recessional velocity of galaxies and their distance from Earth. This result is consistent with a model of the Universe in which every object is moving away from every other object as a result of the expansion of the Universe itself.

The line of best fit $y = mx$ is $v = H_0 d$, where the slope of the graph is Hubble's constant with units of per unit time. The age of the Universe can be determined from the inverse of the slope of the graph, $\frac{1}{H_0}$, where units have been converted to SI units.

Mark breakdown

- 1 mark: a feature of Hubble's observations or conclusion presented, perhaps represented by an unlabelled sketch graph showing a $y = mx$ line
- 2 marks: a sketch graph produced that may not be fully labelled and some attempt to link age of Universe to graph
- 3 marks: valid sketch graph produced with labels on axes. Features of graph, including slope, described and clearly linked to the age of the Universe
- 4 marks: some minor omission or error from the following: valid sketch graph produced with labels on axes; outline of how the linear relationship indicated an expanding Universe. Features of graph, including slope, described and clearly linked to the age of the Universe
- 5 marks: valid sketch graph produced with labels on axes. Outline of how the linear relationship indicated an expanding Universe. Features of graph, including slope, described and clearly linked to the age of the Universe

35 In simple terms, there are a number of similarities and differences that can be identified from the positions of the four stars on the HR diagram. Some predictions about the future stages are somewhat speculative and depend on details of mass that are not obvious from the HR diagram.

Similarities (production of energy):

- All stars (P, Q, X and Y) are emitting energy as a result of converting mass to energy (according to $E = mc^2$) in nuclear fusion (nucleosynthesis) reactions.
- All stars (P, Q, X and Y) are emitting energy as a result of fusing hydrogen to helium: X and Y are both emitting energy as a result of fusing hydrogen to helium in their core; P and Q are both emitting energy as a result of fusing hydrogen to helium in their outermost shell.
- P and Q are both emitting energy as a result of fusing elements heavier than hydrogen in their core.

Differences (production of energy):

- Y is fusing hydrogen to helium predominantly with p–p chain reactions, whereas X is fusing hydrogen to helium predominantly with the CNO cycle.
- P is fusing heavy elements in the core but is likely to only fuse up to carbon or oxygen, whereas Q is able to fuse elements that are even heavier, possibly all the way to iron.
- X and Y are fusing hydrogen to helium in the core, whereas P and Q are fusing heavier elements in the core.

Similarities (remaining stages of evolutionary path):

- When X and Y reach low levels of hydrogen in the core they will become (red) giants.
- P and Q will cease fusion at some point in the future and will no longer be in the giant phase of their evolutionary path.
- X will follow a path similar to Q in the future.

Differences (remaining stages of evolutionary path):

- X will reach giant stage in the relatively near future; Y will take much, much longer to get to that stage.
- P will become a white dwarf after it finishes its giant phase; Q will be large enough to have a supernova followed by subsequent evolutionary stages of either neutron star or black hole.

Mark breakdown

- 1 mark: some relevant information presented
- 2 marks: some relevant information provided about production of energy or evolutionary stages
- 3 marks: sound outline of production of energy and/or evolutionary stages
- 4 marks: sound comparison features at least one similarity and one difference with significant errors/omissions
- 5 marks: thorough comparison features multiple similarities and differences for both the production of energy and evolutionary stages with several errors/omissions
- 6 marks: thorough comparison features multiple similarities and differences for both the production of energy and evolutionary stages with minor errors/omissions
- 7 marks: extensive comparison features multiple similarities and differences for both the production of energy and evolutionary stages

36 Spica

Spica will be on the upper left of the main sequence of the HR diagram where stars are very large, blue and short lived (fast fusing).

- Main sequence star
  - Dominant form of nucleosynthesis is the CNO cycle, in which a carbon-12 nucleus acts as a catalyst. (Note: This carbon is not formed in these stars, but is left over from the nebula dust and gas where the star was initially formed.)
  - Other products of these collisions include gamma photons (3), positrons (2) and the helium-4 nucleus.
  - The overall net equation for this process is $4^{1}_{1}H \rightarrow {}^{4}_{2}He + 2{}^{0}_{+1}e + 2{}^{0}_{0}\nu$ + energy.
  - The carbon-12 is not used up in the process but is recycled so it can take part in further CNO nucleosynthesis.

- Post-main sequence star
  - When hydrogen to helium fusion is finished due to low hydrogen concentrations, the star begins to contract, heating the core where helium fusion will begin. (Note: The triple-alpha reaction can be part of this process.)
  - The star's outer layers will begin to expand and form a red giant/supergiant.
  - A series of heavy element fusion reactions occurs that form elements such as carbon, neon, oxygen, silicon, sulfur and iron. The end point of these fusions is dependent on the star's mass. These stages are accompanied by contractions of the core of the star, forming the hotter, denser conditions required for heavy element fusion. Each time this happens, the star expands a little more.

Note: Elements past iron cannot be synthesised because iron-56 has the peak binding energy per nucleon for fusion.

Barnard's star:

Barnard's star will be found on the lower right of the main sequence group on the Hertzsprung–Russell diagram, where small red stars that are very long-lived and slow fusing are found.

- Main sequence star
  - Dominant form of nucleosynthesis is the p–p chain reaction.
  - Sequence of two particle collisions involving the collision of hydrogen-1 nuclei to form hydrogen-2 nuclei, positrons and neutrinos in the first collision.
  - The hydrogen-2 collides with hydrogen-1 to create hydrogen-3 and gamma photons.
  - The hydrogen-3 nuclei collide to form the helium-4 nucleus and two hydrogen-1 nuclei.
  - The overall reaction of proton–proton nucleosynthesis is $6{}^1_1H \rightarrow {}^4_2He + 2{}^1_1H + 2{}^{0}_{+1}e + 2{}^0_0v + 2{}^0_0\gamma$.
  - As carbon-12 is often present in these stars, CNO nucleosynthesis may occur, but is rare, as hotter and denser conditions are required than found in most smaller main sequence stars.
- Post-main sequence star
  - If the star is large and hot enough, some heavier element fusion will occur as described above for a red giant, but synthesis will stop with lighter elements such as carbon. Iron and other heavier elements will not be formed in these stars.
  - Smaller stars (less than 0.5 solar masses when in main sequence) will not form any elements past helium.

Mark breakdown

- 1 mark: some relevant information presented
- 2–3 marks: limited analysis of one or both stars, missing key details
- 4–5 marks: sound analysis of both stars in main sequence and post-main sequence stages *or* thorough analysis of both stars in main sequence stage *only* with appropriate use of *one* nuclear equation
- 6–7 marks: thorough analysis of both stars in main sequence and post-main sequence stages, including appropriate use of one nuclear equation *or* extensive analysis of both stars in main sequence and post-main sequence stages without appropriate use nuclear equations
- 8 marks: extensive analysis of both stars in main sequence and post-main sequence stages, including appropriate use of *two* nuclear equations

37 Starting from a singularity (an infinitely small point of extreme energy), the Universe began rapidly expanding. During this stage, the four forces (gravity, strong nuclear then electromagnetic and weak nuclear) decoupled.

Next, quarks and leptons, along with their antimatter equivalents, were produced from photons. (Somehow matter particles outnumbered antimatter particles at this stage.)

As the Universe cooled, quarks began to combine to form hadrons, under the influence of the strong nuclear force.

Protons and neutrons then joined by the strong nuclear force to form the first nuclei, as the Universe cooled further.

Much later, the Universe was cool enough for electrons to be brought into orbit around nuclei to form the first atoms, due to the electromagnetic force.

Gravity then began to collect atoms to form gas clouds and then stars and galaxies as the Universe cooled further.

Mark breakdown

- 1 mark: response outlines some processes of Big Bang
- 2 marks: sound outline of processes of Big Bang but with significant errors or omissions
- 3 marks: thorough outline of all key processes of Big Bang with some minor errors or omissions
- 4 marks: thorough outline of all key processes of Big Bang

38 Physical characteristics include details such as temperature, size, colour, density, luminosity. This comparison can be done in a table such as below or in some alternative form.

Characteristic	A	B	C	D
Temperature: A, B and C are similar, and are different from D	Slightly hotter than B, C, cooler than D; main sequence star still fusing hydrogen into helium and generating high heat	Cooler star; core hydrogen fusion has finished, shell fusing hydrogen has expanded and cooled	Cooler star; core hydrogen fusion has finished, shell fusing hydrogen has expanded and cooled	Hottest star; no fusion, but hot remnant core of white dwarf radiates high heat due to small size and high density
Size: all stars different	Mid-size; main sequence stars are relatively compact; this is a relatively small–average main sequence star similar in size to our Sun	Smaller than C, much larger than A and D; star has expanded and cooled, probably formed from A, so is much larger	Largest; star has expanded and cooled, but formed from a larger main sequence star, so larger than B	Smallest; star has finished all fusion, only core remains, gravity is the dominant force; star has contracted massively; approximately Earth-sized
Colour: A, B and C similar and all distinct from D	Yellow-orange: medium main sequence star, so relatively hot, but not very hot	Yellow: cooler stars	Orange: cooler stars	White to blue-white: due to intense temperature
Density: B, C similar, different from both A and D	Medium density; star has not contracted nor expanded as still main sequence star, density high enough for hydrogen fusion to occur	Not very dense; star has cooled and expanded, spreading out the gas	Not very dense; star has cooled and expanded, spreading out the gas	Very dense; all core mass contracted into a very small object
Luminosity: all different from C most luminous through B, A to D (least luminous) (note high luminosity equates to low absolute magnitude values)	Small–average main sequence star, so radiating high heat for its size, but small size means a lower luminosity than B or C	High luminosity due to large surface area, but not producing many of the heavier elements, so not generating as much heat and, therefore, energy as C	Highest luminosity due to huge surface area and heat still being produced by heavier element fusion	Lowest luminosity; hot and dense but extremely small, so low absolute magnitude/ luminosity

Mark breakdown

- 1 mark: identification of at some physical characteristics with some attempt to link to stars *or* attempt at a comparison of one physical characteristic
- 2 marks: identification of a range of physical characteristics with some attempt to expand past identification *or* attempts to compare some physical characteristics
- 3 marks: thorough comparison of some physical characteristics with good descriptions and some explanations *or* sound comparison of a range of physical characteristics with some descriptions or some explanations
- 4 marks: thorough comparison of a range of physical characteristics with good descriptions and some explanations
- 5 marks: extensive comparison of a range of physical characteristics, including descriptions and/or explanations as appropriate

39 Any three of several diagrams could be used to illustrate properties such as:

- Cathode rays cause fluorescence.

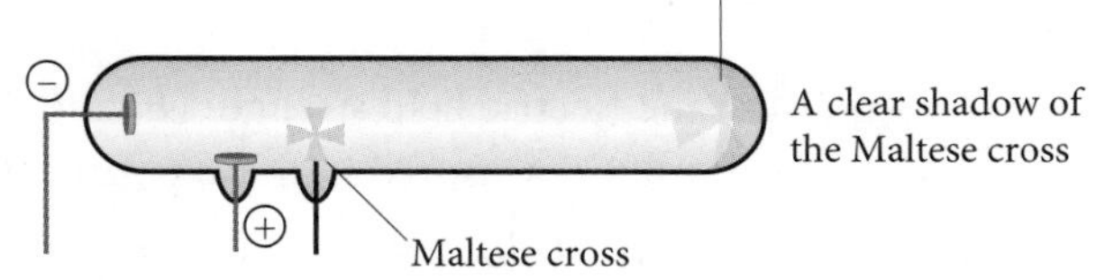

- Cathode rays are deflected by magnetic fields, as shown by the path in this experimental apparatus diagram.

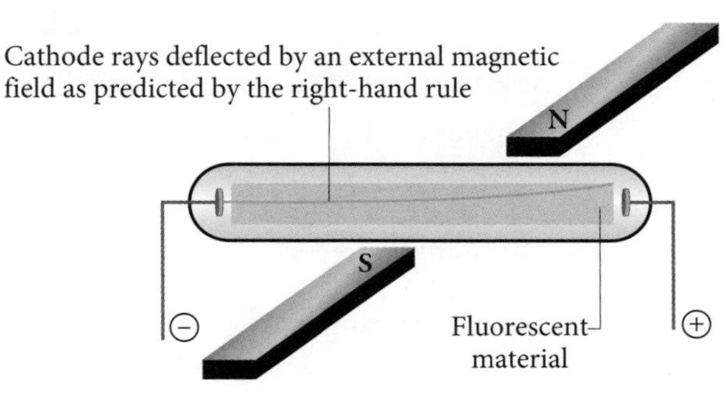

- Cathode rays have momentum, as shown by the effect that they have on a paddle wheel in this experimental apparatus diagram.

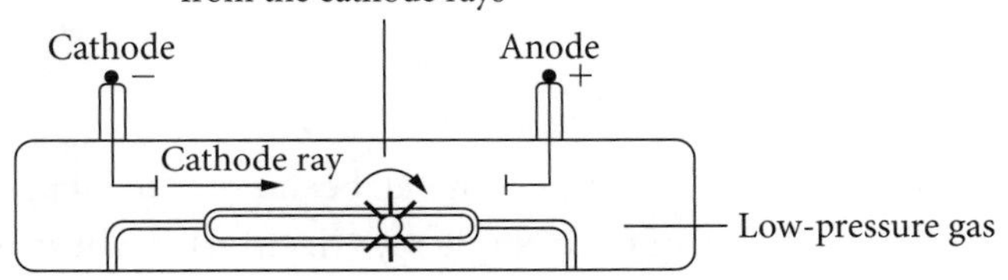

- Cathode rays are deflected by electric fields, as shown by the path in this experimental apparatus diagram.

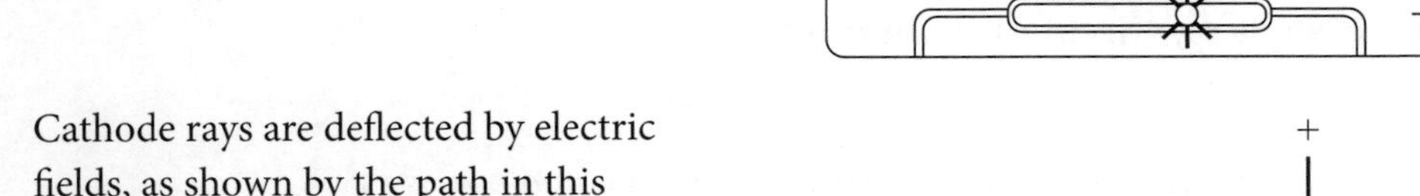

- Cathode rays travel in straight lines and, therefore, cast shadows, as shown by the 'Maltese cross' shadow in this experimental apparatus diagram.

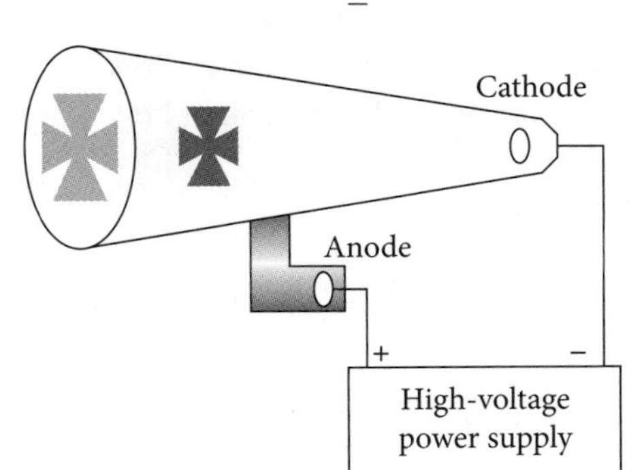

Mark breakdown

- 1 mark: attempt is made to represent at least one cathode ray property by using a diagram
- 2 marks: clearly sketched and labelled diagrams effectively illustrate one experimentally determined property of cathode rays *or* poor-quality diagrams illustrate two properties
- 3 marks: clearly sketched and labelled diagrams effectively illustrate two experimentally determined properties of cathode rays *or* poor-quality diagrams illustrate three properties
- 4 marks: clearly sketched and labelled diagrams effectively illustrate three experimentally determined properties of cathode rays

40 Note: there are some alternative presentations of this process.

Step 1: Balance the forces exerted on the cathode ray beam by the magnetic and electric fields so that the beam is undeflected.

$$F_B = F_E$$
$$\therefore qvB = qE$$

This enables an equation for the velocity of the beam to be determined.

$$v = \frac{E}{B}$$

Both the electric field strength (E) and the magnetic field strength (B) can be determined from values measured in the experiment.

Step 2: The electric field is turned off/removed so that the cathode ray moves in a circular arc under the influence of only the magnetic field.

Measure the radius.

Then it can be concluded that

$$F_B = F_c$$
$$\therefore qvB = \frac{mv^2}{r}$$

which rearranges to

$$\frac{q}{m} = \frac{v}{Br}$$

Since $v = \frac{E}{B}$

Then $\frac{q}{m} = \frac{E}{B^2 r}$

E, B and r have all been experimentally determined, so a value for the charge to mass ratio of the cathode ray (electron) can be determined.

Mark breakdown

- 1 mark: some relevant information about the experimental method provided
- 2 marks: basic response features some details of measurements made at each of the relevant steps or a mathematical description of how q/m is determined in the experiment
- 3 marks: sound response features some details of measurements made at each of the relevant steps, leading to a reasonable description of how q/m is determined in the experiment
- 4 marks: thorough response features most of the correct details of measurements made at each of the relevant steps, leading to a clear description of how q/m is determined in the experiment
- 5 marks: extensive response features details of measurements made at each of the relevant steps, leading to a clear description of how q/m is determined in the experiment

41 Experimental work:

- Convinced that cathode rays were actually particles, Thompson conducted the q/m experiment in which he balanced electric and magnetic fields acting on a cathode ray so that the beam was straight ($qvB = Eq$) and used this to calculate the speed of the electrons. He then turned off the electric field, so the beam turned in circular motion due to the magnetic field.
  - He was able to equate the force causing the cathode ray to move circular motion $\left(\frac{mv^2}{r}\right)$ with the force on a charged particle due to the magnetic field (qvB) to determine a formula to find q/m.
  - This q/m ratio was constant regardless of the material used to create the cathode rays.
- Millikan's experiment used oil drops, charged by exposure to X-rays, and an electric field to measure the charge on the oil drops by balancing the gravitational and electric forces acting on the oil drops. He then used the relationship between the forces ($Eq = mg$) to find the charge on each oil drop.
  - He found that the charges on different oil drops were a multiple of a base charge (approx. 1.6×10^{-19} C). He thus concluded that the charge on the unknown particle causing the oil droplets to be charged was 1.6×10^{-19} C – the quantum of electrical charge.
  - This charge was later confirmed to be the charge on an electron.

Significance:

- Thomson's findings implied that the particles in question were present in all elements, as the q/m ratio was constant regardless of the material used to create the cathode ray beam. This led to the development of an atomic model known as Thomson's plum pudding model. It was later confirmed that all atoms contain electrons that can be removed from the atoms to create cathode beams.
- The combined experimental values determined by Millikan and Thomson were able to determine values for the mass and charge of the electron – thus confirming these properties of the electron.
- The discovery of the electron was significant as it allowed explanation of phenomena seen in electric circuits and machinery, and later allowed for increasingly more accurate models of the atom to be developed with specific detail on the structure and movement of electrons in an atom.

Mark breakdown

- 1 mark: a piece of relevant information
- 2 marks: limited description of the work of either Millikan or Thomson
- 3–4 marks: sound description of the work of Millikan and Thomson
- 5–6 marks: thorough description of the work of Millikan and Thomson with some links to the discovery of the features of electrons
- 7 marks: extensive analysis of the work of Millikan and Thomson with clear links to the discovery of the features of electrons and how this contributed to later scientific work/discoveries.

42 Experiment: (see Figure 4.12 on page 114) The Geiger–Marsden experiment fired alpha particles from a radioactive source in a lead-lined container at a very thin sheet of gold foil in an attempt to verify the Thomson plum pudding model of the atom. A photographic plate arranged to surround the gold foil was able to detect the impact of the alpha particles after their interaction with the gold foil.

Results: A vast number of experimental trials revealed some interesting results.

- The majority of the alpha particles fired passed straight through with little to no deflection. This suggested that most of the atom was empty space.
- Approximately 1 in 8000 particles was deflected back at an angle of greater than 90°, suggesting a relatively small, dense structure within the atoms, since the alpha particles themselves were small and dense.

Both of these observations suggested an atomic model that was very different in structure from the accepted plum pudding model of the time. The plum pudding model was a relatively low-density mass of positive charge, with electrons spread randomly.

Significance: This experiment completely changed the accepted model of the atom at the time from the plum pudding model to one of an atom that was mostly empty space, but featuring a small, dense, central nucleus of positive charge, surrounded by electrons in a known orbit. This was known as a planetary model of the atom.

This led to further advances in the understanding of the atom through experimental work focused on the nucleus (Chadwick and others) and on the orbiting electrons (Bohr, de Broglie, Schrödinger and others) and so can be seen as a most significant experiment in the development of the atomic model.

Mark breakdown

- 1 mark: some relevant information regarding the Geiger–Marsden experiment, its results or how it changed scientific thinking at the time
- 2 marks: thorough description of the Geiger–Marsden experiment *or* its results *or* how it changed scientific thinking at the time *or* basic description of the Geiger–Marsden experiment, its results and how it changed scientific thinking at the time
- 3 marks: sound description of two of the three aspects of response
- 4 marks: sound description of the Geiger–Marsden experiment, its results and how it changed scientific thinking at the time featuring several omissions or errors *or* thorough description of two aspects of the response from the three of the Geiger–Marsden experiment, its results and how it changed scientific thinking at the time
- 5 marks: thorough description of the Geiger–Marsden experiment, its results and how it changed scientific thinking at the time with an omission or error
- 6 marks: thorough description of the Geiger–Marsden experiment, its results and how it changed scientific thinking at the time

43 This response is best supported by a labelled diagram such as that shown.

Alpha particles bombarded a beryllium sample, resulting in protons being detected as they were ejected from an adjacent paraffin sample. Clearly some energy was being transferred from the beryllium to the paraffin.

Experiments revealed that this energy was not deflected by a magnetic field and so it was concluded that it could be gamma ray photons (the only electromagnetic radiation with sufficient energy to dislodge protons from paraffin) or neutral particles.

Chadwick analysed data about the energy of the protons and was able to use principles of conservation of energy and conservation of momentum to determine that the unknown radiation was indeed neutral particles of mass very similar to the proton. These were called neutrons and were surmised to come from the nucleus.

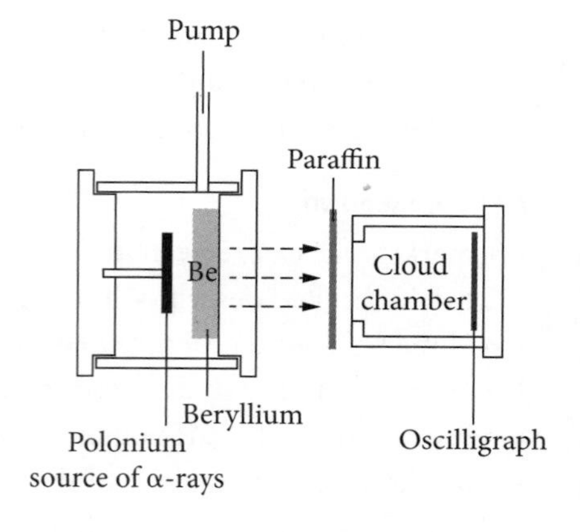

Mark breakdown

- 1 mark: basic description of the experiment, including some detail of materials used, particles detected and principles applied
- 2 marks: sound description of the experiment, including most of the detail of materials used, particles detected and principles applied
- 3 marks: thorough description of the experiment, including detail of materials used, particles detected and principles applied

44 a $\frac{1}{\lambda} = R\left(\frac{1}{n_f^2} - \frac{1}{n_i^2}\right) = 1.097 \times 10^7 \times \left(\frac{1}{2^2} - \frac{1}{5^2}\right) = 2\,303\,700$

$\lambda = 4.3408 \times 10^{-7}$ m

Mark breakdown

- 1 mark: attempt to substitute into correct equation
- 2 marks: values appropriately substituted into correct equation to get correct answer

b The law of conservation of energy states that within a closed system the total amount of energy is constant.

If the Bohr atom and the light it emits or absorbs is considered a closed system, then Bohr's third postulate is a statement of conservation of energy. It states that when an electron moves from one energy level to another the movement is accompanied by an emission or absorption of a photon whose energy is equivalent to the difference in the energy of the energy levels.

So the sum of the electron energy within the atom and the photon energy remains constant throughout the interaction.

Mark breakdown

- 1 mark: basic description has some relevant detail about conservation of energy or Bohr's atomic model
- 2 marks: sound description links conservation of energy with Bohr's atomic model
- 3 marks: thorough description effectively connects conservation of energy with Bohr's atomic model

45 The atomic model developed significantly over a few decades as a consequence of the work of three eminent scientists.

Ernest Rutherford designed an experiment, performed by Geiger and Marsden, that aimed to confirm the atomic model proposed by Thomson. Rutherford's experiment involved directing a beam of the recently discovered alpha particles, which were emitted by some radioisotopes, at a thin gold foil target.

Most alpha particles did pass straight through the gold foil, but some alpha particles (about 1 in 8000) were found to scatter at large angles. Thomson's model of the atom was incompatible with this observation.

Rutherford suggested a new atomic model that featured a small, dense, positively charged centre called the nucleus with electrons orbiting the nucleus, under the influence of an electrostatic attractive force, at a relatively large radius such that the vast majority of the atom was empty.

However, Rutherford's model had limitations.

- Orbiting electrons are accelerating and should, therefore, emit electromagnetic radiation in accordance with Maxwell's theory and consequently spiral into the nucleus, yet they do not.
- It could not explain the internal mechanism of the atom that enabled the production of a characteristic emission spectrum.
- It did not offer any detail of the arrangement of the electrons around the nucleus – simply stating that they were in orbit.
- It did not explain the composition of the nucleus.

In 1913, Neils Bohr, a Danish physicist who had worked with Rutherford, proposed a new model of the atom to explain atomic spectra and overcome the shortcomings of Rutherford's model.

Bohr's model was based on four postulates.

- An electron moves in a circular orbit about the nucleus under the influence of an attractive electrostatic force.
- Electrons can only exist in certain, specific energy levels (shells or orbitals or stationary states). In these specific orbitals, the electrons do not emit electromagnetic radiation as they orbit.

- When electrons move from one energy level to another, a photon is released or absorbed with an energy equivalent to the energy difference of the two levels.
- This is expressed as $\Delta E = E_f - E_i = hf = \frac{hc}{\lambda}$ and links the quantised electron energy to Planck's equation.
  The angular momentum is quantised as indicated by the equation $mvr = \frac{nh}{2\pi}$.

Bohr's atomic model was incomplete and had some significant limitations.

- It only worked for atoms with one electron.
- It could not explain the relative intensities of the spectral lines.
- It could not explain the fine spectral lines that were observed to comprise each spectral line as technology improved (hyperfine lines).
- It could not explain the splitting of lines that occurred when the excited atom was subjected to a magnetic field (Zeeman effect).
- It seemed to be a convenient mixture of classical physics and the new field of quantum physics.

In 1924, Louis de Broglie equated the energy of a photon, as indicated by Planck's equation, with Einstein's mass–energy equivalence equation to derive an equation $\lambda = \frac{h}{mv}$ and in doing so proposed that all objects have a wave–particle nature, rather than just photons.

This matter-wave theory could be used to explain the quantum nature of the electron energy levels that featured in Bohr's atomic model. de Broglie suggested that the stationary states proposed by Bohr existed as the standing waves of electrons positioned around the nucleus. These standing waves would need to have an integer number of wavelengths to fit within the orbital circumference of the electron; otherwise, a complex wave featuring destructive interference would occur. The integers of Bohr's orbitals $n = 1, 2, 3$ etc. corresponded to the number of complete wavelengths of the standing waves within the orbital circumference.

de Broglie was able to:

- explain the quantised electron energy levels proposed, but not explained, by Bohr's model
- explain the lack of emission of electromagnetic radiation that Rutherford was unable to explain, since waves do not emit radiation
- give a reason for the existence of the stationary states of Bohr's model
- rationalise the uncomfortable blend of quantum and classical physics of Bohr's model
- prove Bohr's equation for quantised angular momentum by using standing wave equations.

The concept of wave properties of electrons was tested by Davisson and Germer in 1927 in an experiment in which electrons scattered from the surface of a nickel crystal created a pattern which featured interference maxima. Changes made to the momentum of the electrons, by altering the accelerating voltage, changed the interference pattern in predictable ways. Quantitative analysis of the data found it to be consistent with de Broglie's equation.

Mark breakdown

- 1 mark: response features some relevant aspect of one or more of the three atomic models proposed
- 2–3 marks: basic response analyses a few of the stages of the development of the atomic model through description of the work of one of Rutherford, Bohr and de Broglie
- 4–5 marks: sound response analyses some of the stages of the development of the atomic model through description of the work of Rutherford, Bohr and/or de Broglie, with some major omissions or errors
- 6–7 marks: thorough response clearly and effectively analyses the development of the atomic model through detailed description of the work of Rutherford, Bohr and de Broglie, with some minor omissions or errors
- 8–9 marks: extensive response clearly and effectively analyses the development of the atomic model through detailed description of the work of Rutherford, Bohr and de Broglie

46 Previous atomic models:

Prior to Bohr's work, Rutherford's model of the atom was reluctantly accepted by the scientific community. His model had a small, dense, positive central nucleus (composition undetermined) surrounded at a relatively large radius by orbiting electrons (motion undefined), with the vast majority of the atom being empty space. Rutherford's work had superseded the Thomson plum pudding model, which did not have a central nucleus but an atom consisting of a low-density mass of positive charge.

Limitation of previous model – behaviour of electrons:

Rutherford's model simply stated that electrons were in orbit, but with no indication of the organisation or arrangement – presumably by the force of centripetal acceleration provided by the electrostatic attraction between the positive nucleus and the negative electrons, in a similar manner to orbiting planets and gravitational attraction.

Classical physics (Maxwell's wave theory) predicted that accelerating charged particles such as orbiting electrons would emit electromagnetic radiation. It also predicted that as the electrons lost kinetic energy, their velocity would decrease and they would slowly spiral in to the nucleus.

Of note, Rutherford's model also (i) lacked an explanation for the composition of the nucleus, and it was known at the time that the nuclear mass and charge could not comfortably be reconciled, and (ii) did not explain atomic emission spectra and their quantum nature.

Bohr's model:

Bohr's model addressed the shortcomings of the preceding model by proposing postulates that stated that electrons could only exist in stationary state orbits within which no energy was emitted. These orbits corresponded to different energy levels and movement between these energy levels was accompanied by the emission or absorption of a photon with an energy exactly equivalent to the difference in energy of the starting and finishing electron levels.

This addressed the issues of Rutherford's model by:

- describing the orbital arrangement
- explaining emission spectra
- negating the problem associated with orbiting electrons spiralling into the nucleus as they emit electromagnetic radiation.

Limitations of the Bohr model:

Bohr's model could not explain several observed phenomena such as

- the Zeeman effect in which spectral lines split in two when subjected to a magnetic field
- hyperfine spectral lines when more sensitive observation revealed each spectral line was made up of several finer lines
- relative intensity of spectral lines, which varied
- spectral lines for atoms of more than one electron
- that electrons did not emit electromagnetic radiation in stationary state orbits; it stated this without a reason or justification.

Bohr's model seemed a 'convenient' amalgamation of classical and quantum physics.

Mark breakdown

- 1 mark: some relevant point about Bohr's model or preceding models
- 2–3 marks: basic explanation of how Bohr's model addresses limitations of previous models *and* basic description of limitations of Bohr's model *or* sound explanation of how Bohr's model addresses limitations of previous models *or* sound description of limitations of Bohr's model
- 4 marks: sound explanation of how Bohr's model addresses limitations of previous models *and* sound description of limitations of Bohr's model *or* thorough explanation of how Bohr's model addresses limitations of previous models *or* thorough description of limitations of Bohr's model
- 5 marks: thorough explanation of how Bohr's model addresses limitations of previous models *and* sound description of limitations of Bohr's model *or* extensive addressing of one component of the question with a sound addressing of the other
- 6 marks: extensive explanation of how Bohr's model addresses limitations of previous models *and* extensive description of limitations of Bohr's model with some minor errors or omissions
- 7 marks: extensive explanation of how Bohr's model addresses limitations of previous models *and* extensive description of limitations of Bohr's model

47 a Davisson and Germer observed an electron interference pattern at the collector when the electron beam was directed at the nickel crystal. Measurements of the angles at which electrons were distributed could be used to calculate electron wavelength.

Mark breakdown

- 1 mark: some relevant detail
- 2 marks: an observation or measurement of the Davisson–Germer experiment is clearly identified

b This experiment revealed that a wave property (interference) could be observed for particles and that de Broglie's equation could be used to successfully predict the wavelength of particles of known momentum. It thus confirmed de Broglie's matter-wave theory.

Standing wave analysis and equations could, consequently, be used to show why only certain stationary states existed, why they did not emit electromagnetic radiation, and what the angular momentum and energy of the stationary states was. Rutherford had been unable to provide a model for electron arrangement around the nucleus nor explain why no electromagnetic radiation was emitted by the orbiting electrons. Bohr could not explain why the stationary states existed and were energetically stable.

The model that was confirmed by the Davisson–Germer experiment resolved those issues.

Mark breakdown

- 1 mark: some relevant detail about de Broglie's model or limitations of preceding models included
- 2 marks: sound response includes some detail of limitations of preceding models and how they were addressed by the model confirmed by this experiment
- 3 marks: thorough outline includes detail of limitations of preceding models and how they were addressed by the model confirmed by this experiment

48 In 1924, de Broglie equated the energy of a photon, as indicated by Planck's equation, with Einstein's mass–energy equivalence equation to derive the equation $\lambda = \frac{h}{mv}$ and in doing so proposed that all objects have a wave–particle nature, rather than just photons.

This matter-wave theory could be used to explain the quantum nature of the electron energy levels that featured in Bohr's atomic model. de Broglie suggested that the stationary states proposed by Bohr existed as the standing waves of electrons positioned around the nucleus. These standing waves would need to have an integer number of wavelengths to fit within the orbital circumference of the electron; otherwise, a complex wave featuring destructive interference would occur. The integers of Bohr's orbitals n =1, 2, 3 etc. corresponded to the number of complete wavelengths of the standing waves within the orbital circumference.

de Broglie was able to:

- explain the quantised electron energy levels proposed but not explained by Bohr's model
- explain the lack of emission of electromagnetic radiation that Rutherford was unable to explain, since waves do not emit radiation
- give a reason for the existence of the stationary states of Bohr's model
- overcome the uncomfortable blend of quantum and classical physics of Bohr's model
- prove Bohr's equation for quantised angular momentum using standing wave equations.

Mark breakdown

- 1 mark: response features some relevant detail about de Broglie's theory
- 2 marks: basic response outlines a few features of de Broglie's theory and/or how it addresses limitations of the prevailing model with some omission or error
- 3 marks: sound response outlines most of the key features of de Broglie's theory and how it addresses limitations of the prevailing model with significant omission or error
- 4 marks: thorough response outlines key features of de Broglie's theory and how it addresses limitations of the prevailing model with some omission or error
- 5 marks: extensive response clearly outlines key features of de Broglie's theory and how it addresses limitations of the prevailing model

49 All three physicists accepted that electrons had a specific charge and mass that could be measured, but they had different views about the nature of the electron.

In Bohr's atomic model, electrons orbited a positive nucleus in specific orbitals. For the model to explain the production of the hydrogen emission spectrum, Bohr had to assume that only orbitals with angular momentum (mvr) with specific values that were integer multiples of $\frac{h}{2\pi}$ could exist. His model treated the electron as a simple negatively charged particle that occupied a particular point in space at a particular moment in time as it orbited.

In contrast, de Broglie proposed that electrons had both wave and particle characteristics with an associated wavelength $\left(\lambda = \frac{h}{mv}\right)$. His atomic model explained Bohr's postulate $\left(mvr = \frac{nh}{2\pi}\right)$ by assuming that the wave nature of the electron forced the electrons to only occupy specific orbitals that corresponded to electron standing waves. That is, that the circumference of the orbital was equal to an integer number of electron wavelengths. His matter-wave atomic model envisioned the electron as a particle with wave properties.

Schrödinger's model of the atom treats the electron as a quantum entity. He combined de Broglie's matter-wave equation with classical wave theory to create an equation that describes the probability of finding an electron at any point in the atom. His model shows that the electrons do not move in orbitals and have no specific location in the atom but can only be imagined as existing as an electron cloud in the atom.

Mark breakdown

- 1 mark: some relevant detail about electrons according to at least one of the models
- 2 marks: basic response communicates some valid differences in the nature of the electron as proposed in each of the three models
- 3 marks: sound response communicates most valid differences in the nature of the electron as proposed in each of the three models
- 4 marks: thorough response effectively communicates all key differences in the nature of the electron as proposed in each of the three models, with a minor omission or error
- 5 marks: extensive response effectively communicates detailed, key differences in the nature of the electron as proposed in each of the three models

9780170465304

50

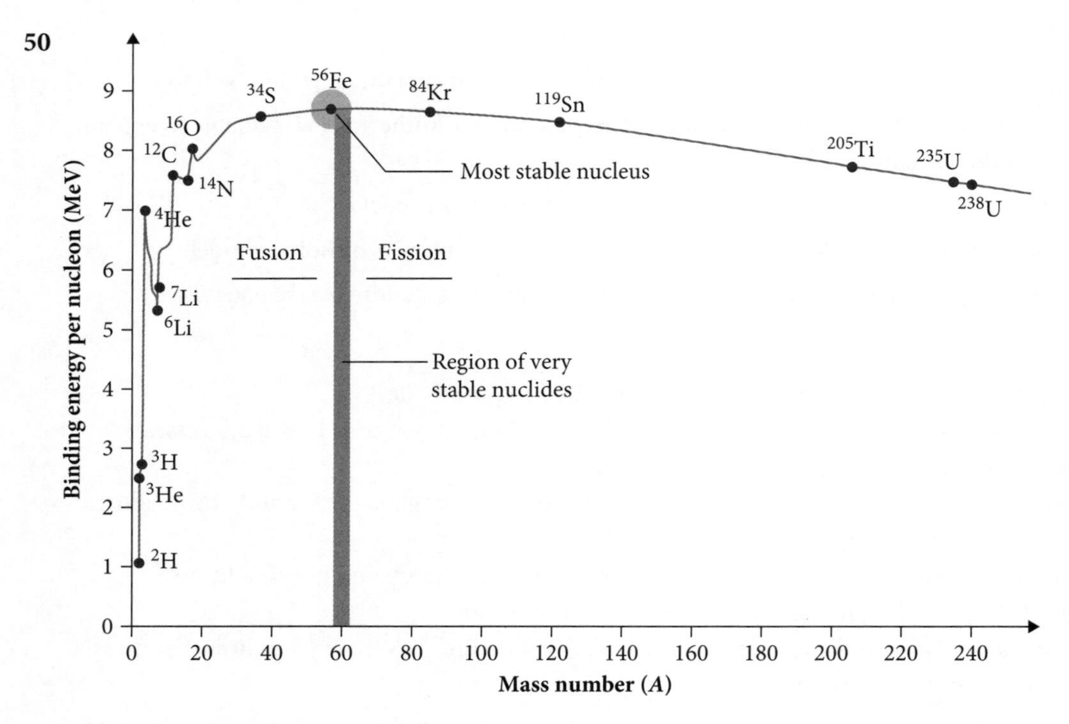

- The graph above shows the trend in binding energy per nucleon vs mass number for a range of elements. A higher binding energy per nucleon indicates a more stable nucleus. It can be seen that iron-56 has the highest binding energy per nucleon of all nuclei.
- Uranium-235 has a relatively low binding energy per nucleon (compared with many other elements, as seen on the right-hand side of the graph) and so is less stable.
- When uranium-235 undergoes fission, each of the fragments will have lower mass numbers (in this example, barium is 141 and krypton is 92) and can be seen to have a higher binding energy per nucleon and thus be more stable than the reactant uranium nucleus.
- Fission is a process in which energy is released and, therefore, product nuclei are more stable than the reactant nucleus.

Mark breakdown

- 1 mark: some information relevant to fission of ^{235}U, stability and energy
- 2 marks: sound description of some aspects of binding energy in uranium fission
- 3 marks: thorough description of binding energy per nucleon in uranium fission using specific isotopes; correct binding energy graph sketched
- 4 marks: extensive description of binding energy per nucleon in uranium fission using relevant specific isotopes; correct binding energy graph sketched and appropriately referred to in answer

51 a Energy is released as a result of mass defect in the alpha decay.

$$\Delta m = m_{\text{reactants}} - m_{\text{products}} = 238.0495 - (234.0409 + 4.0026) = 0.006\,\text{u}$$

Convert this mass to kg by multiplying $0.006 \times 1.661 \times 10^{-27}$.

$$\Delta m = 9.966 \times 10^{-30}\,\text{kg}$$

Use $E = mc^2$.

$$E = 9.966 \times 10^{-30} \times (3 \times 10^8)^2 = 8.9694 \times 10^{-13}\,\text{J}$$

Hence it is shown that the energy released by one reaction is 9.0×10^{-13} J (to two sig. fig.).

Note: there are two alternative ways of reaching a correct answer.

1 Convert all masses to kg before calculating mass defect and proceed as above.

2 Use the 931.5 MeV conversion to calculate the energy directly in MeV and then convert this to energy in joules.

Mark breakdown

- 1 mark: response attempts to find mass defect or converts all atomic mass units to kg or converts atomic mass units
- 2 marks: response successfully works out one of mass defect or energy 'value' of reactants and products
- 3 marks: response clearly shows through conversions, including mass to energy that the energy released by one reaction can be rounded to 9×10^{-13} J

b $\lambda = \frac{\ln 2}{t_{\frac{1}{2}}} = \frac{\ln 2}{87.7} = 0.0079 \text{ year}^{-1}$

$N_t = N_0 e^{-\lambda t}$

$N_{10} = 9 \times 10^{24} \times e^{-0.0079 \times 10} = 8.316 \times 10^{24}$ atoms remaining

$N_{decayed} = N_0 - N_{remaining} = 9 \times 10^{24} - 8.316 \times 10^{24} = 6.84 \times 10^{23}$ atoms

$E_{released} = N_{decayed} \times E_{per\ decay} = 6.84 \times 10^{23} \times 9 \times 10^{-13} = 6.156 \times 10^{11}$ J

Mark breakdown

- 1 mark: response makes some appropriate substitution into a relevant equation
- 2 marks: response completes first two steps successfully or calculates energy from remaining nuclei rather than decayed nuclei
- 3 marks: complete and correct working calculates a correct final energy value within a range that allows for rounding

52 The net energy released in the reaction can be determined from the difference between the total binding energy (BE) of the products and the total binding energy of the reactants.

Total binding energy of reactants = binding energy per nucleon × number of nucleons
= 3 × (7.07 × 4)
= 84.84 MeV

Total binding energy of products = binding energy per nucleon × number of nucleons
= 7.68 × 12
= 92.16 MeV

Energy released = $BE_{products} - BE_{reactants} = 92.16 - 84.84 = 7.32$ MeV

Each gamma ray photon will have $\frac{7.32}{2}$ MeV $= 3.66 \times 10^6$ eV.

So, each gamma ray photon will have $3.66 \times 10^6 \times 1.602 \times 10^{-19} = 5.86332 \times 10^{-13}$ J.

Since $E = hf = \frac{hc}{\lambda}$

$\lambda = \frac{hc}{E} = \frac{6.626 \times 10^{-34} \times 3 \times 10^8}{5.86332 \times 10^{-13}} = 3.39 \times 10^{-13}$ m

Mark breakdown

- 1 mark: response makes an appropriate step in calculations
- 2 marks: response reaches an answer with several errors *or* calculates the energy released from the triple-alpha process correctly
- 3 marks: response reaches a value through a valid process using binding energy per nucleon data to determine energy released to enable calculation of wavelength with some error
- 4 marks: response reaches correct value through a clear and valid process using binding energy per nucleon data to determine energy released to enable calculation of wavelength

53 First determine the mass defect.

$$\text{Mass defect} = \Delta m = m_{\text{reactants}} - m_{\text{products}}$$
$$= 197.999 - (193.988 + 4.00260)$$
$$= 0.0084\,\text{u}$$

Since $\Sigma m_{\text{reactants}} > \Sigma m_{\text{products}}$, energy will be released in this reaction.

Energy released is given by $0.0084 \times 931.5 = 7.8426\,\text{MeV}$.

This is equal to $7.8426 \times 10^6 \times 1.602 \times 10^{-19} = 1.2535 \times 10^{-12}\,\text{J}$.

Considering conservation of energy, the energy of the alpha particle will be given by $E_{\text{alpha}} = E_{\text{total}} - E_{\text{Po}} = 1.2535 \times 10^{-12} - 2.55 \times 10^{-14} = 1.228 \times 10^{-12}\,\text{J}$.

The reason that the kinetic energy of the alpha particle is significantly greater than the energy of the polonium atom is because momentum must be conserved in this interaction as well as energy. Assuming the radon was stationary before the alpha decay (and so initial momentum was zero) then the alpha particle and polonium nucleus will have equal and opposite momenta after the decay.

In terms of magnitude of momentum, then, if:

$$p_{\text{alpha}} = p_{\text{Po}}$$
$$m_{\text{alpha}} v_{\text{alpha}} = m_{\text{Po}} v_{\text{Po}}$$
$$v_{\text{alpha}} = \frac{m_{\text{Po}} v_{\text{Po}}}{m_{\text{alpha}}}$$

The polonium atom is nearly 50 times the mass of the alpha particle and, thus, the velocity of the alpha particle is nearly 50 times the velocity of the polonium atom after the decay.

Substituting into the kinetic energy equation:

$$K_{\text{alpha}} = \frac{1}{2} \times m_{\text{alpha}} (v_{\text{alpha}})^2 = \frac{1}{2} \times \frac{1}{50} \times m_{\text{Po}} \times (50 v_{\text{Po}})^2 = 50 \times \frac{1}{2} \times m_{\text{Po}} \times {v_{\text{Po}}}^2 = 50 \times K_{\text{Po}}$$

The alpha particle will have nearly 50 times the kinetic energy of the polonium atom.

Mark breakdown

- 1 mark: some information relevant to Standard Model of matter or the helium-4 atom
- 2–3 marks: shows some steps in the calculation of kinetic energy of alpha particle *and/or* shows some understanding of the conservation of momentum
- 4–5 marks: shows the main steps in the calculation of kinetic energy of alpha particle *and/or* shows the relevance of the conservation of momentum
- 6 marks: applies correct method to calculate mass defect and kinetic energy of alpha particle, attempts to explain the greater kinetic energy of alpha particle by using conservation of momentum
- 7 marks: applies correct method to calculate mass defect and kinetic energy of alpha particle, explains the greater kinetic energy of alpha particle by using conservation of momentum

54 Using $E = mc^2$, the energy generated by the Sun in one year can be calculated.

$$E = mc^2 = 1.34 \times 10^{17} \times (3 \times 10^8)^2 = 1.206 \times 10^{34}\,\text{J}$$

The power of the Sun is the energy produced per second.

$$P = \frac{E}{t} = \frac{1.206 \times 10^{34}}{365 \times 24 \times 60 \times 60} = 3.824 \times 10^{26}\,\text{W}$$

The mass defect for a single fusion reaction is given by subtracting the mass of the helium nucleus and two positrons from the mass of four protons.

$$4 \times 1.673 \times 10^{-27} - (6.643 \times 10^{-27} + 2 \times 9.109 \times 10^{-31}) = 4.71782 \times 10^{-29}\,\text{kg}$$

If the mass decreases by 1.34×10^{17} kg per year, then there must be

$\dfrac{1.34 \times 10^{17}}{4.71782 \times 10^{-29}}$ reactions occurring each year.

The number of fusion reactions occurring per second is given by $\dfrac{1.34 \times 10^{17}}{4.71782 \times 10^{-29}} \times \dfrac{1}{365 \times 24 \times 60 \times 60} = 9.006515988 \times 10^{37}$.

Mark breakdown

- 1 mark: response attempts to calculate one relevant feature of the Sun's fusion reactions or energy/power
- 2 marks: response attempts working towards determining the Sun's energy output in one year, power, number of reactions per year and number of reactions per second with significant errors
- 3 marks: response attempts to show working to determine the Sun's energy output in one year, power, number of reactions per year and number of reactions per second with an error or omission
- 4 marks: response is clear, complete and correct with working to determine the Sun's energy output in one year, power, number of reactions per year and number of reactions per second

55 a $^{32}_{15}\text{P} \rightarrow {}^{32}_{16}\text{S} + {}^{0}_{-1}\text{e} + \bar{\nu}$

Mark breakdown

- 1 mark: response has some features of beta – decay nuclear equation correct
- 2 marks: response features a full and correct equation for beta-minus decay of ^{32}P

b

$$N_t = N_0 e^{-\lambda t}$$

$$22.64 = 100e^{-30\lambda}$$

$$0.2264 = e^{-30\lambda}$$

$$\ln(0.2264) = -30\lambda$$

$$\lambda = -\frac{\ln(0.2264)}{30} = 0.049\,515\,064\,42 \text{ days}^{-1}$$

$$\lambda = \frac{\ln 2}{t_{\frac{1}{2}}}$$

$$t_{\frac{1}{2}} = \frac{\ln 2}{0.04951506442} = 13.998\,713\,09 = 14 \text{ days to the nearest day}$$

Mark breakdown

- 1 mark: response features some correct working towards decay constant and/or half-life
- 2 marks: response successfully determines decay constant and uses it to calculate half-life

56 The nucleus of the helium-4 atom contains 2 protons and 2 neutrons. According to the Standard Model, each proton comprises of two up quarks and one down quark and each neutron comprises two down quarks and one up quark. The quarks are bound together to form these two different baryons by gluons – the boson particles that mediate the strong nuclear force. Similarly, the baryons (protons and neutrons) are each bound to the adjacent baryons by the gluons of the strong nuclear force.

Since the up quarks have a charge of $+\frac{2}{3}$ and the down quarks have a charge of $-\frac{1}{3}$, the total charge of the proton is +1 and the total charge of the neutron is 0. There will be an electromagnetic repulsive force acting between the two positively charged protons in the nucleus mediated by photons.

Around the nucleus are negatively charged electrons. These are described in the Standard Model as leptons and are attracted to the positively charged protons in the nucleus by the electromagnetic force mediated by photons.

Mark breakdown

- 1 mark: some information relevant to Standard Model of matter or the helium-4 atom
- 2 marks: sound description of helium-4 incorporating several relevant aspects of the Standard Model of matter, such as the role of bosons (both gluon and photon) in providing forces, and the location of leptons (in orbit) and quarks (comprising protons and neutrons)
- 3 marks: thorough description of helium-4 incorporating most relevant aspects of the Standard Model of matter, such as the role of bosons (both gluon and photon) in providing forces, and the location of leptons (in orbit) and quarks (comprising protons and neutrons)
- 4 marks: extensive description of helium-4 incorporating all relevant aspects of the Standard Model of matter, including the role of bosons (both gluon and photon) in providing forces and the location of leptons (in orbit) and quarks (comprising protons and neutrons)

9780170465304

SOLUTIONS

57 a The Standard Model attempts to explain all particles and forces in the Universe through the interaction of elementary particles.

It explains that each of the four forces is mediated by particles called bosons – the strong nuclear force by gluons, the electromagnetic force by photons, the weak nuclear force by W and Z bosons and gravity by gravitons.

Elementary particles of two kinds exist: leptons and quarks. Leptons always exist alone and yet are classified in pairs – one charged and the other neutral. The most stable are the electron and electron-neutrino, less stable the muon and muon-neutrino and least stable the tau and tau-neutrino.

Quarks always exist combined (as hadrons) and form baryons – three quark combinations – and mesons (quark–antiquark combinations). The proton and neutron are examples of baryons.

All elementary particles have antiparticles.

Mark breakdown

- 1 mark: some relevant feature of the Standard Model is included
- 2 marks: sound response identifies some key aspects of the Standard Model
- 3 marks: thorough response identifies most key aspects of the Standard Model
- 4 marks: extensive response identifies all key aspects of the Standard Model

b The Standard Model of matter predicted the existence of several quarks (top and charm) along with the W and Z bosons and the gluon before they had been experimentally observed. Subsequent experiments in particle accelerators such as Fermilab and the Large Hadron Collider have confirmed these predictions.

Mark breakdown

- 1 mark: a prediction is outlined with an error or omission
- 2 marks: a prediction is outlined and some detail of its verification is mentioned

c The Standard Model has predicted the graviton, but it is yet to be observed experimentally. Furthermore, the Standard Model cannot account for dark energy and the disparity between the amount of matter and antimatter created in the Big Bang.

Mark breakdown

- 1 mark: one limitation identified correctly
- 2 marks: two limitations identified correctly

58 Particle accelerators use high potential differences to accelerate charged particles to sufficient kinetic energy so that they can initiate nuclear reactions. By initiating a range of nuclear reactions, characteristics of product particles such as charge, mass, half-life and subsequent products, energy and momentum can be established. This enables a greater understanding of the role and interactions of fundamental and composite particles that comprise the Standard Model of matter. The energies required to initiate these reactions are only possible on Earth in particle accelerators.

Mark breakdown

- 1 mark: basic response includes a feature of particle accelerators or the Standard Model
- 2 marks: response links particle accelerators to the high energy required to investigate aspects of the Standard Model, with some errors or omissions
- 3 marks: extensive response links particle accelerators to the high energy required to investigate aspects of the Standard Model

HIGHER SCHOOL CERTIFICATE EXAMINATION

Data sheet

Charge on electron, q_e	-1.602×10^{-19} C
Mass of electron, m_e	9.109×10^{-31} kg
Mass of neutron, m_n	1.675×10^{-27} kg
Mass of proton, m_p	1.673×10^{-27} kg
Speed of sound in air	$340\ m\ s^{-1}$
Earth's gravitational acceleration, g	$9.8\ m\ s^{-2}$
Speed of light, c	$3.00 \times 10^{8}\ m\ s^{-1}$
Electric permittivity constant, ε_0	$8.854 \times 10^{-12}\ A^2\ s^4\ kg^{-1}\ m^{-3}$
Magnetic permeability constant, μ_0	$4\pi \times 10^{-7}\ N\ A^{-2}$
Universal gravitational constant, G	$6.67 \times 10^{-11}\ N\ m^2\ kg^{-2}$
Mass of Earth, M_E	6.0×10^{24} kg
Radius of Earth, r_E	6.371×10^{6} m
Planck constant, h	6.626×10^{-34} J s
Rydberg constant, R (hydrogen)	$1.097 \times 10^{7}\ m^{-1}$
Atomic mass unit, u	1.661×10^{-27} kg $931.5\ MeV/c^2$
1 eV	1.602×10^{-19} J
Density of water, ρ	$1.00 \times 10^{3}\ kg\ m^{-3}$
Specific heat capacity of water	$4.18 \times 10^{3}\ J\ kg^{-1}\ K^{-1}$
Wien's displacement constant, b	2.898×10^{-3} m K

Formulae sheet

Motion, forces and gravity

$s = ut + \frac{1}{2}at^2$

$v = u + at$

$v^2 = u^2 + 2as$

$\vec{F}_{\text{net}} = m\vec{a}$

$\Delta U = mg\Delta h$

$W = F_{\parallel}s = Fs\cos\theta$

$P = \frac{\Delta E}{\Delta t}$

$K = \frac{1}{2}mv^2$

$\sum\frac{1}{2}mv_{\text{before}}^2 = \sum\frac{1}{2}mv_{\text{after}}^2$

$P = F_{\parallel}v = Fv\cos\theta$

$\Delta\vec{p} = \vec{F}_{\text{net}}\Delta t$

$\Sigma m\vec{v}_{\text{before}} = \Sigma m\vec{v}_{\text{after}}$

$\omega = \frac{\Delta\theta}{t}$

$a_c = \frac{v^2}{r}$

$\tau = r_{\perp}F = rF\sin\theta$

$F_c = \frac{mv^2}{r}$

$v = \frac{2\pi r}{T}$

$F = \frac{GMm}{r^2}$

$U = -\frac{GMm}{r}$

$\frac{r^3}{T^2} = \frac{GM}{4\pi^2}$

Waves and thermodynamics

$v = f\lambda$

$f_{\text{beat}} = |f_2 - f_1|$

$f = \frac{1}{T}$

$f' = f\frac{(v_{\text{wave}} + v_{\text{observer}})}{(v_{\text{wave}} - v_{\text{source}})}$

$d\sin\theta = m\lambda$

$n_1\sin\theta_1 = n_2\sin\theta_2$

$n_x = \frac{c}{v_x}$

$\sin\theta_c = \frac{n_2}{n_1}$

$I = I_{\text{max}}\cos^2\theta$

$I_1r_1^2 = I_2r_2^2$

$Q = mc\Delta T$

$\frac{Q}{t} = \frac{kA\Delta T}{d}$

Electricity and magnetism

$E = \frac{V}{d}$

$V = \frac{\Delta U}{q}$

$W = qV$

$W = qEd$

$B = \frac{\mu_0 I}{2\pi r}$

$B = \frac{\mu_0 N I}{L}$

$\Phi = B_{\parallel} A = BA\cos\theta$

$\varepsilon = -N\frac{\Delta\phi}{\Delta t}$

$\frac{V_p}{V_s} = \frac{N_p}{N_s}$

$\vec{F} = q\vec{E}$

$F = \frac{1}{4\pi\varepsilon_0}\frac{q_1 q_2}{r^2}$

$I = \frac{q}{t}$

$V = IR$

$P = VI$

$F = qv_{\perp}B = qvB\sin\theta$

$F = lI_{\perp}B = lIB\sin\theta$

$\frac{F}{l} = \frac{\mu_0 I_1 I_2}{2\pi r}$

$\tau = nIA_{\perp}B = nIAB\sin\theta$

$V_p I_p = V_s I_s$

Quantum, special relativity and nuclear

$\lambda = \frac{h}{mv}$

$K_{max} = hf - \phi$

$\lambda_{max} = \frac{b}{T}$

$E = mc^2$

$E = hf$

$\frac{1}{\lambda} = R\left(\frac{1}{n_f^2} - \frac{1}{n_i^2}\right)$

$t = \frac{t_0}{\sqrt{\left(1 - \frac{v^2}{c^2}\right)}}$

$l = l_0\sqrt{\left(1 - \frac{v^2}{c^2}\right)}$

$p_v = \frac{m_0 v}{\sqrt{\left(1 - \frac{v^2}{c^2}\right)}}$

$N_t = N_0 e^{-\lambda t}$

$\lambda = \frac{\ln 2}{t_{\frac{1}{2}}}$

9780170465304

Periodic table of the elements

KEY

Atomic Number	79
Symbol	Au
Standard Atomic Weight	197.0
Name	Gold

1 H 1.008 Hydrogen																	2 He 4.003 Helium
3 Li 6.941 Lithium	4 Be 9.012 Beryllium											5 B 10.81 Boron	6 C 12.01 Carbon	7 N 14.01 Nitrogen	8 O 16.00 Oxygen	9 F 19.00 Fluorine	10 Ne 20.18 Neon
11 Na 22.99 Sodium	12 Mg 24.31 Magnesium											13 Al 26.98 Aluminium	14 Si 28.09 Silicon	15 P 30.97 Phosphorus	16 S 32.07 Sulfur	17 Cl 35.45 Chlorine	18 Ar 39.95 Argon
19 K 39.10 Potassium	20 Ca 40.08 Calcium	21 Sc 44.96 Scandium	22 Ti 47.87 Titanium	23 V 50.94 Vanadium	24 Cr 52.00 Chromium	25 Mn 54.94 Manganese	26 Fe 55.85 Iron	27 Co 58.93 Cobalt	28 Ni 58.69 Nickel	29 Cu 63.55 Copper	30 Zn 65.38 Zinc	31 Ga 69.72 Gallium	32 Ge 72.64 Germanium	33 As 74.92 Arsenic	34 Se 78.96 Selenium	35 Br 79.90 Bromine	36 Kr 83.80 Krypton
37 Rb 85.47 Rubidium	38 Sr 87.61 Strontium	39 Y 88.91 Yttrium	40 Zr 91.22 Zirconium	41 Nb 92.91 Niobium	42 Mo 95.96 Molybdenum	43 Tc Technetium	44 Ru 101.1 Ruthenium	45 Rh 102.9 Rhodium	46 Pd 106.4 Palladium	47 Ag 107.9 Silver	48 Cd 112.4 Cadmium	49 In 114.8 Indium	50 Sn 118.7 Tin	51 Sb 121.8 Antimony	52 Te 127.6 Tellurium	53 I 126.9 Iodine	54 Xe 131.3 Xenon
55 Cs 132.9 Caesium	56 Ba 137.3 Barium	57–71 Lanthanoids	72 Hf 178.5 Hafnium	73 Ta 180.9 Tantalum	74 W 183.9 Tungsten	75 Re 186.2 Rhenium	76 Os 190.2 Osmium	77 Ir 192.2 Iridium	78 Pt 195.1 Platinum	79 Au 197.0 Gold	80 Hg 200.6 Mercury	81 Tl 204.4 Thallium	82 Pb 207.2 Lead	83 Bi 209.0 Bismuth	84 Po Polonium	85 At Astatine	86 Rn Radon
87 Fr Francium	88 Ra Radium	89–103 Actinoids	104 Rf Rutherfordium	105 Db Dubnium	106 Sg Seaborgium	107 Bh Bohrium	108 Hs Hassium	109 Mt Meitnerium	110 Ds Darmstadtium	111 Rg Roentgenium	112 Cn Copernicium	113 Nh Nihonium	114 Fl Flerovium	115 Mc Moscovium	116 Lv Livermorium	117 Ts Tennessine	118 Og Oganesson

Lanthanoids

57 La 138.9 Lanthanum	58 Ce 140.1 Cerium	59 Pr 140.9 Praseodymium	60 Nd 144.2 Neodymium	61 Pm Promethium	62 Sm 150.4 Samarium	63 Eu 152.0 Europium	64 Gd 157.3 Gadolium	65 Tb 158.9 Terbium	66 Dy 162.5 Dysprosium	67 Ho 164.9 Holmium	68 Er 167.3 Erbium	69 Tm 168.9 Thulium	70 Yb 173.1 Ytterbium	71 Lu 175.0 Lutetium

Actinoids

89 Ac Actinium	90 Th 232.0 Thorium	91 Pa 231.0 Protactinium	92 U 238.0 Uranium	93 Np Neptunium	94 Pu Plutonium	95 Am Americium	96 Cm Curium	97 Bk Berkelium	98 Cf Californium	99 Es Einsteinium	100 Fm Fermium	101 Md Mendelevium	102 No Nobelium	103 Lr Lawrencium

Standard atomic weights are abridged to four significant figures.
Elements with no reported values in the tables have no stable nuclides.
Information on elements with atomic numbers 113 and above is sourced from the International Union of Pure and Applied Chemistry Periodic Table of Elements (November 2016 version).
The International Union of Pure and Applied Chemistry Periodic Table of the Elements (February 2010 version) is the principal source of all other data. Some data may have been modified.